Nonequilibrium Statistical Mechanics

Basic concepts, models and applications

Online at: https://doi.org/10.1088/978-0-7503-6229-0

Nonequilibrium Statistical Mechanics

Basic concepts, models and applications

Alessandro Sarracino
Dipartimento di Ingegneria, University of Campania 'L. Vanvitelli', Aversa, Italy

Andrea Puglisi
Istituto dei Sistemi Complessi, CNR, Rome, Italy

Angelo Vulpiani
Dipartimento di Fisica, Sapienza University of Rome, Rome, Italy

IOP Publishing, Bristol, UK

ISBN 978-0-7503-6229-0 (ebook)
ISBN 978-0-7503-6227-6 (print)
ISBN 978-0-7503-6230-6 (myPrint)
ISBN 978-0-7503-6228-3 (mobi)

DOI 10.1088/978-0-7503-6229-0

Version: 20250401

IOP ebooks

British Library Cataloguing-in-Publication Data: A catalogue record for this book is available from the British Library.

Published by IOP Publishing, wholly owned by The Institute of Physics, London

IOP Publishing, No.2 The Distillery, Glassfields, Avon Street, Bristol, BS2 0GR, UK

US Office: IOP Publishing, Inc., 190 North Independence Mall West, Suite 601, Philadelphia, PA 19106, USA

To our families

Contents

Preface

> If science ceases to be a rebellion against authority, then it does not deserve the talents of our brightest children.
>
> Freeman Dyson

Equilibrium statistical mechanics is, in many respects, a well-established field. At least on a practical level, all the properties of an equilibrium system are encapsulated in the dependence of its partition function on variables such as temperature, magnetic field, and others. In contrast, the situation for nonequilibrium systems is far more complex. In such cases, there are no simple and elegant prescriptions analogous to those available in equilibrium scenarios.

This raises a fundamental question: what underlies this stark difference? In brief, the (apparent) simplicity of equilibrium statistical mechanics stems from the ability to identify the invariant distribution—namely, the microcanonical distribution—and then use probabilistic methods to construct a robust mathematical framework.

Unfortunately, replicating these steps for nonequilibrium systems is significantly more challenging. In general, there are no established protocols for determining the invariant distribution. Even when strong simplifications are introduced, such as modeling the system using a Fokker–Planck equation, identifying the invariant distribution remains a practical obstacle.

Moreover, even when the stationary distribution is obtained, it typically provides answers to only a limited set of questions, such as mean values. Understanding dynamic features like correlation functions or response functions requires a much deeper mastery of the system under investigation.

Given these inherent difficulties, authors of a book on nonequilibrium statistical mechanics are compelled to make selective choices about the topics to address. Naturally, these choices are shaped by their expertise and personal preferences.

Below, we provide a brief and incomplete list of key milestones in the history of nonequilibrium statistical mechanics. The topics covered in this book will touch upon these milestones, though not necessarily in chronological order:

- 1872: Boltzmann equation and H-theorem;
- 1905–1912: theory and experiments on Brownian motion, stochastic differential equations (Einstein, Perrin, Langevin);
- 1930s: Onsager relations;
- 1950s: Linear response theory (Kubo);
- 1960s: Zwanzig–Mori formalism;
- 1990s: Fluctuation relations (Gallavotti–Cohen);
- 1990s–present: Non-Hamiltonian systems (granular materials, flocks, active matter etc).

The book is divided into two parts. The first part introduces the fundamental concepts of nonequilibrium statistical mechanics. Topics include Brownian motion, the Langevin and Fokker–Planck equations, the Boltzmann equation, Onsager relations, linear response theory, generalized fluctuation–dissipation theorems and fluctuation relations.

The second part explores a selection of advanced topics, such as: constructing effective descriptions (e.g. Langevin equations) from microscopic models, applying fluctuation–dissipation relations to problems of causality and inference in data analysis, with applications including climate science, systems with multiple scales, non-Hamiltonian systems and materials composed of macroscopic grains or self-propelled particles (e.g. micro-swimmers, fishes, insects, birds, or sperm cells), and a miscellany of counterintuitive behaviours in transport phenomena.

The first part of the book can serve as the foundation for a course aimed at master's and PhD students. Its first five chapters provide readers with the essential mathematical tools and physical concepts needed to engage with the advanced topics discussed in the second part, ensuring the book's self-consistency. For the sake of completeness, particularly for students and early-career researchers, we include several appendices. These cover pedagogical aspects and offer suggestions for further reading.

Many other important theoretical and experimental results have been obtained in other recent branches of nonequilibrium statistical mechanics, strictly related to the topics mentioned above, but that will not be treated in this book. In particular, we cite: (i) disordered and glassy systems, that also feature nonstationary (aging) dynamics; (ii) the general approach to the theory of fluctuations developed by G Jona-Lasinio and coworkers; (iii) the rigorous mathematical approach to the Boltzmann equation; (iv) the general field of soft matter, including complex fluids and biological systems.

Acknowledgements

AV has taught a course on nonequilibrium statistical mechanics for several years to master's students in physics at the Sapienza University of Rome. The interactions with numerous students have been invaluable in shaping this book, particularly its first part. The selection of topics and their presentation have also been profoundly influenced by the contributions of many colleagues with whom we have enjoyed intense collaborations over the past decades. We thank A Baldassarri, M Baldovin, O Bénichou, L Caprini, F Cecconi, M Cencini, L Cerino, F Corberi, G Costantini, A Crisanti, A Gnoli, G Gradenigo, P Illien, E Lippiello, D Lucente, U Marini Bettolo Marconi, L Peliti, A Petri, S Pigolotti, A Plati, L Rondoni, M Viale, D Villamaina and M Zannetti.

Author biographies

Alessandro Sarracino

Alessandro Sarracino is Associate Professor of Theoretical Physics at the Engineering Department of the University of Campania 'L. Vanvitelli'.

He graduated in Physics from University of Naples 'Federico II' in 2005; A Coniglio was the supervisor of his thesis. He got his Doctorate in Physics from University of Salerno in 2009, under the supervision of M Zannetti. He held postdoctoral positions at CNR in Rome (2010–13 and 2016–18) and at the LPTMC of Sorbonne University in Paris (2014–15).

His scientific interests include nonequilibrium statistical mechanics, stochastic thermodynamics, granular and active matter, transport phenomena, Ising models, neural networks.

He is author of more than 70 scientific papers.

Andrea Puglisi

Andrea Puglisi is Research Director at the Institute for Complex Systems of the Consiglio Nazionale delle Ricerche (CNR), based at Sapienza University, Rome.

He graduated in Physics in 1998 and got his Doctorate in Physics in 2002, both at Sapienza University of Rome under the supervision of Angelo Vulpiani. He held postdoctoral positions at Sapienza University (2002–03 and 2006–08), Rome and at the LPT in Orsay Paris-Sud (2004–05). Since 2009 he has been a permanent researcher at CNR.

He has studied for many years the statistical mechanics of fluidized granular materials (kinetics, hydrodynamics, structure, nonequilibrium properties). Building upon granular models, he has investigated some problems in the more general field of nonequilibrium statistical mechanics for steady states (fluctuation–response relations, large deviation theory for stochastic processes, fluctuations of currents and entropy production). He has founded a laboratory for experiments with vibrofluidized granular materials. More recent interests have been the fields of stochastic thermodynamics and active matter, where he led experiments with sperm cells.

He has published about 150 scientific papers.

Angelo Vulpiani

Angelo Vulpiani was Full Professor of Theoretical Physics at the Physics Department of the University of Rome Sapienza (until October 2024).

He graduated in Physics from University of Rome Sapienza in 1977; G Jona Lasinio was the supervisor of his thesis. He was CNR Fellow (1978–81), assistant professor at University of Rome (1981–88), associate professor at University of L' Aquila (1988–91) and at the University of Rome (1991–2000), full professor at the University of Rome (2000–24).

His scientific interests include chaos and complexity in dynamical systems, statistical mechanics of nonequilibrium and disordered systems, developed turbulence, phenomena of transport and diffusion and foundations of physics.

He has written about 300 scientific papers and has co-authored several books, in english, italian and french.

He was a visiting fellow in research institutions and universities in France, Belgium, Sweden, Denmark, and the United States.

In 2004 he has been elected Fellow of The Institute of Physics (IOP); he has been professor at *Centro Interdisciplinare B. Segre, Accademia dei Lincei* (2016–19), in addition he is Faculty Member of the Complexity Hub, Vienna (from 2020), Faculty Member of the John Bell Institute (from 2019), Faculty Member of the Gran Sasso Science Institute, L'Aquila (from 2014).

He received the Statistical and Nonlinear Physics Prize of the European Physical Society (2021) and the Richardson Medal of the European Geosciences Union (2023).

Part I

Basic concepts

IOP Publishing

Nonequilibrium Statistical Mechanics
Basic concepts, models and applications
Alessandro Sarracino, Andrea Puglisi and Angelo Vulpiani

Chapter 1

From Brownian motion to the Langevin equation

Multa videbis enim plagis ibi percita caecis // commutare viam retroque repulsa reverti // nunc huc nunc illuc in cunctas undique partis; // scilicet hic a principiis est omnibus rerum[a].

De rerum natura, II 129–132
Lucretius

This chapter focuses on Brownian motion, a phenomenon of immense significance in both physics and mathematics. The theoretical study and the experimental investigation of Brownian motion allowed us to understand the physical reality of atoms. Moreover, the Langevin equation, introduced to describe Brownian motion, laid the foundation for the theory of stochastic processes and it is still one of the pillars for the study of nonequilibrium statistical mechanics.

1.1 From the surprising behaviour of pollen grains to the existence of atoms

In 1827 the botanist Robert Brown, during his study on an Australian plant (*Clarkia Pulchella*), discovered a phenomenon, now called Brownian motion, which had a rather important role in physics [1]. Using a microscope, Brown observed that pollen grains, which are small at macroscopic level, but large (say a few microns) compared

[a] 'For there you will see how many things set in motion by unseen blows change their course and beaten back return back again, now this way, now that way, in all directions. You may be sure that all take their restlessness from the first-beginnings.' Trans. W H D Rouse, revised by Martin F Smith (Harvard University Press: Cambridge, MA and London).

doi:10.1088/978-0-7503-6229-0ch1

with the molecules, suspended in water show a rapid and irregular movement. At the beginning, such a phenomenon was considered only a sort of curiosity; however, after some decades the relevance of Brownian motion and its connection with thermodynamics was realized. For instance, the Italian physicist Cantoni in 1868 claimed that Brownian motion is a *'beautiful and direct experimental demonstration of the fundamental principles of the mechanical theory of heat'* [2]. It is remarkable that the main actors of the building of statistical mechanics (Clausius, Maxwell, Boltzmann and Gibbs) did not show interest in Brownian motion.

The scientist who fully understood the relevance of Brownian motion and was able to develop a theory which constituted the basic point for the acceptance of the atomic structure of matter was Einstein [3]. In his *annus mirabilis* of 1905, Einstein introduced a groundbreaking theory. Essentially, he proposed that the motion of pollen grains in a liquid is governed by two key principles: (a) Stokes' law, which describes the frictional force acting on a body moving through a fluid, and (b) the equipartition of kinetic energy, which ensures that energy is evenly distributed among the system's degrees of freedom, including both the fluid particles and the grain undergoing Brownian motion. The ingenious and bold hypothesis is the validity of both the above assumptions; this in spite of the fact that the mass of the pollen grains, or more generally of a colloidal particle, is much larger that the mass of the molecules.

Once the phenomenon is formalized, it is possible to show that the particle performs an irregular motion whose (Gaussian) statistical features can be computed; namely, one has—for the average over many realizations

$$\langle (x(t) - x(0))^2 \rangle \simeq 2Dt,$$

where $x(t)$ is one component of the position of the particle at time t, and D is the diffusion coefficient whose value is given by the Einstein–Smoluchowski law:

$$D = \frac{k_B T}{6\pi\eta r}, \tag{1.1}$$

where η is the viscosity of the fluid, r the radius of the particle, and k_B the Boltzmann constant, i.e. the ratio of the constant of gases R and the Avogadro number N_A, $k_B = R/N_A$.

In the following we will discuss the great relevance of the Einstein–Smoluchowski relation, which allows us to link the experimentally measurable values of the coefficients D, T, η and r (which are macroscopic quantities), with the Avogadro number.

1.1.1 Langevin's approach

In 1908, a few years after Einstein's paper, Langevin was able to reproduce the results in a simple and elegant way by introducing the first example of the so-called stochastic differential equation [4]. Starting from the basic equation of the mechanics $F = ma$, Langevin, following the ideas of Einstein, split the force in two parts, a systematic one (due to the friction between the colloidal particle and the fluid) and a random one related to the collisions of the (fast) molecules of the liquid with the colloidal particle.

In presenting Langevin's approach, we follow his original treatment which is rather heuristic; later we shall discuss the problems for the building of a consistent mathematical theory of stochastic differential equations. For the sake of simplicity, we consider the one-dimensional case:

$$m\frac{d}{dt}v = f_T, \tag{1.2}$$

where f_T is the force acting on the particle.

In the case of a spherical particle of radius r, with velocity v in a fluid with viscosity η, we have the friction force, given by the Stokes law, $f_{\text{Stokes}} = -6\pi\eta r v$. In addition, we consider the random force given by the collisions of the fluid molecules with the particle, $F_R(t)$ (in modern terms this is white noise). Therefore, we have the evolution equation:

$$m\frac{d}{dt}v = -6\pi\eta r v + F_R(t). \tag{1.3}$$

Let us note that without the random force we have $v(t) = v(0)e^{-\frac{t}{\tau}}$, with

$$\tau = \frac{m}{6\pi\eta r}. \tag{1.4}$$

For colloidal particles of size of the order of a few microns, in a standard fluid we have $\tau \sim 10^{-6}$–10^{-7} s.

Next, let us write equation (1.3) in the form

$$\frac{d^2}{dt^2}x = -\frac{1}{\tau}\frac{d}{dt}x + \frac{1}{m}F_R(t); \tag{1.5}$$

multiplying by x, and using the identities

$$\begin{aligned} x\frac{d}{dt}x &= \frac{1}{2}\frac{d}{dt}x^2, \\ x\frac{d^2}{dt^2}x + \left(\frac{d}{dt}x\right)^2 &= \frac{1}{2}\frac{d^2}{dt^2}x^2, \end{aligned} \tag{1.6}$$

one obtains

$$\frac{d^2}{dt^2}x^2 = -\frac{1}{\tau}\frac{d}{dt}x^2 + \frac{2}{m}xF_R(t) + 2\left(\frac{d}{dt}x\right)^2. \tag{1.7}$$

Let us now perform an average: for instance one can imagine considering a large number of independent colloidal particles, obtaining

$$\frac{d^2}{dt^2}\langle x^2\rangle = -\frac{1}{\tau}\frac{d}{dt}\langle x^2\rangle + \frac{2}{m}\langle xF_R(t)\rangle + 2\left\langle\left(\frac{d}{dt}x\right)^2\right\rangle. \tag{1.8}$$

The term $\langle xF_R(t)\rangle$ is originated by the collisions of the molecules with the particle and it is natural to assume that it is zero. Since we are interested in the long-time behaviour of the particle, we can assume that, as a consequence of the many collisions, the colloidal particle is in thermal equilibrium with the fluid:

$$\frac{1}{2}m\langle \dot{x}^2\rangle = \frac{1}{2}k_B T. \tag{1.9}$$

Therefore, we have

$$\frac{d^2}{dt^2}\langle x^2\rangle = -\frac{1}{\tau}\frac{d}{dt}\langle x^2\rangle + 2\frac{k_B T}{m}, \tag{1.10}$$

whose solution is rather easy: assuming $\langle x^2\rangle|_{t=0} = 0$ and $\frac{d}{dt}\langle x^2\rangle|_{t=0} = 0$[1], we have

$$\langle x(t)^2\rangle = 2\frac{k_B T\tau}{m}\left[t + \tau\left(e^{-\frac{t}{\tau}} - 1\right)\right]. \tag{1.11}$$

For $t \gg \tau$, we have

$$\langle x(t)^2\rangle \simeq 2\frac{k_B T\tau}{m}t, \tag{1.12}$$

and therefore, using (1.4), we obtain the expression of D given in equation (1.1).

1.1.2 The relevance of Brownian motion

It is worth noting that Einstein's interest in Brownian motion was not for its own sake; in his autobiographical notes he wrote: '*My major aim in this was to find facts which would guarantee as much as possible the existence of atoms …*'. To prove (or disprove) the existence of atoms was indeed Einstein's purpose, as clearly stated in his paper on Brownian motion: '*If the movement discussed here can actually be observed (together with the laws relating to it that one would expect to find), then classical thermodynamics can no longer be looked upon as applicable with precision to bodies even of dimensions distinguishable in a microscope: an exact determination of actual atomic dimensions is then possible. On the other hand, had the prediction of this movement proven to be incorrect, a weighty argument would be provided against the molecular-kinetic conception of heat*' [3].

After the theoretical work of Einstein (and Smoluchowski) and some experiments on the diffusion of colloidal particles performed by Svedberg, the conclusive experimental contribution was given by Perrin with his study on the sedimentation and the diffusion [5], see figure 1.1. Perrin was able to determine N_A from the measurement of D: the agreement of N_A with values obtained independently definitively closed the heated controversy about the physical existence of atoms

[1] This is not really important in the asymptotic limit, however, it comes from the physical fact that the motion of the Brownian particle at very short time ($t \ll \tau$) is ballistic, i.e. it is straight with some finite velocity $x(t \to 0) \approx v_0 t$; therefore, one expects $\langle x^2\rangle|_{t\to 0} \approx \langle v_0^2\rangle t^2$. The solution in equation (1.11) shows that $\langle v_0^2\rangle = \frac{k_B T}{m}$.

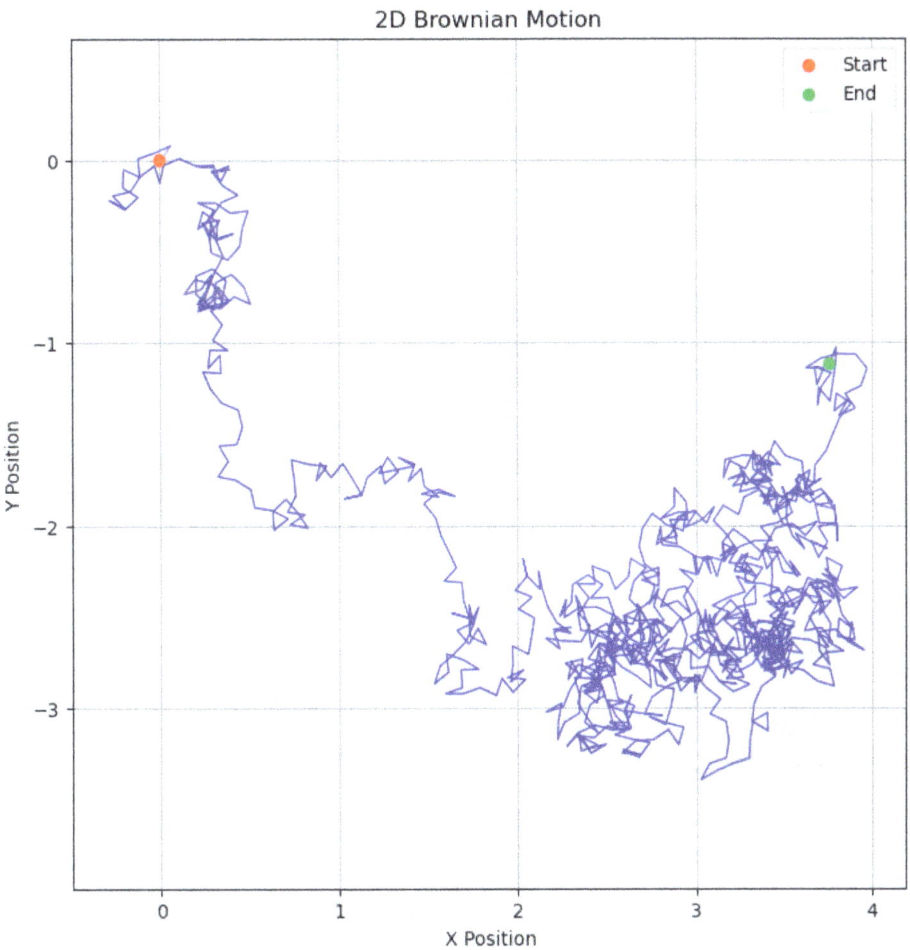

Figure 1.1. A simple example of Brownian motion, not very different from what was observed by Perrin in his renowned experiment. The figure has been generated numerically: it is actually a kind of random walk in two dimensions, where each step has a fixed length and a uniformly distributed (in $(0, 2\pi)$) random angle.

between Boltzmann, on one side, and Mach and Ostwald, on the other, who regarded atoms as useful, but merely as a mathematical tool for constructing a coherent description of Nature.

In a lecture in Paris, in 1911, Arrhenius, summarizing the works of Einstein and Perrin, declared: '*after this, it does not seem possible to doubt that the molecular theory entertained by the philosophers of antiquity, Leucippus and Democritus, has attained the truth at least in essentials*'.

1.2 Toward a mathematical theory

The reader can have the feeling that in the previous discussion some non-trivial aspects have been hidden: actually, we just repeated the straightforward argument in the original Langevin's paper. Now we shall discuss the delicate aspects in the mathematical treatment, introducing the Fokker–Planck equation and its link with the stochastic differential equations.

In order to convince the reader that such a careful mathematical treatment is necessary, we show what happens if one tries to compute the mean squared velocity of the Brownian particle using the same argument as Langevin used for the mean squared displacement. In fact, an equation for the mean squared velocity is immediately obtained by multiplying by v the Langevin equation (1.3), and taking an average, which results in:

$$\frac{1}{2}m\frac{d}{dt}\langle v^2\rangle = -6\pi\eta r\langle v^2\rangle + \langle F_R(t)v(t)\rangle. \tag{1.13}$$

Now, if one assumes that $\langle F_R(t)v(t)\rangle = 0$ as Langevin assumed for $\langle x(t)F_R(t)\rangle$, the result $\langle v^2\rangle \to 0$ for $t \gg \tau$ is obtained, which is in paradoxical contradiction with Einstein and Langevin assumptions of energy equipartition. Actually, the force $F_R(t)$ is a subtle mathematical function whose properties have been clarified in the first decades of the 20th century, giving rise to the theory of stochastic processes.

1.2.1 A discrete time version of the Langevin equation

In order to understand the mathematical difficulties in the treatment of stochastic differential equations, it can be useful, at least at a pedagogical level, to study a discrete time version, due to Lorentz, see [6], of the Langevin equation:

$$v_{n+1} = av_n + bw_n, \tag{1.14}$$

where n is a non-negative integer, $v_n = v(t_n)$, with $t_n = n\Delta t$, $a = 1 - \frac{\Delta t}{\tau}$ and b is a constant whose value will be discussed in the following; the $\{w_n\}$ are independent identically distributed Gaussian variables with zero mean and unitary variance which are indicated with $\mathcal{N}(0, 1)$. Clearly the terms bw_n and av_n correspond to the random force and the Stokes friction, respectively.

Let us assume that at the initial time the probability density of v_0 is Gaussian with mean value $\langle v_0\rangle$ and variance σ_0^2, i.e. $\mathcal{N}(\langle v_0\rangle, \sigma_0^2)$. Since the w_n are independent of v_n, using the well known result that a linear combination of Gaussian variables is Gaussian, we have that at time n the probability density is $\mathcal{N}(\langle v_n\rangle, \sigma_n^2)$. It is quite easy, from (1.14), to show that

$$\begin{aligned}\langle v_n\rangle &= a^n\langle v_0\rangle,\\ \langle v_{n+1}^2\rangle &= a^2\langle v_n^2\rangle + b^2.\end{aligned} \tag{1.15}$$

Noting that $a < 1$, we have that $\langle v_n\rangle = a^n\langle v_0\rangle$ tends to zero for $n \to \infty$. For the variance $\sigma_n^2 = \langle v_n^2\rangle - \langle v_n\rangle^2$, we have

$$\sigma_{n+1}^2 = a^2\sigma_n^2 + b^2, \tag{1.16}$$

whose solution is:

$$\sigma_n^2 = \langle v^2 \rangle + a^{2n}(\sigma_0^2 - \langle v^2 \rangle), \tag{1.17}$$

where $\langle v^2 \rangle = \frac{b^2}{1-a^2}$; therefore, for $n \to \infty$, $\sigma_n^2 \to \sigma_\infty^2 = \langle v^2 \rangle$. So we have that starting with an arbitrary Gaussian distribution after a short transient we have always a Gaussian with zero mean and variance $\langle v^2 \rangle$.

Let us now consider the position $x_n = x(n\Delta t)$ which evolves with the rule

$$x_{n+1} = x_n + v_n \Delta t.$$

It is easy to show that for large n one has

$$\langle (x_n - x_0)^2 \rangle = n\Delta t^2 \langle v^2 \rangle + 2n\Delta t^2 \sum_{j=1}^{\infty} \langle v_j v_0 \rangle.$$

Since $\langle v_j v_0 \rangle = \langle v^2 \rangle a^j$, we can determine the diffusion coefficient as the limit for $n \to \infty$ and $\Delta t \to 0$, of the ratio

$$\frac{\langle (x_n - x_0)^2 \rangle}{2n\Delta t},$$

obtaining the result $D = \langle v^2 \rangle \tau$, i.e. equation (1.1).

Let us consider again σ_∞^2; for small Δt, we have

$$\sigma_\infty^2 = \langle v^2 \rangle = \frac{b^2}{1-a^2} \simeq \frac{b^2}{2\Delta t}\tau,$$

and therefore

$$b \simeq \sqrt{\frac{2\langle v^2 \rangle}{\tau}} \sqrt{\Delta t}. \tag{1.18}$$

This allows us to understand the mathematical issue of the continuous limit; writing

$$\frac{v_{n+1} - v_n}{\Delta t} = -\frac{1}{\tau} v_n + \sqrt{\frac{2\langle v^2 \rangle}{\tau}} \frac{1}{\sqrt{\Delta t}} w_n, \tag{1.19}$$

and taking the limit $\Delta t \to 0$, we have the formal expression

$$\frac{d}{dt} v = -\frac{v}{\tau} + \sqrt{\frac{2\langle v^2 \rangle}{\tau}} \xi, \tag{1.20}$$

where $\xi = \lim_{\Delta t \to 0} (w_n / \sqrt{\Delta t})$ is a divergent quantity!

From a mathematical point of view it is better to avoid the derivative and it is more appropriate to write

$$dv = -\frac{v}{\tau} dt + \sqrt{\frac{2\langle v^2 \rangle}{\tau}} dW, \tag{1.21}$$

where W is the so-called Wiener process, i.e. a Gaussian process with $\langle W(t)\rangle = 0$, $\langle dW(t)\rangle = 0$ and $\langle dW(t)^2\rangle = dt$ (see the following for more details)[2].

1.2.2 The Fokker–Planck equation

The Langevin equation involves a noisy term whose treatment can appear ambiguous; for such a reason it is desirable to introduce an equation for the time evolution of the probability distribution of the stochastic process associated with the Langevin equation [7–9].

Let us briefly describe how to obtain the Fokker–Planck equation ruling the evolution of the probability density. We start recalling the evolution rule in a Markov chain for the probability to be in the state i at time $t+1$:

$$P_i(t+1) = \sum_j P_j(t) P_{j\to i},$$

where $P_j(t)$ is the probability to be in the state j at time t and $P_{j\to i}$ is the transition probability form j to i. In the case of continuous states we have

$$P(x, t+\Delta t) = \int P(y, t) W(y, t \to x, t+\Delta t)\, dy, \tag{1.22}$$

where now $P(x, t)$ is the probability density at time t, and $W(y, t \to x, t+\Delta t)$ is the transition rate starting from y at time t to be in x at $t+\Delta t$.

Let us introduce the increment $\Delta x(\Delta t) = z = x - y$ and consider the following conditions: (a) small Δt, and (b) small Δx; the precise meaning will be clear in the following. We consider the case where W is function of x, z and Δt, i.e. $W(y, t \to x, t+\Delta t) = \psi(x-z, z, \Delta t)$. Equation (1.22) can be written in the form

$$P(x, t+\Delta t) = \int P(x-z, t)\psi(x-z, z, \Delta t)dz.$$

For small z we can expand $P(x-z, z, \Delta t)\psi(x-z, z, \Delta t)$ around x:

$$\begin{aligned} P(x, t+\Delta t) &= \int P(x-z, t)\psi(x-z, z, \Delta t)dz \\ &= P(x, t)\int \psi(x, z, \Delta t)dz - \frac{\partial}{\partial x}\left[P(x, t)\int \psi(x, z, \Delta t)zdz\right] \\ &+ \frac{1}{2}\frac{\partial^2}{\partial x^2}\left[P(x, t)\int \psi(x, z, \Delta t)z^2 dz\right] - \cdots \end{aligned} \tag{1.23}$$

Let us note that $\int \psi(x, z, \Delta t)dz = 1$, while $\int \psi(x, z, \Delta t)zdz = E(\Delta x|x)$ is the mean value of Δx starting from x, in a time interval Δt; in a similar way

[2] Some readers may wonder what is the meaning of an equation with differentials instead of derivatives. Naively, writing $dx = a(t)df(t)$ seems identical to writing $x(t) - x(t_0) = \int_{t_0}^{t} a(t')df(t')$. The meaning of the latter is a so-called Riemann–Stieltjes integral, which is—roughly speaking—the continuous limit ($N \to \infty$) of an integral sum $\sum_{i=0}^{N-1} a_i[f(t_i + dt) - f(t_i)]$ where $t_i = t_0 + idt$, $dt = (t - t_0)/N$ and a_i is $a(t)$ evaluated at some time $t \in [t_i, t_{i+1}]$. For the differential of the Wiener process, dW, however, this notion must be carefully adapted for two reasons: (1) it is a stochastic function, (2) even if continuous, it is non-differentiable anywhere. This point will be briefly discussed in the following.

$\int \psi(x, z, \Delta t)z^2dz = E(\Delta x^2|x)$ and so on. Therefore, since $P(x, t + \Delta t) = P(x, t) + \partial_t P(x, t)\Delta t$, we have:

$$\begin{aligned}\frac{\partial}{\partial t}P(x, t)\Delta t = &- \frac{\partial}{\partial x}[P(x, t)E(\Delta x(\Delta t)|x)] \\ &+ \frac{1}{2}\frac{\partial^2}{\partial x^2}[P(x, t)E(\Delta x^2(\Delta t)|x)] - \cdots\end{aligned} \tag{1.24}$$

Let us now assume that

$$\lim_{\Delta t \to 0} \frac{E(\Delta x|x)}{\Delta t} = a(x), \ \lim_{\Delta t \to 0} \frac{E(\Delta x^2|x)}{\Delta t} = b(x), \tag{1.25}$$

$$\lim_{\Delta t \to 0} \frac{E(\Delta x^n|x)}{\Delta t} = 0 \text{ for } n \geqslant 3. \tag{1.26}$$

The above assumptions are the mathematical relations to express the condition that Δx is small when Δt is small. With the above assumptions we have the Fokker–Planck equation:

$$\frac{\partial}{\partial t}P(x, t) = -\frac{\partial}{\partial x}[a(x)P(x, t)] + \frac{1}{2}\frac{\partial^2}{\partial x^2}[b(x)P(x, t)]. \tag{1.27}$$

Let us now consider the case in N dimensions:

$$\mathbf{x} = (x_1, x_2, \ldots, x_N) \in \mathbb{R}^N. \tag{1.28}$$

Repeating the same argument presented for the one-dimensional case we have the multi-dimensional Fokker–Planck equation:

$$\frac{\partial P(\mathbf{x}, t)}{\partial t} = -\sum_{n=1}^{N}\frac{\partial}{\partial x_n}[a_n(\mathbf{x})P(\mathbf{x}, t)] + \frac{1}{2}\sum_{n,n'}\frac{\partial^2}{\partial x_n \partial x_{n'}}[b_{nn'}(\mathbf{x})P(\mathbf{x}, t)], \tag{1.29}$$

where

$$\begin{aligned} a_n(\mathbf{x}) &= \lim_{\Delta t \to 0} \frac{\langle \Delta x_n|\mathbf{x}\rangle}{\Delta t}, \\ b_{nn'}(\mathbf{x}) &= \lim_{\Delta t \to 0} \frac{\langle \Delta x_n \Delta x_{n'}|\mathbf{x}\rangle}{\Delta t}, \\ \lim_{\Delta t \to 0} \frac{\langle \Delta x_n^k \Delta x_{n'}^{k'}|\mathbf{x}\rangle}{\Delta t} &= 0 \quad \text{if } k + k' > 2. \end{aligned} \tag{1.30}$$

1.2.3 Relation between Fokker–Planck and Langevin equations

The procedure to obtain the Fokker–Planck equation can appear a sort of dirty game, a truncation of the Taylor series performed with an ad hoc assumption. So it is rather natural to wonder whether the conditions (equations 1.25, 1.26) hold for some interesting stochastic processes. Consider the Langevin equation

$$dx = a(x)dt + \sqrt{b(x)}\,dW, \tag{1.31}$$

where W is the so-called Wiener process, introduced above, i.e. a Gaussian process with the following properties (assuming $W(t_0) = w_0$)

$$\langle W(t)\rangle = w_0,\ \langle W(t)W(t')\rangle = \min\{t - t_0, t' - t_0\} + w_0^2,\ \langle [W(t) - w_0]^2\rangle = t - t_0,$$

which imply for the infinitesimal increment $dW = W(t + dt) - W(t)$:

$$\langle dW\rangle = 0,\ \langle dW^2\rangle = dt.$$

It is easy to show the validity of the following relation

$$W(t) = \int_0^t \xi(t')dt', \tag{1.32}$$

where $\xi(t)$ is the so-called white noise, i.e. a Gaussian process with

$$\langle \xi(t)\rangle = 0,\ \langle \xi(t)\xi(t')\rangle = \delta(t - t').$$

In the physics literature often one finds the equation

$$\frac{d}{dt}x = a(x) + \sqrt{b(x)}\,\xi, \tag{1.33}$$

which is formally equivalent to (1.31) once one identifies ξ with dW/dt. On the other hand W is not differentiable, and therefore, strictly speaking, the expression $\xi = dW/dt$ is meaningless; however, following an established tradition, sometimes we shall write the stochastic differential equations in the form (1.33).

Now, for simplicity, we consider only the case of constant b, postponing the discussion of the general case with $b(x)$ to the following. We have

$$\Delta x(\Delta t) = x(t + \Delta t) - x(t) = \int_t^{t+\Delta t} a[x(t')]dt' + \sqrt{b}\int_t^{t+\Delta t} \xi(t')dt'; \tag{1.34}$$

noting that the mean value of the second integral is zero, for small Δt one has

$$E(\Delta x|x) = \int_t^{t+\Delta t} a[x(t')]dt' = a[x(t)]\Delta t \implies \lim_{\Delta t\to 0}\frac{E(\Delta x|x)}{\Delta t} = a(x). \tag{1.35}$$

Repeating the computation for $\Delta x(\Delta t)^2$ we obtain

$$\left(\int_t^{t+\Delta t} a[x(t')]dt'\right)^2 + b\iint_t^{t+\Delta t} \xi(t')\xi(t'')dt'dt'' + 2\sqrt{b}\iint_t^{t+\Delta t} a[x(t')]\xi(t'')dt'dt''.$$

The average of the third integral is zero[3], the average of the second one can be easily computed using the property $\langle \xi(t')\xi(t'')\rangle = \delta(t - t')$, and therefore we have

[3] This is actually one of the most delicate points of the theory. The result of this integral depends upon the particular choice of stochastic integration, as discussed in the following [8]. In the Ito integration scheme the average of a stochastic integral, working only with the so-called 'non-anticipating functions', $\int a(t')dW(t')$ is always zero.

$$E(\Delta x^2|x) = a^2(x)\Delta t^2 + b\Delta t \implies \lim_{\Delta t \to 0} \frac{E(\Delta x^2|x)}{\Delta t} = \lim_{\Delta t \to 0} a^2(x)\Delta t + b = b. \quad (1.36)$$

In a similar way it is easy to verify that

$$\lim_{\Delta t \to 0} \frac{E(\Delta x^n|x)}{\Delta t} = 0 \text{ for } n > 2.$$

Therefore, we have that there exists a stochastic process, the Langevin equation, whose time evolution is given by the Fokker–Planck equation.

1.2.3.1 A mathematical digression

Let us open a brief mathematical parenthesis, for simplicity of notation in the one-dimensional case. We can write the solution of equation (1.31) as

$$x(t) = x(0) + \int_0^t a[x(t')]\, dt' + \int_0^t \sqrt{b(x(t'))}\, dW(t').$$

So we have to treat integrals of the form

$$I(t) = \int_0^t f[x(t')]\, dW(t'),$$

where f depends on W through $x(t)$ which solves the stochastic differential equation where $W(t)$ appears. At variance with the usual integrals, such a problem is rather delicate. Following the procedure of the theory of the Riemann–Stieltjes integral, it seems natural to introduce $\Delta t = t/N$, $x_n = x(n\Delta t)$, $W_n = W(n\Delta t)$ and define $I(t)$ as the limit for $N \to \infty$ of

$$I_I^N = \sum_{n=0}^{N-1} f(x_n)(W_{n+1} - W_n). \quad (1.37)$$

Another possibility is

$$I_S^N = \sum_{n=0}^{N-1} f\left(\frac{x_{n+1} + x_n}{2}\right)(W_{n+1} - W_n). \quad (1.38)$$

The first procedure is called Ito stochastic integral, and the second Stratonovich stochastic integral; actually the two protocols are not equivalent. This can be seen with an explicit example: in the case $f = W$, a simple computation shows that

$$I_S^N = \frac{W_N - W_0}{2} \to \langle I_S^N \rangle = \frac{t}{2},$$

while

$$\langle I_I^N \rangle = 0.$$

This means that when we consider the formula (1.31) in the case where $b(x)$ is not constant we have to specify how we are interpreting the stochastic integral; once this has been specified we have a well-defined mathematical theory, in particular we can associate with a Langevin equation its correspondent Fokker–Planck equation. Of course there are infinitely many other interpretations of the stochastic integral, since the function $f(x)$ can be computed with other prescriptions. Stratonovich and Ito prescriptions are the most used, since each of them has an important advantage which we recall here, briefly, as a rule of thumb to decide between one and the other. The Stratonovich prescription has the advantage of having the usual differentiation rule $dy(x) = y'(x)dx$ but the disadvantage of having (in general) a complicated result for $\langle f[x(t')]dW(t')\rangle$, since $f(x)$ is computed in x evaluated in the middle of the time interval and therefore could be entangled with the Wiener increment. The Ito prescription, in contrast, has the advantage of always giving $\langle f[x(t')]dW(t')\rangle = 0$, since—in the set of non-anticipating functions, which is sufficient for physical situations— is computed at the beginning of the time interval and is independent of the successive Wiener increment; however, it has the disadvantage of having an unusual differentiation rule, i.e. in general $dy(x) \neq y'(x)dx$. More precisely, in one dimension, if $dx = adt + bdW$, one has

$$\begin{aligned} dy(x, t) &= \frac{\partial y}{\partial t}dt + \frac{\partial y}{\partial x}dx + \frac{1}{2}\frac{\partial^2 y}{\partial x^2}dx^2 \\ &= \frac{\partial y}{\partial t}dt + \frac{\partial y}{\partial x}(adt + bdW) + \frac{1}{2}\frac{\partial^2 y}{\partial x^2}b^2 dt, \end{aligned} \tag{1.39}$$

where in the last passage we have kept for dx^2 only the term b^2dt, while the others are of order $O(dt^{3/2})$ or higher.

Note that with the Ito prescription one can solve the conundrum of the mean squared velocity mentioned at the beginning of this section. In fact, starting from the correct writing of the Langevin equation, equation (1.3) in differential form

$$dv = -\frac{1}{\tau}vdt + bdW, \tag{1.40}$$

with $\tau = m/(6\pi\eta r)$ and some noise amplitude b which has to be determined, if one calls $y(v) = v^2$, then the Ito prescription says that

$$\langle dy\rangle = \langle 2vdv + dv^2\rangle = \left\langle 2v\left[-\frac{v}{\tau}dt + bdW\right] + b^2dt\right\rangle = \left(b^2 - 2\frac{y}{\tau}\right)dt, \tag{1.41}$$

after using the Ito property $\langle vdW\rangle = 0$. The final equation leads at large times to $\langle v^2\rangle \to \tau b^2/2$ and this immediately sets $b^2 = 2\frac{k_B T}{m\tau}$. In contrast, if the Stratonovich prescription is used, the term dv^2 should not appear, but the term $\langle vdW\rangle$ could not be put to zero: anyway, after more complicate calculations, the result of the two integration schemes would be the same.

A final remark on the Ito–Stratonovich prescription is how to go from one scheme to the other. Let us forget for a moment the deterministic part of the Langevin equation and consider the equation $dx = b(x)dW$. If interpreted according to Stratonovich, it means

$$dx = b\left(\frac{x(t) + x(t+dt)}{2}\right)dW(t) = b\left(x(t) + \frac{dx}{2}\right)dW(t)$$
$$= \left\{b[x(t)] + \frac{1}{2}\frac{\partial b}{\partial x}dx\right\}dW(t) \tag{1.42}$$
$$= b[x(t)]dW(t) + \frac{1}{2}\frac{\partial b}{\partial x}b[x(t)]dt,$$

which is now to be integrated with the Ito scheme, since b is computed in $x(t)$. In the last passage we have again used the fact that $dxdW = b(x)dWdW \approx b(x)dt$. In general, therefore, these two stochastic equations—the first integrated with the Stratonovich prescription, the second with the Ito one,

$$dx_S = adt + bdW, \tag{1.43a}$$

$$dx_I = \left(a + \frac{1}{2}\frac{\partial b}{\partial x}b\right)dt + bdW, \tag{1.43b}$$

have the same solution, i.e. $dx_S = dx_I$. The knowledge of this transformation rule makes the Fokker–Planck equation for the Stratonovich equation easy to derive. We discuss this point in the general N-dimensional case below.

In N dimensions one has

$$dx_n = a_n(\mathbf{x})dt + \sum_j \sigma_{nj}(\mathbf{x})\, dW_j(t), \tag{1.44}$$

where the $\{W_j\}$ are independent Wiener processes, i.e. $\langle W_j W_i(t')\rangle = 0$ for $i \neq j$. Using the Ito rule we have the following Fokker–Planck equation

$$\frac{\partial P(\mathbf{x}, t)}{\partial t} = -\sum_{n=1}^{N}\frac{\partial}{\partial x_n}[a_n(\mathbf{x})P(\mathbf{x}, t)] + \frac{1}{2}\sum_{n,n'}\frac{\partial^2}{\partial x_n \partial x_{n'}}[D_{nn'}(\mathbf{x})P(\mathbf{x}, t)] \tag{1.45}$$

where $D_{nn'}(\mathbf{x}) = \sum_k \sigma_{nk}(\mathbf{x})\sigma_{n'k}(\mathbf{x})$.

In order to get the corresponding Stratonovich equation we can exploit the generalization to N dimensions for the Ito–Stratonovich connection shown in equation (1.43):

$$dx_n^S = a_n(\mathbf{x})dt + \sum_j \sigma_{nj}(\mathbf{x})\, dW_j(t), \tag{1.46a}$$

$$dx_n^I = \left[a_n(\mathbf{x}) + \frac{1}{2}\sum_{jk}\frac{\partial b_{nj}}{\partial x_k}b_{kj}\right]dt + \sum_j \sigma_{nj}(\mathbf{x})\, dW_j(t), \tag{1.46b}$$

which leads to the following Fokker–Planck equation for the Stratonovich equation (1.46)

$$\frac{\partial P(\mathbf{x}, t)}{\partial t} = -\sum_{n=1}^{N}\frac{\partial}{\partial x_n}[a_n(\mathbf{x})P(\mathbf{x}, t)] + \frac{1}{2}\sum_{k,j,i}\frac{\partial}{\partial x_i}\left[\sigma_{ik}(\mathbf{x})\frac{\partial}{\partial x_j}(\sigma_{jk}(\mathbf{x})P(\mathbf{x}, t))\right]. \tag{1.47}$$

Let us note that in the case of constant σ_{ik} we obtain the same Fokker–Planck equation.

We remark that when equation (1.44) is a model of a physical system, the integration prescription (e.g. Ito or Stratonovich) is part of the model, i.e. it cannot be chosen independently, otherwise one would have the paradox of a model that describes a single physical system but predicts infinite different behaviours. In contrast, a Fokker–Planck equation is unambiguous and therefore can be the description of a physical system. Once a Fokker–Planck is understood as a correct model for the system, then for the sake of calculations, it can be useful to put it in the corresponding Langevin form, and the exact form depends upon the choice of the integration prescription.

1.2.4 Beyond the Fokker–Planck equation

We can now wonder why to take into account only two terms in the Taylor expansion, i.e. why equation (1.26) holds. Actually one can consider the Kramers–Moyal expansion [8], containing all the terms:

$$\frac{\partial}{\partial t}P(x, t) = \sum_{n=1}^{\infty}\frac{\partial^n}{\partial x^n}[D_n(x)P(x, t)],$$

where

$$D_n(x) = \frac{(-1)^n}{n!}\lim_{\Delta t\to 0}\frac{E(\Delta x^n|x)}{\Delta t}.$$

The Fokker–Planck equation is recovered in the case where $D_n(x) = 0$ for $n \geqslant 3$. In particular, one can show a theorem due to Pawula [9] which states that either the sequence becomes zero at the third term, or all its even terms are positive. In physical situations, the possibility to truncate the expansion to $n = 2$ is related to the existence of a small parameter that guarantees the rapid convergence of the series and the fact that the sum of the terms for $n \geqslant 3$ is really negligible. A fundamental example is the so-called van Kampen size expansion which takes as small parameter the ratio between the size (or mass) of the molecules and the size (mass) of the Brownian particle [7, 10]. It becomes reasonable, then, that in the limit of a very large (massive) particle, its dynamics is described by observables which are continuous in time and therefore obey a stochastic differential equation. This programme is described in section 6.3.2.

More in general, if one considers non-Gaussian noises, instead of the Fokker–Planck equation we have that the evolution of the probability distribution contains also an integral part. Let us discuss for notation simplicity a one-dimensional example

$$\frac{dx}{dt} = a(x)dt + f(t),$$

where now the noise f is

$$f(t) = \sqrt{2D_2}\xi(t) + \zeta(t),$$

where ξ is a Gaussian white noise, and $\zeta(t)$ a Poissonian noise

$$\zeta(t) = \sum_j z_j \delta(t - t_j).$$

The z_j are independent random variables distributed according to a symmetric $\rho(z)$, while the intervals $(t_j - t_{j-1})$ are distributed according to a Poissonian distribution with parameter λ. In such a stochastic process the evolution equation for the probability density is

$$\frac{\partial}{\partial t}P(x, t) = \sum_{n=1}^{2}\frac{\partial^n}{\partial x^n}[D_n(x)P(x, t)] + \lambda\int (P(x - z, t) - P(x, t))\rho(z)\,dz.$$

In the above equation the Fokker–Planck part corresponds to the contribution of the Gaussian part of the noise, while the integral part includes the contribution of the Kramers–Moyal expansion for $n > 2$ and is given by the non-Gaussian noise terms. The integral part shown above is a particular case of the integral that appears in the so-called master equations, describing discontinuous Markov processes (e.g. in a discrete space or in continuous space but with *jumps* [8]), and which are discussed in detail in section 2.1.1.

1.3 Again about the Langevin equation

1.3.1 The Langevin equation from the microscopic dynamics

The Langevin equation for the dynamics of the colloidal particle originally had been introduced in a phenomenological way, but a derivation of Brownian motion from the first principles is an interesting general problem. In chapter 6 we shall discuss this formidable task, which can be accomplished in a rigorous way only in few cases. Here, we briefly discuss the general idea of the procedure to derive the Langevin equation from molecular dynamics.

Consider a system of colloidal particles suspended in a liquid (or in a gas). At the microscopic level we have the canonical coordinates $(\mathbf{Q}_i, \mathbf{P}_i)$ and $(\mathbf{q}_n, \mathbf{p}_n)$ of colloidal particles and solvent molecules, respectively; m is the mass of a solvent molecule, M is the mass of a colloidal particle, and we assume $M \gg m$.

The evolution of such a system is ruled by the Hamiltonian equations; if we are interested in the colloidal subsystem alone, we must remove somehow the degrees of freedom of the solvent particles. Since, in comparison with the solvent molecules, the colloidal particles have a much larger mass, they have a much slower evolution. As a consequence of this timescale separation between the two subsystems, and because of the huge number of the solvent particles, we can conjecture that the fast solvent dynamics can be consistently decoupled from the slow colloid dynamics, by approximating its effects on the big suspended particles by means of an effective force. This latter may be decomposed into a systematic part, of viscous type, and a truly stochastic fluctuating part, described by white noise. In such a limit we recover a stochastic differential equation for the colloidal subsystem.

1.3.2 Langevin equation as a tool for numerical computations

If we are interested in a numerical study of $P(\mathbf{x}, t)$ we can follow two paths. The first one is rather obvious: one can study the Fokker–Planck equation using some numerical methods for partial differential equations. Another possibility is to study the Langevin equation looking at the time evolution of a large number of trajectories and then determine $P(\mathbf{x}, t)$. The procedure is the following:

1. at $t = 0$ we consider $\mathcal{N} \gg 1$ trajectories with initial conditions
$$\mathbf{x}^{(1)}(0), \cdots, \mathbf{x}^{(\mathcal{N})}(0),$$
distributed according to some initial condition $P(\mathbf{x}, 0)$;
2. we follow each trajectory which evolves according to the Langevin equation, where for each particle one has to use independent noisy terms;
3. from the $\mathbf{x}^{(1)}(t),\ldots, \mathbf{x}^{(\mathcal{N})}(t)$ we obtain a proxy of $P(\mathbf{x}, t)$, which approaches the correct value as $\mathcal{N} \to \infty$.

The above procedure has several advantages: at variance with the numerical treatment of the Fokker–Planck equation which can be rather difficult, with several troubles such as numerical instabilities, the approach based on the Langevin equation is quite easy, because it is enough to have a method to solve numerically the Langevin equation. There exist efficient algorithms, based on the generalization to the stochastic case of the usual approach: time is discretized and then one obtains $\mathbf{x}(t + \Delta t)$ starting from $\mathbf{x}(t)$, using a stochastic version of the Euler or Runge–Kutta method. It is easy to see that the Lorentz model, equation (1.14) for small Δt, can be interpreted as a numerical algorithm for the Langevin equation for the velocity.

In order to clarify the difference between the Fokker–Planck or the Langevin approach to the study of a stochastic process, we can mention the problem of the spreading of a pollutant in a steady velocity field $\mathbf{u}(\mathbf{x})$. The pollutant concentration $C(\mathbf{x}, t)$ evolves according to

$$\partial_t C(\mathbf{x}, t) = -\nabla \cdot (\mathbf{u}(\mathbf{x})C(\mathbf{x}, t)) + D\Delta C(\mathbf{x}, t),$$

where Δ denotes the Laplacian, which is nothing but the Fokker–Planck equation associated with the equation for the evolution of a test particle advected by the velocity field and molecular noise:

$$\frac{d}{dt}\mathbf{x} = \mathbf{u}(\mathbf{x}) + \sqrt{2D}\xi.$$

For this problem the approach in terms of Fokker–Planck equation corresponds to the Eulerian point of view, while the Langevin equation is the analogous of the Lagrangian description.

1.4 Further remarks

We conclude this chapter with some digressions which we consider useful for students as well as for curious scientists.

1.4.1 A pedagogical parenthesis on the Lorentz model

Let us consider again the Lorentz model (1.14), where now w_n are independent variables distributed according to a generic $P_w(w)$ which is regular enough, i.e. its cumulants exist. We can show that starting from an arbitrary probability distribution for v_0, $P_v(v; 0)$, then $P_v(v; n) \to P^{(\infty)}(v)$, where $P^{(\infty)}(v)$ can be determined in terms of $P_w(w)$; in particular $P^{(\infty)}(v)$ is Gaussian if and only if $P_w(w)$ is Gaussian.

Let us recall the definition of the characteristic function of a density probability

$$\Phi_x(t) = \langle e^{itx} \rangle = \sum_{k=0}^{\infty} \frac{(it)^k}{k!} \langle x^k \rangle, \tag{1.48}$$

with

$$\ln \Phi_x(t) = \sum_{k=1}^{\infty} \frac{(it)^k}{k!} c_k \tag{1.49}$$

where $c_1, c_2, \ldots$ are the cumulants of the distribution; we recall that the first cumulants can be written in terms of the moments as:

$$c_1 = \langle x \rangle, \quad c_2 = \langle x^2 \rangle - \langle x \rangle^2, \quad c_3 = \langle (x - \langle x \rangle)^3 \rangle, \quad c_4 = \langle (x - \langle x \rangle)^4 \rangle - 3c_2^2.$$

For a Gaussian distribution the cumulants c_k with $k > 2$ are zero.

Now let us introduce

$$\Phi_{v_n}(t) = \langle e^{iv_n t} \rangle, \quad \Phi_{w_n}(t) = \langle e^{iw_n t} \rangle.$$

Reminding that v_n and w_n are independent, for $v_{n+1} = av_n + bw_n$ we have

$$\Phi_{v_{n+1}}(t) = \langle e^{iv_{n+1}t} \rangle = \langle e^{iav_n t} e^{ibw_n t} \rangle = \langle e^{iav_n t} \rangle \langle e^{ibw_n t} \rangle = \Phi_{v_n}(at) \Phi_{w_n}(bt), \tag{1.50}$$

and taking the logarithm

$$\ln \Phi_{v_{n+1}}(t) = \ln \Phi_{v_n}(at) + \ln \Phi_{bw_n}(t). \tag{1.51}$$

Denoting by $C_k^{(n)}$ and c_k the k-th cumulant of v_n and w_n, respectively, we have

$$\sum_{k=1}^{\infty} \frac{(it)^k}{k!} C_k^{(n+1)} = \sum_{k=1}^{\infty} \frac{a^k (it)^k}{k!} C_k^{(n)} + \sum_{k=1}^{\infty} \frac{b^k (it)^k}{k!} c_k. \tag{1.52}$$

This allows us to find a recursive rule for each k:

$$C_k^{(n+1)} = a^k C_k^{(n)} + b^k c_k. \tag{1.53}$$

Repeating the analysis previously done for σ_n in section 1.2.1, we have

$$C_k^{\infty} = \frac{b^k c_k}{1 - a^k}. \tag{1.54}$$

Therefore, if $P_w(w)$ is Gaussian we have

$$C_1^{(n)} \to \frac{b \langle w \rangle}{1 - a} \qquad C_2^{(n)} \to \frac{b^2 \sigma_w^2}{1 - a^2} \qquad C_{k \geqslant 3}^{(n)} \to 0,$$

and therefore $P^{(\infty)}(v)$ must be Gaussian. In the case where $P_w(w)$ is not Gaussian, we have $P(v; n) \to P^{(\infty)}(v)$ which is not Gaussian in general, but, using (1.54), it can be determined by $P_w(w)$. In physical models a and b should scale with Δt and therefore one may have cases where a Gaussian limit distribution is recovered. For instance, the case considered in section 1.2.1, where $a = 1 - \Delta t/\tau$ and $b \sim \Delta t^{1/2}$ (assuming $\langle w \rangle = 0$), corresponds to $C_k^\infty \to 0$ for $\Delta t \to 0$ when $k > 2$, that yields a Gaussian distribution for $P^{(\infty)}(v)$. A general discussion of the continuous-time limit of discrete stochastic processes can be found in [8]. A particularly interesting case of this problem is analysed in section 6.3.2.

1.4.2 Einstein's approach to Brownian motion

Let us briefly recall the original idea of Einstein who based his theory on:

- the Stokes law for the friction force acting on a sphere of radius r with velocity v in a fluid with viscosity η; only for notation simplicity we consider the one-dimensional case: $F_S = -6\pi r\eta v$;
- the Van't Hoff law for the osmotic pressure.

As already mentioned, the simple and ingenious idea was to assume that the statistical mechanics holds even for macroscopic objects whose size is on the order of a few microns. Let us consider particles of mass M in a liquid in the presence of a constant gravity field along the z-axis. From the barotropic formula (as well as the Van't Hoff law), we have:

$$\rho(z) = \rho(0)e^{-Mgz/k_B T}, \tag{1.55}$$

where ρ is the density. Each particle feels the force

$$F_S - Mg,$$

which is zero for

$$v = v^* = -\frac{Mg}{6\pi\eta r}.$$

At the height z we have a down forward current density

$$j_-(z) = v^*\rho(z) = -\frac{Mg}{6\pi\eta r}\rho(z). \tag{1.56}$$

Since the density is not constant we have an up forward current density $j_+(z)$ due to the motion of particles from a region of larger density to a region of lower density:

$$j_+(z) = -D\frac{\partial}{\partial z}\rho(z, t).$$

From equation (1.55) one has

$$j_+(z) = D\frac{Mg}{k_B T}\rho(z, t), \tag{1.57}$$

and in the case of stationarity one has a balance of the two fluxes, i.e. $j_+ + j_- = 0$. Therefore, using equations (1.56) and (1.57) one has

$$D = \frac{k_B T}{6\pi\eta r}.$$

It is straightforward to check that the result does not change if instead of the gravity force $-Mg$ one considers colloidal particles in a generic potential $U(z)$ and therefore a force $f(z) = -dU(z)/dz$.

1.4.3 Remarks on Perrin's experiments

Jean Baptiste Perrin was a sophisticated experimental physicist, and had the great merit to verify in a very clever way the theoretical prediction of Einstein, and determine the value of the Avogadro number from the diffusion coefficient of the Brownian motion. In his laboratory he was able to overcome enormous practical difficulties, in particular, at that time it is was not so easy to have little spheres of a given mass and/or radius. Studying, with the help of a microscope, the statistical features of grains with diameters in the range [0.2–5.5] microns, after a long series of experiments he was able to verify the relation (1.1), and was awarded the Nobel Prize in Physics in 1926, the official motivation being *'for his work on the discontinuous structure of matter'*.

Once one assumes, according to Einstein, that the same statistical laws are valid for molecules as well as for colloidal particles, one has a powerful method to determine the Avogadro number. For instance Perrin verified the validity of the barotropic formula for grains of size of the order of micron and mass M (order 10^{-12}–10^{-11} g) in water at temperature T, studying the density of the grains at varying height with a microscope. In the case of high dilution, the interaction among the colloidal particles can be neglected and we can assume the barotropic formula for perfect gases in an external field, equation (1.55). Let us note that the value of the quantity $\zeta = k_B T/(Mg)$ decreases with the mass M; for colloidal particles ζ is small enough so that the behaviour of $\rho(z)$ can be observed in the laboratory. The experimental results of Perrin gave further evidence of Einstein's idea about the validity of the statistical mechanics of macroscopic objects, as well as another way to determine the Avogadro number from the value of ζ.

1.4.4 The legacy of Brownian motion

Brownian motion is not only an important historical and technical aspect of modern physics, but it has been also the starting point for the development of the mathematical theory of stochastic processes; in addition, Brownian motion has played a role in the recent progress in biophysics [11] and finance [12].

The Langevin equation had been the first non-trivial example of stochastic process. Such a mathematical field, now largely used in physics, chemistry, biology and applied sciences, was developed in a systematic way in the 30s of the last century, mainly by Kolmogorov with the formalization of the Fokker–Planck and master equations for the continuous-time Markov processes [13]. A seminal work, which started from the mathematical description of Brownian motion, is due to Wiener who introduced the idea of path integral which plays an essential role for the Feynman formulation of quantum mechanics [14].

A few years before the paper of Einstein, the French mathematician Louis Jean-Baptiste Alphonse Bachelier in his doctoral dissertation *Théorie de la Spéculation* (1900) proposed that the price of a stock behaves as a particle performing Brownian motion. Recently, the stochastic processes, in particular the differential stochastic equations, have received vast attention from the financial community. The celebrated Black and Scholes theory [15] for the option pricing is, from a mathematical point of view, nothing but an application of the Langevin equation. It is interesting to note that, despite their completely different origins, the phenomenon of Brownian motion and the option pricing in finance are described by the same mathematical formalism.

At first glance the explanation of Brownian motion from kinetic theory can appear a perfect example of the success of reductionism: an explanation of complicated visible phenomena in terms of simple invisible elements. Actually, this point of view cannot be considered correct: Brownian motion theory was derived by Einstein without him being really familiar with the details of the experiments that had been performed for several decades, and the real interest of Einstein was not to describe Brownian motion from the detailed knowledge of the motion of atoms. He started from the bold assumption that particles, even if of macroscopic size, had to obey the laws of statistical mechanics, and then he found an expression for the diffusion coefficient which can be compared with the experiments. In the Einstein approach to Brownian motion, we do not find a path from the microscopic to the macroscopic world; on the contrary, different levels of description were mixed up several times. The force acting on pollen grains was assumed to be made up of two contributions: one of macroscopic character and one of microscopic origin; in particular, the macroscopic term was given by the Stokes law, assumed as a phenomenological result and not derived from the microscopic realm.

It is interesting to note that, sometimes, Brownian motion, one of the major chapters of modern physics, has been so poorly understood, and even misconstrued as incompatible with the second law of thermodynamics, by leading philosophers of science, for instance Popper [16] and Feyerabend [17], who stated that Brownian motion is a serious problem for the second law.

We conclude with a few words about present times: today modern technologies allow scientists easy access to the study of thermal fluctuations even in small systems such as colloidal suspensions. This kind of study is rather important in nanoscience, and fluctuations can have a crucial role in noise-assisted transport mechanisms, also called Brownian motors [18] (see chapter 8). As an example, molecular motors in biophysics use chemical energy to lock in the fluctuations due to thermal noise, and move the molecular machine forward.

Therefore, we can say that the paper of Einstein on Brownian motion, which sometimes had been considered the less relevant among those of the *annus mirabilis*, with its huge impact on the physics of colloidal particles, other forms of soft matter, and of biophysical systems, had its revenge on the subatomic world.

References

[1] Robert B 1827 A brief account of microscopical observations *Phil. Mag.* **4** 161

[2] Cantoni G 1867 Su alcune condizioni fisiche dell'affinità e sul moto Browniano *Il Nuovo Cimento (1855–1868)* **27** 156–67

[3] Einstein A 1905 Über die von der molekularkinetischen theorie der wärme geforderte bewegung von in ruhenden flüssigkeiten suspendierten teilchen *Ann. Phys., Lpz.* **4** 549–60

[4] Langevin P 1908 Sur la théorie du mouvement Brownien *Compt. Rendus* **146** 530–3

[5] Perrin J 1909 Mouvement Brownien et réalité moléculaire *Ann. Chim. Phys.* **18** 1–114

[6] Pauli W 1973 *Pauli Lectures on Physics, Volume 4: Statistical Mechanics* (Cambridge: Cambridge University Press)

[7] van Kampen N G 1992 *Stochastic Processes in Physics and Chemistry* (Amsterdam: Elsevier)

[8] Gardiner C 2009 *Stochastic Methods* (Berlin: Springer)

[9] Risken H 1996 *Fokker-Planck equation* (Berlin: Springer)

[10] van Kampen N G 1961 A power series expansion of the master equation *Can. J. Phys.* **39** 551–67

[11] Frey E and Kroy K 2005 Brownian motion: a paradigm of soft matter and biological physics *Ann. Phys., Lpz.* **517** 20–50

[12] Karatzas I, Shreve S E, Karatzas I and Shreve S E 1998 *Methods of Mathematical Finance* (Berlin: Springer)

[13] Kolmogoroff A 1931 Über die analytischen methoden in der wahrscheinlichkeitsrechnung *Math. Ann.* **104** 415–58

[14] Wiener N 1976 *Collected Works* **vol 1** (Cambridge, MA: MIT Press)

[15] Black F and Scholes M 1974 From theory to a new financial product *J. Finance* **29** 399–412

[16] Popper K R 1957 Irreversibility; or, entropy since 1905 *Br. J. Phil. Sci.* **8** 151–5

[17] Feyerabend P K 2010 *Against Method* (London: Verso Books)

[18] Reimann P 2002 Brownian motors: noisy transport far from equilibrium *Phys. Rep.* **361** 57–265

Chapter 2

The Boltzmann equation

> Since in the differential equations of mechanics themselves there is absolutely nothing analogous to the Second Law of thermodynamics, the latter can be mechanically represented only by means of assumptions regarding initial conditions.
>
> Ludwig Boltzmann

A few decades before the reality of atoms was confirmed by the work of Einstein and Perrin (see chapter 1), Ludwig Boltzmann, one of the founding figures of theoretical physics, laid the bases of statistical mechanics to explain the behaviour of macroscopic matter [1]. Boltzmann's framework rested on the erratic and unpredictable motion of countless atoms and molecules, which, while governed by the laws of mechanics, could only be studied through probabilistic methods. He grasped the molecular basis of entropy, a concept already central to thermodynamics, and provided a 'statistical mechanics proof' for its increase in isolated systems. This groundbreaking insight was encapsulated in his development of the Boltzmann equation, which describes the evolution of the single-particle probability distribution for molecules in a gas. As a cornerstone of nonequilibrium statistical mechanics, this equation captures the dynamics of probabilistic descriptions of systems, making it more broadly applicable than equilibrium statistical mechanics. In this chapter, we will present these ideas in modern terms.

First we give a succinct description of the essence of the Boltzmann equation and the H-theorem, which stems from the microscopic time reversal invariance, concluding with a brief discussion of the H-theorem for Markov processes. Then we sketch the usual derivation of the Boltzmann equation from Hamiltonian mechanics, underlining its most delicate passages. In the third section we discuss the two main objections that have been put forward against the consistency between classical mechanics and the Boltzmann equation and its main consequence, that is the H-theorem: we show that both arguments do not constitute real obstacles for the validity of the theory. In the

doi:10.1088/978-0-7503-6229-0ch2

fourth section we discuss a few kinetic models that widen the application of the Boltzmann equation to other conservative (e.g. finance or systems with negative temperatures) and non-conservative (e.g. granular) systems. In the final fifth section we discuss further remarks about irreversibility and typicality of entropy growth, with an emphasis on the Ehrenfest urn model that clarifies several issues.

2.1 Boltzmann equation in a nutshell

Perhaps the simplest way to grasp the meaning of the Boltzmann equation is to consider the case where the states of the system are discrete. Each state can represent, for instance, a range of the velocities that one molecule of a gas can assume. Let us call $P_i(t)$ the probability of observing a state i at time t. Another simplifying assumption is that interactions are pairwise, i.e. a molecule changes its state (its velocity) because it interacts with another molecule. Pairwise interactions require the introduction of another probability, $P_{ij}(t)$, that is the joint probability of observing at time t two molecules in the states i and j, respectively. Formally, we can write for the evolution in time of $P_i(t)$

$$\frac{dP_i(t)}{dt} = \sum_{k,l,j} \left[W^{(2)}_{(k,\,l)\to(i,\,j)} P_{kl}(t) - W^{(2)}_{(i,\,j)\to(k,\,l)} P_{ij}(t) \right]. \tag{2.1}$$

Such an equation contains on the right-hand side two terms that contribute to the variation in time of $P_i(t)$: the first one is the so-called *gain* term, the second one is the *loss* term. In both terms the function $W^{(2)}_{(k,\,l)\to(i,\,j)}$ represents the rate (per unit time) of transition—for a couple of molecules—from states k, l to states i, j, due to the physical interaction (e.g. collisions). The above equation is quite straightforward as it only expresses the trivial fact that, in the presence of only pairwise interactions, if a collision transforms the states k, l of two interacting molecules in states i, j then the relative number of molecules in state i increases. At the same time, if a couple of molecules in states i, j collide, the relative number of molecules in state i decreases. Of course there are collisions that could transform one state of a molecule in the same state (for instance in a collision that perfectly exchanges states), but these events perfectly cancel between the gain and loss terms.

The above 'simple' equation becomes the Boltzmann equation under the assumption of molecular chaos, which is a fundamental physical idea introduced by Boltzmann. He conjectured that under conditions of strong diluteness, the molecules of a gas collide as if they never collided before and therefore cannot be correlated. In probabilistic terms this is equivalent to saying that $P_{ij}(t) = P_i(t)P_j(t)$. Assuming the validity of the molecular chaos assumption, the evolution equation becomes the following Boltzmann equation:

$$\frac{dP_i(t)}{dt} = \sum_{k,l,j} \left[W^{(2)}_{(k,\,l)\to(i,\,j)} P_k(t)P_l(t) - W^{(2)}_{(i,\,j)\to(k,\,l)} P_i(t)P_j(t) \right]. \tag{2.2}$$

The above equation is closed in the sense that, given the transition rates, it only contains the probability $P_i(t)$. In principle it should be solved by plugging initial conditions and looking at the evolution and asymptotic fate of the probability. Anyway, it represents—in general—a hard problem to be solved, as it is nonlinear in $P_i(t)$, exactly as a consequence of the molecular chaos assumption. The precise form of the transition rates $W^{(2)}_{(k,\,l)\to(i,\,j)}$ depends upon the particular kind of interaction, for instance upon the pairwise potential that represents the reciprocal forces between molecules. In the following we will sketch a derivation of the Boltzmann equation from Hamiltonian equations, and in the case of hard spheres we will show the particular form of the transition rates. Anyway, for the following discussion, the explicit form of transition rates is irrelevant. Only a fundamental physical property is necessary, which is a consequence of the invariance under time reversal, and it reads

$$W^{(2)}_{(k,\,l)\to(i,\,j)} = W^{(2)}_{(i,\,j)\to(k,\,l)}. \tag{2.3}$$

Thanks to this assumption, equation (2.2) takes the form

$$\frac{dP_i(t)}{dt} = \sum_{k,l,j} W^{(2)}_{(k,\,l)\to(i,\,j)}[P_k(t)P_l(t) - P_i(t)P_j(t)]. \tag{2.4}$$

Using such a structure it is easy to show the H-theorem, i.e. that the $H(t)$ function

$$H(t) = \sum_i P_i(t)\ln P_i(t) \tag{2.5}$$

is monotonically decreasing:

$$\frac{dH(t)}{dt} \leqslant 0, \tag{2.6}$$

and reaches its minimum, i.e. $\frac{dH(t)}{dt} = 0$, when $P_i(t) = P_i^{\infty}$, where the P_i^{∞} corresponds to the Maxwell–Boltzmann distribution.

It is useful, to slightly simplify the notation, to introduce the variable $A_{i,j,k,l} = W^{(2)}_{(i,\,j)\to(k,\,l)}$. Then we have, from equation (2.4)

$$\begin{aligned}\frac{dH(t)}{dt} &= \sum_i \frac{dP_i(t)}{dt}\ln P_i(t) + \sum_i P_i(t)\frac{dP_i(t)}{dt}\frac{1}{P_i(t)}\\ &= \sum_{i,k,l,j} A_{k,l,i,j}[P_k(t)P_l(t) - P_i(t)P_j(t)]\ln P_i(t),\end{aligned} \tag{2.7}$$

where we have used the conservation of the normalization $\sum_i P_i(t) = 1$ to remove the second sum in the second equality.

The symmetry between collisions $(k, l) \to (i, j)$ and $(k, l) \to (j, i)$ implies the following identities $A_{k,l,i,j} = A_{i,j,k,l} = A_{l,k,j,i} = A_{l,k,i,j}$, so that

$$\frac{dH(t)}{dt} = \frac{1}{2}\sum\nolimits_{i,j,k,l} A_{i,j,k,l}\{P_k(t)P_l(t) - P_i(t)P_j(t)\}\ln[P_i(t)P_j(t)],$$

and by means of the symmetry between $(k, l) \to (i, j)$ and $(i, j) \to (k, l)$ one gets

$$\frac{dH(t)}{dt} = -\frac{1}{4}\sum_{i,j,k,l} A_{i,j,k,l}\{P_i(t)P_j(t) - P_k(t)P_l(t)\}\{\ln[P_i(t)P_j(t)] - \ln[P_k(t)P_l(t)]\}.$$

We notice that $A_{i,j,k,l} > 0$. Furthermore, since $(\ln x - \ln y)(x - y) \geqslant 0$ for every $x > 0$ and $y > 0$, we finally get $\frac{dH(t)}{dt} \leqslant 0$. The equality holds only when $P_i(t) \to P_i^\infty$ with the condition $P_i^\infty P_j^\infty = P_k^\infty P_l^\infty$ where $i, j \to k, l$ is a physically allowed collision. Taking the logarithm of this equality and denoting as $\phi_i(t) = \ln P_i^\infty$, we get the stationarity condition

$$\phi_i + \phi_j = \phi_k + \phi_l; \tag{2.8}$$

therefore ϕ_i is a conserved quantity in the collisions. The collision conserves energy $m_i v_i^2 + m_j v_j^2$ (where m_i is the mass of the ith particle) and the momentum $m_i v_i + m_j v_j$; this last conservation law is not relevant since in a closed system the total momentum is zero. The stationary distribution is therefore the Maxwell–Boltzmann distribution

$$P_i^\infty(v) = \mathcal{N} e^{-\beta \frac{m_i v^2}{2}}, \tag{2.9}$$

where β is the inverse temperature and $\mathcal{N}$ is a normalization factor.

The above proof is the standard one contained in any textbook and is due to Ludwig Boltzmann.

2.1.1 Master equations

Boltzmann equations can be written down also for mixtures of particles with different properties, e.g. different masses. For instance, for the evolution of the single-particle distribution $P_i(t)$ for a massive tracer particle in a gas of smaller particles whose single-particle distribution is now denoted as $p_i(t)$, one can write

$$\frac{dP_i(t)}{dt} = \sum_{k,l,j}\Big[W^{(\text{tr-gas})}_{(k,l)\to(i,j)}P_k(t)p_l(t) - W^{(\text{tr-gas})}_{(i,j)\to(k,l)}P_i(t)p_j(t)\Big], \tag{2.10}$$

$$\frac{dp_i(t)}{dt} = \sum_{k,l,j}\Big[W^{(2)}_{(k,l)\to(i,j)}p_k(t)p_l(t) - W^{(2)}_{(i,j)\to(k,l)}p_i(t)p_j(t)\Big] + \tag{2.11}$$

$$\sum_{k,l,j}\Big[W^{(\text{gas-tr})}_{(k,l)\to(i,j)}p_k(t)P_l(t) - W^{(\text{gas-tr})}_{(i,j)\to(k,l)}p_i(t)P_j(t)\Big]. \tag{2.12}$$

The first equation concerns the tracer particle which only feels collisions with the gas particles, ruled by the rate $W^{(\text{tr-gas})}_{(i,j)\to(k,l)}$. The second equation governs the evolution of the gas distribution, which merges the effect of the collisions among gas particles (ruled by $W^{(2)}_{(i,j)\to(k,l)}$) as well as of the collisions among gas particles and the tracer (with transition rate $W^{(\text{gas-tr})}_{(i,j)\to(k,l)}$). Assuming that the dynamics of the tracer probability distribution is much slower than the gas probability distribution and that the tracer's influence upon the gas particles is negligible, one can assume that the gas probability is stationary, $p_i(t) = p_i^\infty$, and cast the first equation (2.10) into a linear uncoupled equation which goes under the name of master equation:

$$\frac{dP_i(t)}{dt} = \sum_{k\neq i}\left[W^{(1)}_{k\to i}P_k(t) - W^{(1)}_{i\to k}P_i(t)\right], \tag{2.13}$$

where

$$W^{(1)}_{i\to k} = \sum_{l,j} W^{(\text{tr}-\text{gas})}_{(i,\,j)\to(k,\,l)}p_j(t). \tag{2.14}$$

In the first chapter of this book we discussed stochastic processes with the Markovian property, whose probability are generally described by the same master equation. The Fokker–Planck equation, describing processes which are continuous in space and time, is a particular form of master equation. Its precise form can be deduced from the continuous space version of equation (2.13)

$$\frac{dP(x,\,t)}{dt} = \int dx'[W^{(1)}(x' \to x)P(x',\,t) - W^{(1)}(x \to x')P(x,\,t)], \tag{2.15}$$

through a perturbative procedure which consists of two steps: a first formal rewriting into the so-called Kramers–Moyal expansion, already seen in section 1.2.4, that is a series with infinite terms with growing power of the jump distance $x' - x$; a second physical step where the tracer particle is assumed to have a much larger mass M with respect to the mass of the gas particles $m \ll M$ and therefore each term of the Kramers–Moyal series is expanded in powers of m/M (see section 6.3.2 and chapter 8 for some examples). In the limit $m/M \to 0$ then one can retain the two most meaningful terms of the expansion and the Fokker–Planck equation is obtained, see section 1.2.2 in chapter 1.

2.1.1.1 H-*Theorem for master equations*

It is clear that the master equation has a structure which resembles the structure of the Boltzmann equation, with a gain and a loss term, however it is *linear*, rather than nonlinear as the Boltzmann equation is. An H-theorem also holds under rather general hypothesis for processes governed by a master equation. However, the H-function is different from the Boltzmann one and must be informed about the stationary solution. In this case indicating with $\{\Pi_j\}$ the invariant probabilities, which are the solutions of the following equation:

$$\Pi_i = \frac{1}{\gamma_i}\sum_{k\neq i} W^{(1)}_{k\to i}\Pi_k,$$

where

$$\gamma_i = \sum_{k\neq i} W^{(1)}_{i\to k},$$

one has that the function

$$H_K(t) = \sum_i P_i(t) \ln \frac{P_i(t)}{\Pi_i}$$

is non-increasing, i.e.

$$\frac{dH_K(t)}{dt} \leqslant 0, \tag{2.16}$$

and attains its minimum when $P_j = \Pi_j$, where $H_K = 0$. Let us note that $-H_K$ is also known as Kullback–Leibler or relative entropy between P_i and Π_i.

Let us introduce the variables $f_i(t) = P_i(t)/\Pi_i$ and $B_{ik} = W^{(1)}_{k\to i}\Pi_k$. It is also useful to define the function $F(x) = x \ln x$, so that $F'(x) = 1 + \ln x$. It is easy to verify that

$$\frac{dH_K}{dt} = \sum_{i,k} F'(f_i)(B_{ik}f_k - B_{ki}f_i) = \sum_{i,k} B_{ik}[f_k F'(f_i) - f_k F'(f_k)], \tag{2.17}$$

where we have exchanged the indexes in the second part of the sum, in order to collect B_{ik}. Then one notices that for a set of numbers ψ_i the following relation holds

$$\sum_{i,k} B_{ik}(\psi_i - \psi_k) = 0, \tag{2.18}$$

which is a consequence of stationarity for Π_i. Adding equation (2.18) to equation (2.17), and choosing $\psi_n = F(f_n) - f_n F'(f_n)$, one gets

$$\frac{dH_K}{dt} = \sum_{i,k} B_{ik}[(f_k - f_i)F'(f_i) + F(f_i) - F(f_k)]. \tag{2.19}$$

Now, since $F''(x) > 0$, it is immediate to see that

$$[(f_k - f_i)F'(f_i) + F(f_i) - F(f_k)] \leqslant 0, \tag{2.20}$$

and therefore $\frac{dH_K}{dt} \leqslant 0$.

It is interesting to notice that in the above proof we have not used any assumption of micro-reversibility, such as the one used in the proof of the H-theorem for the Boltzmann equation.

In the case of continuous variables, where the probability $P_i(t)$ is replaced by the probability density $P(x, t)$ (where x is a state in a continuous space, e.g. velocity of a particle) and the sums with the integrals, the previously introduced entropy functionals become

$$H(t) = \int P(x, t) \ln P(x, t) dx, \tag{2.21}$$

$$H_K(t) = \int P(x, t) \ln \frac{P(x, t)}{\Pi(x)} dx. \tag{2.22}$$

Let us notice that H_K has an intrinsic property: switching to a new variable $y(x)$, if the transformation between x and y is invertible, H_K is invariant, whereas H is not.

Finally, one may wonder if H_K is non-increasing also in the case of the Boltzmann equation, equation (2.4). It is immediate to verify that this is true, indeed:

$$H_K(t) = H(t) - \sum_i P_i(t) \ln \Pi_i,$$

and $\ln \Pi_i = \ln P_i^\infty = \phi_i$ being a linear combination of conserved quantities, the second term of the right-hand side is constant, so that $dH_K/dt = dH/dt$.

2.2 Boltzmann equation for hard spheres

Parts of this section have been reprinted from [21] copyright (2015), with permission from Springer Nature.

In this section we rapidly sketch the main passages needed to go from the Hamiltonian mechanics of hard spheres to the Boltzmann equation. This is necessary, for instance, when one needs the explicit form of the scattering rates $W^{(2)}_{(i,j)\to(k,l)}$. Of course this is not a rigorous derivation, which would be a lengthy mathematical procedure, not particularly pedagogical from a physical point of view, but we attempt to stress the fundamental assumptions that underly the construction of a probabilistic description from a deterministic one [2].

We consider a system of N identical hard spheres (of diameter σ and mass m): the phase space is therefore a $6N$-dimensional space where the coordinates are the $3N$ components of the N position vectors of the sphere centres $\mathbf{r}_i$ and the $3N$ components of the N velocities $\mathbf{v}_i$. The state of the system is represented by a point in this space, $\mathbf{z}$. If the positions $\mathbf{r}_i$ of the spheres are restricted in a space region Ω, then the full phase space $\mathbf{D}$ is given by the product $\Omega^N \times \mathfrak{R}^{3N}$. If the state is not known with absolute accuracy, or if a distribution of initial conditions is considered, then one must introduce a probability density $P(z, t)$ which is defined by

$$\mathrm{Prob}(z \in \mathbf{D} \text{ at time } t) = \int_{\mathbf{D}} P(\mathbf{z}, t) d\mathbf{z}, \tag{2.23}$$

where $d\mathbf{z}$ is the Lebesgue measure in phase space; therefore, we assume that the probability is a measure absolutely continuous with respect to it.

In a general system (not necessarily made of hard spheres) with conservative and additive interactions, the force between the particles (ij), at distance vector $\mathbf{r}_{ij}$ with modulus r_{ij}, is $\mathbf{F}_{ij} = -\partial U(r_{ij})/\partial \mathbf{r}_{ij}$, so that the time evolution of the probability in phase space is dictated by the Liouville equation

$$\frac{\partial}{\partial t} P(\mathbf{z}, t) = \left(-\sum_i L_i^0 + \sum_{i<j} \Theta(ij) \right) P(\mathbf{z}, t), \tag{2.24}$$

which is an expression of the incompressibility of the flow in phase space, with

$$L_i^0 = \mathbf{v}_i \cdot \frac{\partial}{\partial \mathbf{r}_i}, \tag{2.25a}$$

$$\Theta(ij) = \frac{1}{m}\frac{\partial U(r_{ij})}{\partial \mathbf{r}_{ij}} \cdot \left(\frac{\partial}{\partial \mathbf{v}_i} - \frac{\partial}{\partial \mathbf{v}_j}\right). \tag{2.25b}$$

Two hard spheres in 3D (hard disks in 2D, hard rods in 1D) of diameters σ_1 and σ_2 interact by means of a discontinuous potential $U(r)$ of the form:

$$U(r) = 0 \quad (r > \sigma_{12}) \tag{2.26a}$$

$$U(r) = \infty \quad (r < \sigma_{12}), \tag{2.26b}$$

where $\sigma_{12} = (\sigma_1 + \sigma_2)/2 = r_m$ is the distance of the centres of the spheres at contact. The potential in equation (2.26) can be taken as a definition of hard spheres systems. In this case it can be shown that there is no contraction of phase space at collision, i.e.

$$P(\mathbf{z}', t) = P(\mathbf{z}, t), \tag{2.27}$$

where $\mathbf{z}'$ and $\mathbf{z}$ are the phase space points before and after a collision. It is important to stress that $\mathbf{z}' \neq \mathbf{z}$: a collision represents a time discontinuity in the velocity section of phase space. At this discontinuity the total momentum and the total energy are conserved. From the technical point of view, in the derivation of the Boltzmann equation, the collisions events $\mathbf{z} \to \mathbf{z}'$ are considered as boundary conditions for the Liouville equation (2.24).

The first key passage to go from the Liouville equation to the Boltzmann equation is to introduce the reduced (marginal) probability densities P_s:

$$\begin{aligned} &P_s(\mathbf{r}_1, \mathbf{v}_1, \mathbf{r}_2, \mathbf{v}_2, \ldots, \mathbf{r}_s, \mathbf{v}_s, t) = \\ &\int_{\Omega^{N-s} \times \Re^{3(N-s)}} P(\mathbf{r}_1, \mathbf{v}_1, \mathbf{r}_2, \mathbf{v}_2, \ldots, \mathbf{r}_N, \mathbf{v}_N, t) \prod_{j=s+1}^{N} d\mathbf{r}_j d\mathbf{v}_j. \end{aligned} \tag{2.28}$$

Some computation that takes into account the collisional boundary conditions allows to derive—from the Liouville equation—the equation for the reduced probability:

$$\frac{\partial P_s}{\partial t} + \sum_{i=1}^{s} \mathbf{v}_i \cdot \frac{\partial P_s}{\partial \mathbf{r}_i} = (N - s)\sum_{i=1}^{s} \int P_{s+1} \mathbf{V}_i \cdot \hat{\mathbf{n}}_i d\sigma_i d\mathbf{v}_*, \tag{2.29}$$

where $\mathbf{V}_i = \mathbf{v}_i - \mathbf{v}_*$, $\hat{\mathbf{n}}_i = (\mathbf{r}_i - \mathbf{r}_*)/\sigma$ and the arguments of P_{s+1} are $(\mathbf{r}_1, \mathbf{v}_1, \mathbf{r}_2, \mathbf{v}_2, \ldots, \mathbf{r}_s, \mathbf{v}_s, \mathbf{r}_*, \mathbf{v}_*, t)$. Integrations in equation (2.29) are performed over the 1-particle velocity space $\mathbb{R}^3$ and over the sphere S^i (given by the condition $|\mathbf{r}_i - \mathbf{r}_*| = \sigma$) with surface elements $d\sigma_i$. Equation (2.29) is complemented by reflecting boundary conditions on the reduced boundary surface Λ_s which extends to the entire space $\mathbb{R}^{3(N-s)}$ for the velocity variables, and to Ω^{N-s} deprived of the spheres $|\mathbf{r}_i - \mathbf{r}_j| < \sigma$ $(i = 1, \ldots, N, i \neq j)$ with respect to the position variables. Equation (2.29) states that the evolution of the reduced probability density P_s is governed by the free evolution operator of the s-particle dynamics, which appears in the left hand side, with corrections due to the effect of the interaction with the remaining $(N - s)$

particles. The effect of this interaction is described by the right-hand side of this equation.

Usually equation (2.29) is written in a different form, obtained using some symmetries of the problem,

$$\frac{\partial P_s}{\partial t} + \sum_{i=1}^{s} \mathbf{v}_i \cdot \frac{\partial P_s}{\partial \mathbf{r}_i} = (N-s)\sigma^2 \sum_{i=1}^{s} \int_{\mathbb{R}^3} \int_{S_+} (P'_{s+1} - P_{s+1}) |\mathbf{V}_i \cdot \hat{\mathbf{n}}| d\hat{\mathbf{n}} d\mathbf{v}_*, \tag{2.30}$$

where S^+ is the positive hemisphere and

$$P'_{s+1} = P_{s+1}(\mathbf{r}_1, \mathbf{v}_1, \dots, \mathbf{r}_i, \mathbf{v}_i - \hat{\mathbf{n}}_i(\hat{\mathbf{n}}_i \cdot \mathbf{V}_i), \dots \mathbf{r}_s, \mathbf{v}_s, \mathbf{r}_i - \sigma\hat{\mathbf{n}}_i, \mathbf{v}_* + \hat{\mathbf{n}}_i(\hat{\mathbf{n}}_i \cdot \mathbf{V}_i)). \tag{2.31}$$

The system of equations (2.30) is usually called the BBGKY hierarchy for the hard-sphere gas (from Bogoliubov, Born, Green, Kirkwood and Yvon, sometimes called simply Bogoliubov hierarchy). The hierarchy is closed only when the number of molecules N is finite, and in such a case it is nothing but another way to write the Liouville equation.

In a rarefied gas N is a very large number and σ is very small; let us say, to fix ideas, that we have a box whose volume is 1 cm^3 at room temperature and atmospheric pressure. Then $N \simeq 10^{20}$ and $\sigma \simeq 10^{-8} cm$ and (from equation (2.30)) for small s we have $(N-s)\sigma^2 \simeq N\sigma^2 \simeq 1m^2$; at the same time the difference between $\mathbf{r}_i$ and $\mathbf{r}_i + \sigma\hat{\mathbf{n}}$ can be neglected and the volume occupied by the particles ($N\sigma^3 \simeq 10^{-4}$ cm^3) is very small so that the collision between two selected particles is a rather rare event. In this spirit, the Boltzmann–Grad limit has been suggested as a procedure to obtain a closure for equation (2.30): $N \to \infty$ and $\sigma \to 0$ in such a way that $N\sigma^2$ remains finite. The total number of collisions in the unit of time (for volume and typical velocities both of order 1) is proportional to the total scattering cross-section multiplied by N, which for a system of hard spheres gives $N\pi\sigma^2$. The Boltzmann–Grad limit, therefore, states that the single-particle collision probability must vanish, but the total number of collisions remains of order 1. Within this limit, the BBGKY hierarchy reads:

$$\frac{\partial P_s}{\partial t} + \sum_{i=1}^{s} \mathbf{v}_i \cdot \frac{\partial P_s}{\partial \mathbf{r}_i} = N\sigma^2 \sum_{i=1}^{s} \int_{\mathbb{R}^3} \int_{S_+} (P'_{s+1} - P_{s+1}) |\mathbf{V}_i \cdot \hat{\mathbf{n}}| d\hat{\mathbf{n}} d\mathbf{v}_*, \tag{2.32}$$

where the arguments of P'_{s+1} and of P_{s+1} are the same as above, except that the position of the $(s+1)$-th particle ($\mathbf{r}'_*$ and $\mathbf{r}_*$) is equal to $\mathbf{r}_i$ (as $\sigma \to 0$). Equation (2.32) gives a complete description of the time evolution of a Boltzmann gas (i.e. the ideal gas obtained in the Boltzmann–Grad limit), usually called the Boltzmann hierarchy.

Finally, the Boltzmann equation is obtained if the molecular chaos assumption is taken into account:

$$P_2(\mathbf{r}_1, \mathbf{v}_1, \mathbf{r}_2, \mathbf{v}_2, t) = P_1(\mathbf{r}_1, \mathbf{v}_1, t) P_1(\mathbf{r}_2, \mathbf{v}_2, t), \tag{2.33}$$

for particles that are about to collide (that is, when $\mathbf{r}_2 = \mathbf{r}_1 - \sigma\hat{\mathbf{n}}$ and $\mathbf{V}_{12} \cdot \hat{\mathbf{n}} < 0$). This assumption naturally stems from the Boltzmann–Grad limit, as it is reasonable that, in the limit of vanishing single-particle collision rate, two colliding particles are uncorrelated. The lack of correlation of colliding particles is the essence of the molecular chaos assumption. We underline that nothing is said about correlation of particles that have just collided. With the assumption (2.33) one can rewrite the first equation of the hierarchy (2.32), omitting the $_1$ subscript (and obvious time dependence) for simplicity:

$$\frac{\partial P(\mathbf{r}, \mathbf{v})}{\partial t} + \mathbf{v} \cdot \frac{\partial P(\mathbf{r}, \mathbf{v})}{\partial \mathbf{r}} = \\ N\sigma^2 \int_{\mathbb{R}^3} \int_{S_+} (P(\mathbf{r}, \mathbf{v}')P(\mathbf{r}, \mathbf{v}_*') - P(\mathbf{r}, \mathbf{v})P(\mathbf{r}, \mathbf{v}_*))|\mathbf{V} \cdot \hat{\mathbf{n}}| d\mathbf{v}_* d\hat{\mathbf{n}}, \tag{2.34}$$

with $\mathbf{v}' = \mathbf{v} - \hat{\mathbf{n}}(\mathbf{V} \cdot \hat{\mathbf{n}})$, $\mathbf{v}_*' = \mathbf{v}_* + \hat{\mathbf{n}}(\mathbf{V} \cdot \hat{\mathbf{n}})$, $\mathbf{V} = \mathbf{v} - \mathbf{v}_*$. This represents the Boltzmann equation for hard spheres. The integral in equation (2.34) is extended to the hemisphere S_+ but could be equivalently extended to the entire sphere S provided a factor 1/2 is inserted in front of the integral itself, as changing $\hat{\mathbf{n}} \to -\hat{\mathbf{n}}$ does not change the integrand.

From a rigorous point of view, the molecular chaos has to be assumed and cannot be proved. However, it has been demonstrated that if the Boltzmann hierarchy has a unique solution for data that satisfies for $t = 0$ a generalized form of chaos assumption:

$$P_s(\mathbf{r}_1, \mathbf{v}_1, \ldots, \mathbf{r}_s, \mathbf{v}_s, t) = \prod_{j=1}^{s} P_1(\mathbf{r}_j, \mathbf{v}_j, t), \tag{2.35}$$

then equation (2.35) holds at any time and therefore the Boltzmann equation is fully justified. Otherwise it has also been proved that if equation (2.35) is satisfied at $t = 0$ and the Boltzmann equation (2.34) admits a solution for the given initial data, then the Boltzmann hierarchy (2.32) has at least a solution which satisfies (2.35) at any time t [3].

It is not difficult to recognize that equation (2.34)—when space coordinates are neglected—is a continuous version of the Boltzmann equation discussed in equation (2.2) where the state index i is replaced by the combination of velocity $\mathbf{v}_*$ and unit vector $\hat{\mathbf{n}}$ and the transition rate is replaced by $N\sigma^2|\mathbf{V} \cdot \hat{\mathbf{n}}|$ together with the collisional condition (a Dirac delta) the connects $\mathbf{v}'$, $\mathbf{v}_*'$ to $\mathbf{v}$, $\mathbf{v}_*$. In fact, the derivation of the H-theorem for the Boltzmann equation for hard spheres follows the same conceptual passages already discussed above.

2.3 The Loschmidt and Zermelo objections

Two conceptually important objections to the Boltzmann equation and the consequent H-theorem were put forward by Loschmidt in 1876 and Zermelo in 1896, both clearly discussed in a celebrated paper by Paul and Tatiana Ehrenfest in 1907. The first one is usually known as *Umkehreinwand* or inversion objection. It recalls that the mechanics underlying the Boltzmann equation is Hamiltonian and therefore if a trajectory of the phase space is allowed then the time-reversed trajectory is also allowed, where all molecules trace back the forward trajectory with inverted velocities. In this trajectory the H-functional would increase, with an apparent violation of the theorem itself. The second objection, usually named *Wiederkehreinwand* or return objection, is based upon a theorem demonstrated by Poincaré which states that an isolated mechanical system with a bounded phase space has a 'quasi-return' behaviour. Namely, we have that a point in phase space representing the system, which at a given time t_0 is in $\mathbf{z}_0$, after some time t_r can be found in a point $\mathbf{z}_r$ as closely as wanted to the initial point z_0. Such a *quasi-return* would imply that the H-functional should come back at time t_r as close as possible to its value at time t_0: therefore, if it grows for some time then it should decrease. Both objections suggested a problem in the H-theorem or in the Boltzmann equation. At the heart of both objections there is the fact, very important, that the Boltzmann equation is an equation for a system made of a very large number of particles, and it treats the evolution of the single-particle probability: this suggests that the single-particle probability is obtained from the many molecules of a large single system (rather than from the evolution of an ensemble of systems prepared in an ensemble of initial conditions). Therefore, if in the single system a reverted or recurrent trajectory appears, all single molecules would do the same and the single probability itself would do the same, violating the H-theorem.

Both objections are based upon the confusion between possibilities and probabilities. Trajectories (such as the reverted one or the returning one) are mechanically *possible*—the returning one is actually necessary—but both types are statistically *unlikely*. Their unlikeliness is—in fact—overwhelming when the number of molecules is large, and makes them almost unobservable. If a system is prepared far from equilibrium, one has that it starts with a large value of H with respect to its minimum, and—when N is large—the majority of the microscopic configurations corresponding to the macroscopic parameters of the initial condition satisfy molecular chaos and therefore H decreases. Of course an infinitely precise manipulation of the molecules of the system could prepare the gas in a situation which violates molecular chaos, and this explains the citation at the beginning of this chapter. Note that at a time t where molecular chaos holds, the H-function must have a local maximum since inverting time leads the Boltzmann equation to have exactly the opposite behaviour and therefore $dH/dt < 0$ when t grows while $dH/dt > 0$ when t decreases. It could seem counterintuitive that there are many local maxima even if H on average decreases, but it is sufficient to imagine a very irregular curve with several local maxima (and minima) and an average decreasing behaviour, to reconcile with this concept. In section 2.5 we discuss a simplified stochastic model where this behaviour is quantitatively demonstrated and the presence of very frequent local maxima is shown to be perfectly compatible with an overall decrease of the entropy.

Note that the unlikeliness of the reverted trajectory stems from the dynamical instability which makes the preparation of reverted initial conditions impossible: the exact reverted condition is physically unachievable unless one has infinite resolution in observation and preparation of the system's molecule coordinates; a condition which is close to the reverted one is perhaps achievable but would rapidly lead to a new trajectory, totally different from the reverted one. The unlikeliness of the recurrent trajectory—on the contrary—is due to the fact that the recurrence time is extremely long when a system is made of a large number of molecules. This was made quantitatively clear by a later lemma due to Kac, which extends a heuristic argument of Boltzmann, by noting that the average return time is inversely proportional to the measure in phase space of the set of points of the initial conditions where the points are expected to return. Since this measure is inversely proportional to the exponential of the number of variables of the phase space, the recurrent time goes as $\sim e^{\alpha N}$, with $\alpha > 0$.

2.4 Other kinetic models

2.4.1 Granular gases

Parts of this subsection have been reprinted from [21] copyright (2015), with permission from Springer Nature.

A granular gas is a gas made of *macroscopic* particles, e.g. grains or spheres with diameters of a few millimetres, such as sand or cereals. We discuss this fascinating topic in detail in chapter 8, however, we briefly mention the basic idea here. Granular particles collide dissipating the kinetic energy of their relative motion [4]. This is due to the macroscopic nature of the grains: during the interaction, irreversible processes happen inside the grain and energy is dissipated in the form of heat. In a collision between two free particles, these processes conserve momentum, so that the velocity of the centre of mass of the two grains is not modified. The most used model in granular gas literature is also the simplest: the gas of inelastic smooth hard spheres, with fixed restitution coefficient. It is given by the following prescriptions for the collision of two particles of mass m_1 and m_2:

$$m_1\mathbf{v}_1' + m_2\mathbf{v}_2' = m_1\mathbf{v}_1 + m_2\mathbf{v}_2 \tag{2.36a}$$

$$(\mathbf{v}_1' - \mathbf{v}_2') \cdot \hat{\mathbf{n}} = -r(\mathbf{v}_1 - \mathbf{v}_2) \cdot \hat{\mathbf{n}}, \tag{2.36b}$$

where the primes denote the postcollisional velocities, $\hat{\mathbf{n}}$ is the unity vector in the direction joining the centres of the grains, and $0 \leqslant r \leqslant 1$ is the so-called restitution coefficient. When $r = 1$ the gas is elastic and the rule coincides with hard spheres. When $r = 0$ the gas is perfectly inelastic, that is, the particles exit from the collision with no relative velocity in the $\hat{\mathbf{n}}$ direction.

Variants of this model have been largely used in the literature. The importance of tangential frictional forces acting on the grains at contact may be studied taking into account the rotational degree of freedom of the particles, i.e. adding a variable ω_i to each grain. The simplest model which takes into account the rotational degree of freedom of particles is the rough hard spheres gas. In this model the postcollisional translational and angular velocities are given by instantaneous rules that conserve total translational and angular momentum, while kinetic energy is partly transferred between the translational and angular degrees of freedom and partly lost in dissipation.

Even if rigorous results such as those for hard spheres have not been proved for this kind of system, one may assume that in the dilute limit a molecular chaos assumption holds, and under this assumption a Boltzmann equation for granular gases is obtained [5] (see section 8.2):

$$\left(\frac{\partial}{\partial t} + L_1^0\right)P(\mathbf{r}_1, \mathbf{v}_1, t) = N\sigma^2 Q(P, P) \tag{2.37}$$

$$Q(P, P) = \int d\mathbf{v}_2 \int_{\mathbf{V}_{12}\cdot\hat{\mathbf{n}}>0} d\hat{\mathbf{n}}|\mathbf{V}_{12}\cdot\hat{\mathbf{n}}|\left[\frac{1}{r^2}P(\mathbf{r}_1, \mathbf{v}_1', t)P(\mathbf{r}_1, \mathbf{v}_2', t) + \right. \tag{2.38}$$

$$\left. P(\mathbf{r}_1, \mathbf{v}_1, t)P(\mathbf{r}_1, \mathbf{v}_2, t)\right], \tag{2.39}$$

where the primed velocities are the so-called pre-collisional velocities. A major difference with respect to the elastic case is the presence of the factor $1/r^2$ in front of the gain collisional term. This term is the main source of unbalance between gain and loss, and is at the basis of the violation of time reversal symmetry and of the H-theorem.

In fact the task of finding something similar to an H-theorem for granular gases is made more complicated by dissipation: an isolated granular system cannot be stationary unless there are no more collisions. Usually it is in a cooling state of some sort and for a long, possibly infinite, time. A possibility could be looking for rescaled H-theorems, i.e. assuming some asymptotic form of the single-particle distribution where time is entirely contained in the evolution of one or a few parameters (e.g. kinetic temperature), and the H-theorem is studied for a probability distribution with some of its moments which are rescaled, in order to look for an asymptotic 'stationary' state. A more physical alternative is to introduce in the model some mechanism of energy injection, as would happen in real experiments, for instance in a constantly vibrated box. A discussion of the fate of the H-theorem for such kinds of model can be found in [6], briefly summarised in section 8.2. There are strong indications that an H-theorem of the kind of that for master equations (where the H-functional is H_K and not H) holds.

2.4.2 Maxwell models

Parts of this section have been reprinted from [21] copyright (2015), with permission from Springer Nature.

The collisional integral of the Boltzmann equation for hard spheres, equation (2.34), contains a term $g = |\mathbf{V}\cdot\hat{\mathbf{n}}|$ which multiplies the probabilities of particles entering or coming out from a collision. In general, the collisional integral is proportional to the differential collision rate for a particle coming at a certain relative velocity, which may be expressed in terms of the scattering cross-section s, which depends strongly on the kind of interaction between the molecules of the gas. For power law repulsive interaction potential $U(r) \sim r^{-(a-1)}$, the scattering cross-section can be calculated and it is found that in dimension d, when $a = 1 + 2(d - 1)$ (i.e. $a = 5$ for $d = 3$ and $a = 3$ for $d = 2$) the collision rate *does not depend upon g*. This property defines the so-called Maxwell molecules [7]. Interaction with $a < 1 + 2(d - 1)$ are called soft interactions (e.g. the electrostatic or gravitational interaction). Interactions with $a > 1 + 2(d - 1)$ are called hard interactions. Hard spheres ($a \to \infty$) belongs to this set of interactions, with the collision rate $\sim g$. The Very Hard Particles model has also been studied, which is characterized by a collision rate $\sim g^2$, which is not attainable with an inverse power potential, as it requires an interaction harder than the hard-sphere interaction.

The advantage of Maxwell molecules is that the Boltzmann equation is greatly simplified, as g does not appear in the collision integral, which therefore takes a form similar to a convolution integral. A further simplification of the Boltzmann equation came from Krook and Wu, who studied the Boltzmann equation of Maxwell molecules with an isotropic scattering cross-section. A great deal of literature exists for linear and nonlinear model-Boltzmann equations (for a review see [7]). The importance of the Maxwell molecules model relies on the possibility of obtaining solutions for it: the general method (extended to other model-Boltzmann equations) is to obtain an expansion in orthogonal polynomials where the coefficients are polynomial moments of the solution distribution function. For Maxwell molecules the moments satisfy a recursive system of differential equations that can be solved sequentially. Given an initial distribution, one can solve the problem if the series expansion converges. The Maxwell molecules model has also been a subject of study in the framework of the kinetic theory of granular gases.

2.4.3 Redistribution models

A plethora of kinetic models exist where the rules of collisions are not directly connected to some mechanical model, including some which are inspired by phenomena in Nature or in social systems, for instance in finance or in epidemiology. In these models the collisions are events where some quantity is redistributed and/or varied according to some rules which have some resemblance with the collisions among particles.

One of the most popular models in this class was introduced by S Ulam at the Los Alamos laboratories [8]. He performed numerical simulations for such a kinetic model and conjectured that it admits a stationary state, which was proven a few years later [9]. This process has been used to model biological processes, (e.g. cell mutations or the velocity exchange in flocks) and applied to finance, replacing the distribution of energy with that of wealth in closed economic systems. Recently an H-theorem for the Ulam model was demonstrated [10]. A fascinating similar result was also obtained in [11], where the authors showed that the Gini coefficient (a measure which quantifies the degree of inequalities in a given wealth distribution) acts as the H-function in the finance version of the Ulam model.

The model can be briefly described in the following way. Imagine a system consisting of N particles and let $E_n(t) > 0$ be the energy of the nth particle at time t. The system evolves through exchange of energy between random pairs of particles, i.e.

$$\begin{aligned} E_i(t+1) &= \alpha(E_i(t) + E_j(t)), \\ E_j(t+1) &= (1-\alpha)(E_i(t) + E_j(t)), \end{aligned} \tag{2.40}$$

where the indices i and j are drawn from the uniform distribution among all possible pairs (i, j), while $\alpha \in [0, 1]$ is a random number with a symmetric distribution, $p(\alpha) = p(1 - \alpha)$. For $N \gg 1$ this dynamics is equivalent to the following nonlinear transformation which corresponds to a Boltzmann equation in discrete time and continuous space:

$$\begin{aligned} \rho_{t+1}(E) &= \int_0^1 \mathrm{d}\alpha\, \mathrm{p}(\alpha) \int_0^\infty \mathrm{dE}_1 \int_0^\infty \mathrm{dE}_2 \rho_t(\mathrm{E}_1)\rho_t(\mathrm{E}_2)\delta(\mathrm{E} - \alpha(\mathrm{E}_1 + \mathrm{E}_2)) \\ &= \int \mathrm{dq} \frac{\mathrm{e}^{iqE}}{2\pi} \int_0^1 \mathrm{d}\alpha\, \mathrm{p}(\alpha)\hat{\rho}_t^2(\alpha \mathrm{q}), \end{aligned} \tag{2.41}$$

where $\hat{\rho}$ denotes the characteristic function of ρ. As conjectured by Ulam and proved in [9], asymptotically the system reaches a stationary state whose distribution $\rho_\infty(E)$ depends on the choice of $p(\alpha)$. In the special cases where $p(\alpha) = 1$ or $p(\alpha) \propto \alpha^a(1 - \alpha)^a$ one has $\rho_\infty(E) \propto e^{-\beta E}$ and $\rho_\infty(E) \propto E^a e^{-\beta E}$ with β fixing the mean energy. Unfortunately, for a generic $p(\alpha)$ the shape of $\rho_\infty(E)$ is unknown. Later it has been shown that the entropy $S(\xi) = -\int_E \xi(E)\log(\xi(E))dE$, $\xi(E)$ being a generic function, increases during the evolution when $p(\alpha) = 1$. If α is distributed according to a Beta distribution, a rigorous proof exists that S is no longer monotone. However, both the relative entropy $K(\xi) = \int_E \xi(E)\log(\frac{\xi(E)}{\xi_\infty(E)})$ and the functional $G(\xi) = S(\xi) - a\int_E \xi(E)\log(E)$ increase at each iteration. It is not clear if in all cases at least the relative entropy K is monotone.

2.5 Some remarks about irreversibility

We conclude this chapter with few technical and conceptual remarks about the problem of irreversibility [12].

2.5.1 About ensembles and entropies

Parts of this subsection have been reprinted from [22] copyright (2022), with permission from Springer Nature.

Surely the ensembles are among the cornerstones of statistical mechanics. On the other hand, they are often introduced in rather unfortunate fashions: for instance, one can find obscure statements, such as *'an ensemble is an infinite collection of identical systems'*. Mathematically such a collection is nothing but an intuitive way to introduce a probability distribution of points in the phase space.

On the other hand in the real activity of a laboratory, and in practice even in numerical computations, we deal with just a single system, so it is important to understand the physical relevance of a computation based on the ensembles [13]. In our opinion, the main issue to be addressed, for such a question, is the justification of typicality, i.e. the fact that 'good' quantities of a single macroscopic object, with probability almost one, are close to the corresponding averages computed with a suitable ensemble.

We believe that such an attempt is not a hopeless project, and the basic idea is one of the best propositions used to link probability and physics, the Cournot's principle: *An event with very small probability will not happen.* Actually, this statement may be associated with a celebrated sentence in the Jakob Bernoulli's book *Ars Conjectandi* (1713), which reads: *Something is morally certain if its probability is so close to certainty that shortfall is imperceptible.* We do not enter into the debate about the validity of such a principle; however, we recall that eminent mathematicians, such as P Lévy, J Hadamard, and A N Kolmogorov, considered the Cournot's principle as the only sensible connection between probability and the empirical world [14], and we share their opinion[1].

Khinchin [16] made a significant contribution to equilibrium statistical mechanics by demonstrating that, for a certain class of observables, the time average of a single system's evolution closely approximates the ensemble average, with the exception of a set of initial conditions that have a very low probability. In the following, we will explore both analytical and numerical results showing that even in nonequilibrium conditions, the behaviour of observables in a single macroscopic system remains close to their mean values. The occurrence of 'atypical' behaviours is extremely rare and can be considered to have vanishing probability, aligning with the deterministic nature of thermodynamics, which effectively rules out such exceptions.

Consider the probability distribution at time t, of the microscopic state $\mathbf{X}$ in the phase space, $\rho(\mathbf{X}, t)$; it is quite evident that this quantity cannot be really computed in systems of interest for the statistical mechanics, i.e. with many degrees of freedom, and its practical relevance is restricted to the context of low-dimensional dynamical systems. Let us discuss the properties of the so-called Gibbs entropy, defined as:

$$S_G(t) = -k_B \int \rho(\mathbf{X}, t) \ln \rho(\mathbf{X}, t)\, d\mathbf{X}. \tag{2.42}$$

Introducing the semigroup $\mathcal{S}^t$ which gives the time evolution of the Hamilton equations, i.e. $\mathbf{X}(t) = \mathcal{S}^t\mathbf{X}(0)$, and denoting by $\mathbf{Y}$ the vector such that $\mathbf{X} = \mathcal{S}^t\mathbf{Y}$, from the Liouville theorem one has $\rho(\mathbf{X}, t) = \rho(\mathbf{Y}, 0)$, so it is easy to show that the Gibbs entropy S_G is constant in time: $S_G(t) = S_G(0)$. The Gibbs entropy correctly yields the equilibrium thermodynamic entropy of the system, but it cannot represent the growing entropy of an isolated nonequilibrium system: the Gibbs entropy S_G is not a suitable counterpart in statistical mechanics of the thermodynamic entropy.

A possible way to overcome this difficulty is with the introduction of a coarse-graining of the phase space, i.e. a partition made of cells of given small size, say ε.

[1] We quote from the incipit of the book by Gnedenko and Kolmogorov [15]: 'all epistemologic value of the theory of probability is based on this: that large-scale random phenomena in their collective action create strict, nonrandom regularity. The very concept of mathematical *probability* would be fruitless if it did not find its realization in the *frequency* of occurrence of events under large-scale repetition of uniform conditions.'

In such a way one has the coarse-grained version of the probability density, i.e. the probability for the microscopic phase to lie in the ith cell at time t is expressed by:

$$P_\varepsilon(i, t) = \int_{\Lambda_\varepsilon(i)} \rho(\mathbf{X}, t)\, d\mathbf{X}, \tag{2.43}$$

where $\Lambda_\varepsilon(i)$ is the ith cell of size ε, and the corresponding coarse-grained Gibbs entropy:

$$S_{G,\varepsilon}(t) = -k_{\mathrm{B}} \sum_i P_\varepsilon(i, t) \ln P_\varepsilon(i, t). \tag{2.44}$$

Then, unlike S_G, the quantity $S_{G,\varepsilon}$ can grow in time; however, this success cannot be considered completely satisfying because the evolution of $S_{G,\varepsilon}$ depends on ε; we do not discuss such a topic whose interest is restricted to dynamical systems.

In a system of weakly interacting particles, at the equilibrium for the stationary distribution we have $\rho_s(\mathbf{X}) \simeq \prod_n f_s(\mathbf{q}_n, \mathbf{p}_n)$, and therefore $S_G \simeq NS_B$, where $S_B = -k_{\mathrm{B}} H$ is the Boltzmann entropy. However, the Gibbs and the Boltzmann definitions have a very different conceptual and physical meaning: the Gibbs entropy is defined in terms of the very abstract notion of phase space probability density, while the Boltzmann entropy can be derived from a very material property, i.e. the number of particles of one concrete system occupying a given small volume in the μ-space.

Let us consider again the entropy S_B used for the H-theorem. We note that if the number of particles N is very large, then, the one-particle distribution function $f(\mathbf{q}, \mathbf{p}, t)$, that is, the main theoretical object in the theory, can be seen as an empirical distribution function, concerning the positions $\{\mathbf{q}_n(t)\}$ and momenta $\{\mathbf{p}_n(t)\}$ of the N particles, formally expressed by:

$$f(\mathbf{q}, \mathbf{p}, t) = \frac{1}{N} \sum_{n=1}^{N} \delta(\mathbf{q} - \mathbf{q}_n(t)) \delta(\mathbf{p} - \mathbf{p}_n(t)). \tag{2.45}$$

Therefore, at variance with $\rho(\mathbf{X}, t)$, the one-particle distribution $f(\mathbf{q}, \mathbf{p}, t)$ can be measured for a single system, and for $S_B(t)$ there is no need to refer to the statistical ensembles.

Let us consider again the Zermelo paradox: perhaps the most important mathematical contribution in the understanding of this problem is due to Lanford [3], who was able to show that the microscopic Hamiltonian, reversible, dynamics is not incompatible with the H-theorem. Indeed, given a hard-sphere system, consider the Boltzmann–Grad limit, i.e. $N \to \infty$, $\sigma \to 0$ so that $N\sigma^2 \to$ constant, where σ is the diameter of the particles; starting from an initial condition in a 'good set', one obtains that with good approximation $f(\mathbf{q}, \mathbf{p}, t)$ evolves as prescribed by the Boltzmann equation: hence the H-theorem holds. In other words, Lanford proved that, a part for very atypical initial conditions, we have

$$f(\mathbf{q}, \mathbf{p}, t) \simeq f_B(\mathbf{q}, \mathbf{p}, t), \tag{2.46}$$

where $f_B(\mathbf{q}, \mathbf{p}, t)$ is the solution of the Boltzmann equation. This result has been proven in a rigorous way only for a short time, namely a fraction of the mean collision time, for $N \gg 1$, and a typical $\mathbf{X}(0)$; but this is enough to prove rigorously that Hamiltonian dynamics, in the proper limit, does not violate the Boltzmann equation, and one shows that it is possible to obtain an irreversible behaviour from a microscopic reversible dynamics. Beyond the lack of a rigorous proof for a large time, there is overwhelming evidence that the Boltzmann equation is able to describe with good accuracy the time evolution of $f(\mathbf{q}, \mathbf{p}, t)$ of a diluted gas.

Let us now briefly discuss again the Loschmidt paradox looking at velocity inversion for the time evolution of a system in computer simulations, e.g. a two-dimensional diluted gas of 100 hard disks in a periodic box. One can compute the quantity H via the function $f(\mathbf{q}, \mathbf{p}, t)$ that is obtained by making histograms of the positions and velocities of the disks. Starting from a nonequilibrium configuration, one follows the time-behaviour of H in the direct evolution and in evolutions with the velocity inversion at certain times. If very small errors are introduced into the initial data after the velocity inversion, in a chaotic system, one does not observe the exact antikinetic behaviour; increasing the error the antikinetic behaviour is suppressed more or less completely, as shown in figure 2.1. Roughly speaking, we can say that an inversion of the velocity with some error produces a state into the 'good' ensemble of Lanford's theorem.

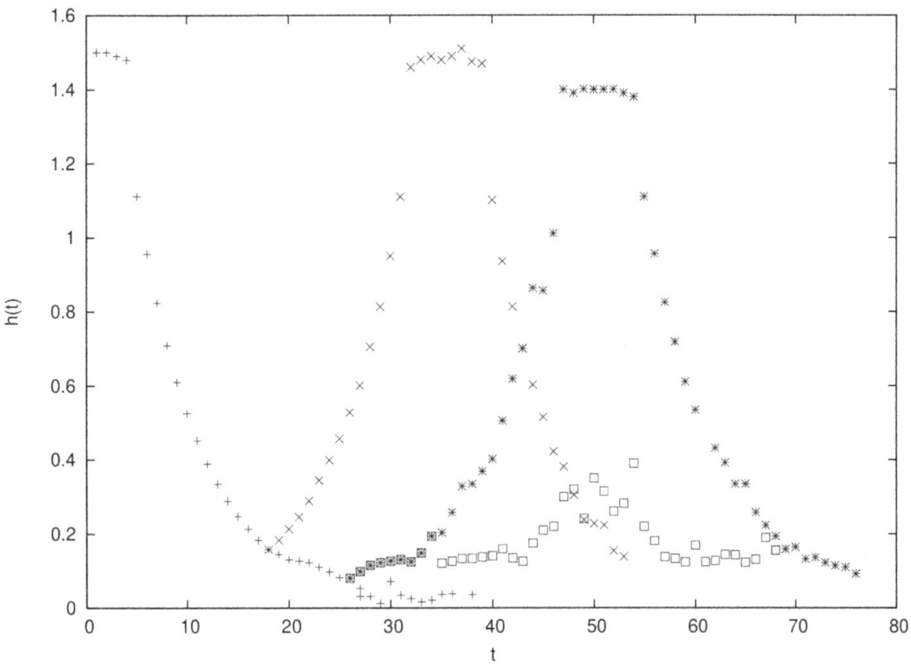

Figure 2.1. The effect of velocity inversion on $h(t) = H(t) - H_{\min}$. The pluses correspond to direct evolution; the crosses, the stars and the squares are obtained by reversing the velocities after 18 collision with very small errors, after 26 collisions with small errors and after 26 collisions with (relatively) large errors, respectively. This picture shows in a schematic way the computer experiment by Orban and Bellemans [17]. Reprinted with permission from [12]. Copyright Cambridge University Press.

2.5.2 Ehrenfest model and typicality

The work of Lanford is rather technical; in order to show the basic physical aspects, avoiding the mathematical difficulties, we discuss a popular model whose simplicity allows a neat discussion of irreversibility: the well known Ehrenfest model. Although such a model is nothing but a Markov chain, it is (according to Kac) *probably one of the most instructive models in the whole of physics.* We have N 'particles', each of which can either be in a box called A, or in another box called B, and the state of the system at time t is identified by the number n_t of particles in A; the evolution is ruled by the transition probabilities

$$P_{n\to n-1} = \frac{n}{N}, \; P_{n\to n+1} = 1 - \frac{n}{N},$$

dictating how the state n_t changes in one time step to become $n_{t+1} = n_t \pm 1$. The state n_t can be seen as the 'macroscopic' state of the system and the equilibrium state is $n_{eq} = N/2$.

Let us note that such a system has the property of detailed balance, which is somehow analogous in the stochastic realm of the reversibility in deterministic systems: denoting with $P_s(n)$ the stationary probability to be in n we have

$$P_s(n)P_{n\to m} = P_s(m)P_{m\to n};$$

the relevance of the detailed balance in statistical mechanics will be discussed in detail in chapter 3. It is interesting to stress that in such a model an important intuition of Boltzmann on the prevalence of the configuration for which the molecular chaos holds, can be explicitly proved. Consider the configuration

$$C_A(m) = \{n_{t-1} = m - 1, \, n_t = m, \, n_{t+1} = m - 1\},$$

which corresponds to the molecular chaos, i.e. it is the analogous of a maximum of H; an easy direct computation shows that if m is very far from the equilibrium value $N/2$, the $C_A(m)$ are dominant with respect to the other sequences

$$C_B(m) = \{n_{t-1} = m - 1, \, n_t = m, \, n_{t+1} = m + 1\}$$

$$C_C(m) = \{n_{t-1} = m + 1, \, n_t = m, \, n_{t+1} = m + 1\}$$

$$C_D(m) = \{n_{t-1} = m + 1, \, n_t = m, \, n_{t+1} = m - 1\},$$

see figure 2.2. Let us compute the probability of $C_A(m)$ under the condition $n_t = m$:

$$P(C_A(m)) = P(n_{t-1} = m - 1, \, n_t = m, \, n_{t+1} = m - 1 | n_t = m);$$

since the model is ruled by a Markov chain one has

$$P(C_A(m)) = \frac{P_s(m-1)P_{m-1\to m}P_{m\to m-1}}{P_s(m)} = \frac{m^2}{N^2},$$

where we used the expression for the stationary probability distribution:

$$P_s(m) = \frac{N!}{m!(N-m)!}2^{-N}.$$

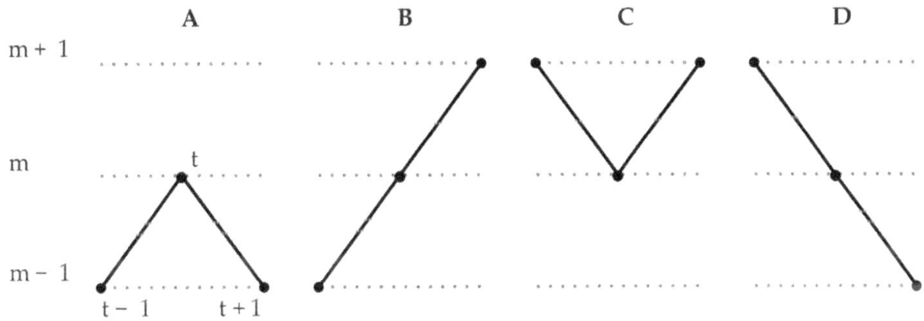

Figure 2.2. The configuration A corresponds to the molecular chaos and, if m is close to N, is much more frequent than the other configurations.

So we have that $P(C_A(m))$ is close to 1 for $m \simeq N$, and therefore it is much larger than $P(C_B(m))$, $P(C_C(m))$ and $P(C_D(m))$.

The evolution of an ensemble of initial conditions starting from a given state n_0 can be described computing the mean population $\langle n_t \rangle$ and its variance $\sigma_t^2 = \langle n_t^2 \rangle - \langle n_t \rangle^2$; one obtains

$$\langle n_t \rangle = \frac{N}{2} + \left(1 - \frac{2}{N}\right)^t \Delta_0, \; \sigma_t^2 = \frac{N}{4} + \left(1 - \frac{4}{N}\right)^t \left(\Delta_0^2 - \frac{N}{4}\right) + \left(1 - \frac{2}{N}\right)^{2t} \Delta_0^2, \quad (2.47)$$

where $\Delta_0 = n_0 - N/2$. One has $\langle n_t \rangle \to n_{eq} = N/2$, exponentially fast with a characteristic time $\tau_c = -1/\ln(1 - 2/N) \simeq N/2$ and the standard deviation σ_t reaches its equilibrium value $\sqrt{N}/2$ with a characteristic time $O(N)$.

The monotonic relaxation of $\langle n_t \rangle$ to its equilibrium value, at first glance, can appear as the analogoue of the second law. On the other hand, in real life we are interested in the behaviour of a unique (large) system, not to an ensemble of systems, therefore, a natural question is: what about the time evolution of a single macroscopic object, i.e. a single realisation of the process? Figure 2.3 illustrates the result of numerical simulations, suggesting that for large N, the single object behaves 'typically', i.e. a macroscopic variable is close to its mean value, apart from cases with a very small probability. In mathematical terms it is possible to show [18] that

$$\text{Prob}(n_t \simeq <n_t> \text{ for any } t \in [0, T]) \simeq 1 \text{ where } T = O(N), \quad (2.48)$$

i.e. each single realization of n_t stays 'close' to the time dependent average $\langle n_t \rangle$. Actually one has:

$$\text{Prob}\left(\frac{|n_t - \langle n_t \rangle|}{N} < \varepsilon_N \text{ for any } t \in [0, T]\right) \geqslant 1 - a_N, \quad (2.49)$$

with the quantities $\varepsilon_N \to 0$, and $a_N \to 0$ as $N \to \infty$. Taking $\varepsilon_N \sim N^{-B}$ with $0 < B < 1/3$, one has $a_N \sim N^{-A}$ with $A > 0$: for instance, $B = 0.2$ implies $A \geqslant 0.2$.

The above exact result neatly shows the notion of typicality, i.e. a macroscopic quantity of the system has a very low probability of resulting sensibly different from the expected value [19]. Of course one can wonder whether the above scenario is still

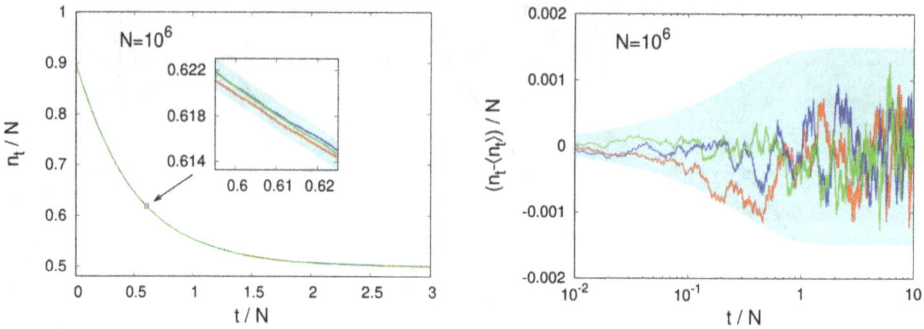

Figure 2.3. Several realisations of the time evolution of the state of the Ehrenfest flea model, n_t, for $N = 10^6$. The coloured region corresponds to three standard deviations from the running mean: $\langle n_t \rangle - 3\sigma_t < n_t < \langle n_t \rangle + 3\sigma_t$. Reprinted from reference [18], Copyright (2019), with permission from Elsevier.

valid in the more realist case of systems with many particles evolving with deterministic evolution. The gap between the stochastic and the deterministic realm can, however, be bridged by numerical simulations.

Consider, for instance, a channel containing N particles of mass m, closed by a fixed vertical wall on the left, and by a frictionless piston of mass M on the right. Denoting by $x_n(t)$ the horizontal coordinate of the nth particle and with $X(t)$ the position of the piston, we have $0 \leqslant x_n \leqslant X$; a constant force F acts on the piston, and in addition there are the collisions with particles inside the channel, so that the Hamiltonian of this system reads:

$$H = \frac{P^2}{2M} + \sum_i \frac{p_i^2}{2m} + \sum_{i<j} U(|\mathbf{q}_i - \mathbf{q}_j|) + U_w(\mathbf{q}_1, \ldots, \mathbf{q}_N, X) + FX,$$

where U is the interacting potential among the particles, and U_w denotes the interaction of the particles with the wall. In the case of non-interacting particles, $U = 0$, and with U_w describing elastic collisions, the dynamics is not chaotic, and it is easy to find the 'equilibrium' position of the piston, $\langle X \rangle$, and its variance σ_X^2. In the presence of interactions, e.g. for interaction potentials like $U(r) = U_0/r^{12}$ and $U_w = U_0 \sum_n |x_n - X|^{-12}$, it is not possible to determine analytically the equilibrium statistical properties, but one can study numerically the problem.

At variance with the non-interacting case, numerical computations show that in the system with interacting particles one has a positive Lyapunov exponent, i.e. chaos. Both for the chaotic and non-chaotic case, the initial conditions are far from equilibrium: in particular, the positions of the particles are initially distributed uniformly in the interval $[0, X(0)]$, while the velocities initially follow a Maxwell–Boltzmann distribution at a temperature T different from the equilibrium temperature T_{eq}, and $|X(0) - X_{\text{eq}}| \gg \sigma_{\text{eq}}$, where the subscript eq refers to the equilibrium state.

The main outcome of simulations of the model, shown in figure 2.4, is that the single trajectories are typical both for the chaotic and the non-chaotic case: although far from equilibrium, fluctuations are small compared to the corresponding

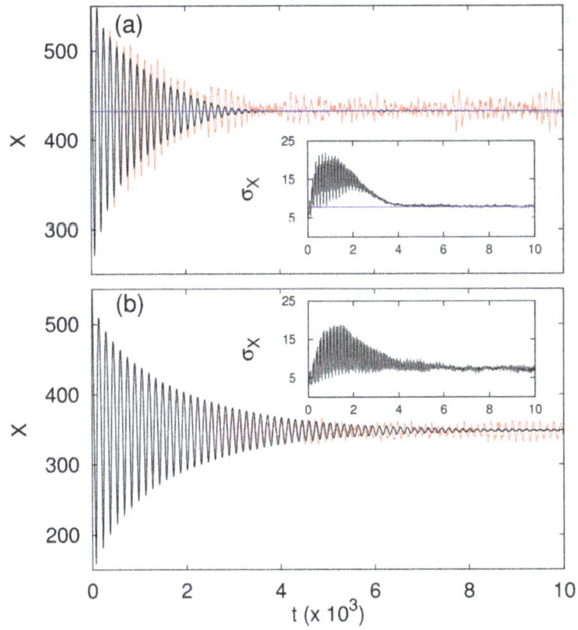

Figure 2.4. $X(t)$ versus t for $N = 1024$ and $X(0) = X_{\text{eq}} + 10\sigma_{\text{eq}}$ in a chaotic piston (a), and in a non-chaotic piston (b). Red lines represent $X(t)$ for a single realization; black lines refer to the ensemble average $\langle X(t)\rangle$. Reprinted from reference [20], Copyright (2016), with permission from Elsevier.

averages, as in the case of the stochastic Ehrenfest model. In the limit $N \gg 1$, there is no need to average over ensembles of independent evolving systems: in the case one really wants to perform such an average, the result will be very close to what one observes in a single trajectory. Summarizing, we have that numerical simulations support the scenario that for a macroscopic observable M, one has

$$\text{Prob}(M(t) \simeq \langle M(t)\rangle) \simeq 1. \tag{2.50}$$

2.5.3 Again on chaos and irreversibility

The previous results for the piston model suggest that in the limit of $N \gg 1$, chaos is not a relevant ingredient for the irreversible behaviour. Since there is still a persistent confusion concerning the role of chaos in statistical mechanics, we believe it is appropriate to clarify the situation.

Consider a mixing system: one has a relaxation to the invariant measure, i.e. independently of the initial density distribution $\rho(\mathbf{X}, 0)$, for large t one has

$$\rho(\mathbf{X}, t) \rightarrow \rho_{inv}(\mathbf{X}). \tag{2.51}$$

Such a property at first glance can appear analogous of irreversibility, but it is not so. The relaxation to the invariant measure is a property of the ensemble of the initial conditions, and it is rather different from the irreversibility of real life, which concerns a unique (large) system.

We underline that the property (2.51) can hold even for low-dimensional systems. In this case it is not possible to define macroscopic variables, and therefore there is not a distinction micro/macro: it is impossible to speak of irreversible behaviour for a single system. In order to clarify such a point we consider as an example of a chaotic system a bidimensional area preserving map, the celebrated Arnold cat:

$$x_{t+1} = x_t + y_t \bmod 1, \quad y_{t+1} = x_{t+1} + y_t \bmod 1.$$

In spite of the fact that the above system is chaotic, and mixing, i.e. that equation (2.51) holds, at variance with the observed behaviour of the piston in figure 2.4, we cannot observe any qualitative difference between a direct trajectory

$$\mathbf{x}_0, \mathbf{x}_1, \ldots, \mathbf{x}_{t-1}, \mathbf{x}_t,$$

and its time reversal

$$\mathbf{x}_t, \mathbf{x}_{t-1}, \ldots, \mathbf{x}_1, \mathbf{x}_0.$$

The above remark allows us to appreciate the difference between the physical irreversibility in a single macroscopic object, and the relaxation of a phase-space probability distribution $\rho(\mathbf{X}, t)$ to an invariant distribution.

References

[1] Cercignani C 1988 *The Boltzmann equation* (Berlin: Springer)

[2] Huang K 2008 *Statistical Mechanics* (New York: Wiley)

[3] Lanford O E III 1981 The hard sphere gas in the Boltzmann-Grad limit *Physica A: Stat. Mech. Appl.* **106** 70–6

[4] Brilliantov N V and Pöschel T 2004 *Kinetic Theory of Granular Gases* (Oxford: Oxford University Press)

[5] Brey J J, Ruiz-Montero M J and Cubero D 1996 Homogeneous cooling state of a low-density granular flow *Phys. Rev.* E **54** 3664

[6] Marini Bettolo Marconi U, Puglisi A and Vulpiani A 2013 About an H-theorem for systems with non-conservative interactions *J. Stat. Mech. Theory Exp.* **2013** P08003

[7] Ernst M H 1981 Nonlinear model-Boltzmann equations and exact solutions *Phys. Rep.* **78** 1–171

[8] Ulam S 1980 On the operations of pair production, transmutations, and generalized random walk *Adv. Appl. Math.* **1** 7–21

[9] Blackwell D and Daniel Mauldin R 1985 Ulam's redistribution of energy problem: collision transformations *Lett. Math. Phys.* **10** 149–53

[10] Apenko S M 2013 Monotonic entropy growth for a nonlinear model of random exchanges *Phys. Rev.* E **87** 024101

[11] Boghosian B M, Johnson M and Marcq J A 2015 An *H* theorem for Boltzmann's equation for the Yard–Sale model of asset exchange: the Gini coefficient as an H functional *J. Stat. Phys.* **161** 1339–50

[12] Castiglione P, Falcioni M, Lesne A and Vulpiani A 2008 *Chaos and Coarse Graining in Statistical Mechanics* (Cambridge: Cambridge University Press)

[13] Zurek W H 2018 Eliminating ensembles from equilibrium statistical physics: Maxwell's demon, Szilard's engine, and thermodynamics via entanglement *Phys. Rep.* **755** 1–21

[14] Shafer G and Vovk V 2005 *Probability and Finance: It's Only a Game!* (New York: Wiley)
[15] Gnedenko B V and Kolmogorov A N 1968 *Limit Distributions for Sums of Independent Random Variables* (Reading, MA: Addison-Wesley)
[16] Khinchin A Y 2013 *Mathematical Foundations of Information Theory* (North Chelmsford, MA: Courier Corporation)
[17] Orban J and Bellemans A 1967 Velocity-inversion and irreversibility in a dilute gas of hard disks *Phys. Lett.* **24** 620–1
[18] Baldovin M, Caprini L and Vulpiani A 2019 Irreversibility and typicality: a simple analytical result for the Ehrenfest model *Physica A: Stat. Mech. Appl.* **524** 422–9
[19] Goldstein S 2012 Typicality and notions of probability in physics ed Y Ben-Menahem and M Hemmo *Probability in Physics* (Berlin: Springer) pp 59–71
[20] Cerino L, Cecconi F, Cencini M and Vulpiani A 2016 The role of the number of degrees of freedom and chaos in macroscopic irreversibility *Physica A: Stat. Mech. Appl.* **442** 486–97
[21] Puglisi A 2015 *Transport and Fluctuations in Granular Fluids Springer Briefs in Physics* (Springer)
[22] Chibbaro S, Rondoni L and Vulpiani A 2022 *Probability, Typicality and Emergence in Statistical Mechanics The Frontiers Collection, From Electrons to Elephants and Elections* (Springer) pp 339–60

IOP Publishing

Nonequilibrium Statistical Mechanics
Basic concepts, models and applications
Alessandro Sarracino, Andrea Puglisi and Angelo Vulpiani

Chapter 3

Statistical and dynamical features of fluctuations

Never build your muscle and then go hunt for a problem. Build your muscle while solving the problem. That is the key to good research.

Lars Onsager

3.1 Fluctuations are small but relevant

It is well known that in macroscopic systems with N particles, fluctuations are relatively small. For example, consider the familiar result for energy fluctuations:

$$\langle (E - \langle E \rangle)^2 \rangle = N c_v k_B T^2, \tag{3.1}$$

where c_v is the specific heat per particle. The relative fluctuations of an extensive quantity are of the order $O(N^{-1/2})$, making them insignificantly small at large N. This naturally leads to the tempting conclusion that such fluctuations are negligible and that the statistical descriptions of macroscopic systems can be firmly grounded in the law of large numbers.

Even some of the great pioneers of statistical mechanics shared this view. For instance, Boltzmann wrote: *'In the molecular theory we assume that the laws of the phenomena found in nature do not essentially deviate from the limits that they would approach in the case of an infinite number of infinitely small molecules'*. Similarly, Gibbs expressed: *'[the fluctuations] would be in general vanishing quantities, since such experience would not be wide enough to embrace the more considerable divergences from the mean values'*.

We know that the above point of view is not correct: this was understood by Einstein who realized the central role played by fluctuations; e.g. he noted how from equation (3.1) we can obtain the Avogadro number *'if it were possible to determine the average of the square of the energy fluctuations of the system; this is however not possible according to our present knowledge'*. Although for macroscopic objects equation (3.1) in practice cannot be used for the determination of the Avogadro number, Einstein's intuition was correct and, as discussed in chapter 1, it is possible

doi:10.1088/978-0-7503-6229-0ch3

to relate the Avogadro number to macroscopic quantities such as the diffusion coefficient, obtained by observing the fluctuations of Brownian particles.

The relation (3.1) can be generalized, and remarkably, fluctuations, although very small, are detectable and are important ingredients of statistical mechanics. Let us assume that the macroscopic state of a system is described by m variables $(\alpha_1,\ldots, \alpha_m)$, functions of the microscopic state $\mathbf{X}$, $\alpha_j = g_j(\mathbf{X})$ with $j = 1,\ldots, m$. Denote by $\mathscr{P}$ the parameters which determine the probability density of the microscopic state $\mathbf{X}$, e.g. in the canonical ensemble $\mathscr{P} = (T, N, V)$, and in the grand canonical ensemble $\mathscr{P} = (T, \mu, V)$. The probability density of the $\{\alpha_j\}$ is

$$p(\alpha_1,\ldots, \alpha_m) = \int \rho(\mathbf{X}, \mathscr{P}) \prod_{j=1}^{m} \delta(\alpha_j - g_j(\mathbf{X}))d\mathbf{X},$$

where $\rho(\mathbf{X}, \mathscr{P})$ is the probability density of $\mathbf{X}$ in the ensemble with parameters $\mathscr{P}$. It is straightforward to have

$$p(\alpha_1,\ldots, \alpha_m) = \exp\{-\beta[F(\alpha_1,\ldots, \alpha_m|\mathscr{P}) - F(\mathscr{P})]\}, \tag{3.2}$$

where $F(\mathscr{P})$ is the free energy of the system with parameters $\mathscr{P}$ and $F(\alpha_1,\ldots, \alpha_m|\mathscr{P})$ is the free energy of the system where one has the parameters $\mathscr{P}$ and the macroscopic variables $\alpha_1,\ldots, \alpha_m$. For instance in the canonical ensemble

$$F(\alpha_1,\ldots, \alpha_m|\mathscr{P}) = -k_B T \ln \int \prod_{j=1}^{m} \delta(\alpha_j - g_j(\mathbf{X}))e^{-\beta H(\mathbf{X})}d\mathbf{X},$$

$$F(\mathscr{P}) = -k_B T \ln \int e^{-\beta H(\mathbf{X})}d\mathbf{X}.$$

We can identify the quantity

$$\delta S(\alpha_1,\ldots, \alpha_m) = -\frac{1}{T}[F(\alpha_1,\ldots, \alpha_m|\mathscr{P}) - F(\mathscr{P})],$$

as the difference of the entropy between that states with parameters $(\alpha_1,\ldots, \alpha_m, \mathscr{P})$ and that one with $\mathscr{P}$. From equation (3.2), one can write

$$p(\alpha_1,\ldots, \alpha_m) = \exp\left[\frac{\delta S(\alpha_1,\ldots, \alpha_m)}{k_B}\right]. \tag{3.3}$$

The above relation is called Boltzmann–Einstein principle, due to the fact that equation (3.3) is (at least formally) similar to the equation $S = k_B \ln W$ (assuming $W \propto p$).

In macroscopic systems, the fluctuations, with respect to the thermodynamic equilibrium, of extensive variables are small (compared with the corresponding mean values); therefore, it is sensible to expand $\delta S(\alpha_1,\ldots, \alpha_n)$ around the mean values $\{\langle\alpha_j\rangle\}$, i.e. the values at the equilibrium $\{\alpha_j^*\}$:

$$\delta S(\alpha_1,\ldots, \alpha_n) \simeq -\frac{1}{2}\sum_{i,j}\delta\alpha_i G_{ij}\delta\alpha_j, \tag{3.4}$$

where

$$\delta\alpha_i = \alpha_i - \alpha_i^*, \;\; G_{ij} = -\frac{\partial S}{\partial\alpha_i\partial\alpha_j}\bigg|_{\alpha*}.$$

Since δS must be negative, the matrix $\mathcal{G}$, whose elements are G_{ij} (matrices will be indicated by calligraphic symbols in the following), has positive eigenvalues, and the small fluctuations are well described by a multivariate Gaussian distribution:

$$p(\alpha_1,\dots,\alpha_m) = \sqrt{\frac{\det\mathcal{G}}{(2\pi k)^m}}\exp\left[-\frac{1}{2k}\sum_{i,j}\delta\alpha_i G_{ij}\delta\alpha_j\right]. \tag{3.5}$$

Let us now introduce the affinities (also called conjugated variables)

$$X_n = \frac{\partial\delta S}{\partial\delta\alpha_n} = -\sum_j G_{nj}\delta\alpha_j;$$

using equation (3.5) and integrating by parts we have

$$\langle\delta\alpha_j X_n\rangle = \int\delta\alpha_j X_n p(\alpha_1,\dots,\alpha_m)\,d\alpha_1\cdots d\alpha_m = -k_{\mathrm{B}}\delta_{jn}, \tag{3.6}$$

so that

$$\langle\delta\alpha_i\delta\alpha_j\rangle = k_{\mathrm{B}}[\mathcal{G}^{-1}]_{ij}. \tag{3.7}$$

The matrix elements G_{ij} are computed in equilibrium, and therefore from the knowledge of the entropy at the thermodynamic equilibrium it is possible to determine the fluctuations. The result in equation (3.7) is the core of the Einstein theory of small fluctuations; although obtained in a very simple way from a Gaussian approximation, it is rather interesting at a conceptual level: the fluctuations, even if small, can be detected from the equilibrium thermodynamics.

3.1.1 Fluctuations and responses

Note that equation (3.1) is a particular case of equation (3.7); another similar relation can be derived for the fluctuations of the magnetization m:

$$\langle(\delta m)^2\rangle = \frac{k_{\mathrm{B}}}{N}T\chi, \tag{3.8}$$

where χ is the magnetic susceptibility. In general, the variance of the fluctuations of a variable is proportional to a suitable response function: in equation (3.1) one has the specific heat $c_v = \partial\langle e\rangle/\partial T$ which determines the variation of the mean energy per particle at varying the temperature; analogously, in equation (3.8) the susceptibility $\chi = \partial\langle m\rangle/\partial h$ gives the magnetization change as a function of the magnetic field h.

It is possible to show that in general there exists a link between fluctuations and (suitable) responses. Consider a small perturbation of the Hamiltonian:

$$H_0(\mathbf{X}) \to H_0(\mathbf{X}) - \lambda U(\mathbf{X});$$

in order to understand the effect of the perturbation $-\lambda U(\mathbf{X})$ on an observable A, we can compute

$$\langle \delta A \rangle = \langle A \rangle_\lambda - \langle A \rangle_0,$$

where

$$\langle A \rangle_\lambda = \frac{\int A\, e^{-\beta(H_0 - \lambda U)} d\mathbf{X}}{\int e^{-\beta(H_0 - \lambda U)} d\mathbf{X}}.$$

A simple computation shows that for small λ, one has

$$\langle A \rangle_\lambda \simeq \langle A \rangle_\lambda + \lambda\beta(\langle AU \rangle_0 - \langle A \rangle_0 \langle U \rangle_0),$$

and therefore

$$\langle \delta A \rangle_\lambda = \lambda\beta\langle (A - \langle A \rangle_0)(U - \langle U \rangle_0) \rangle_0. \tag{3.9}$$

It is simple to realize that equation (3.8) corresponds to the case where $\lambda = h$ and U is the magnetization, while for $U = E$ and $\lambda = \delta T/T$ one has equation (3.1).

The relation (3.9) is a particular aspect of the fluctuation–dissipation theorem, which will be discussed in detail in chapter 4, and shows how fluctuations, although small, can be computed in terms of responses.

3.1.2 A digression about temperature fluctuations

We conclude this section with a brief discussion on the 'temperature fluctuations'. Using in a blind way equation (3.7) one can derive misleading results, which, surprisingly, appear also in well-known textbooks. For instance if we express S as a function of T and V, we have

$$\delta S = -\frac{C_V}{2T^2}\delta T^2 + \frac{1}{2T}\frac{\partial P}{\partial V}\bigg|_T \delta V^2, \tag{3.10}$$

which represents an explicit form of equation (3.4) with $\alpha_1 = T$, $\alpha_2 = V$, $G_{11} = C_V/T^2$, $G_{22} = \partial P/\partial V\,|_T/T$, $G_{12} = G_{21} = 0$, where $C_V = Nc_V$ is the heat capacity at constant volume. Now using (3.10) and (3.7), we obtain

$$\langle \delta T^2 \rangle = \frac{k_B T^2}{c_v}. \tag{3.11}$$

Let us note that the expression of δS in equation (3.10) is a thermodynamic relation, whereas the quantity δS appearing in equation (3.3) has a rather different status: it is a function of mechanical variables. Therefore, the derivation of equation (3.11) is just formal in the sense of the mere manipulation of symbols.

From equations (3.11) and (3.1) we have $\langle \delta T^2 \rangle \langle \delta E^2 \rangle = k_B T^4$, which is equivalent to

$$\langle \delta\beta^2 \rangle \langle \delta E^2 \rangle = 1. \tag{3.12}$$

The above result has been interpreted by some authors as a 'thermodynamic uncertainty relation' [1] formally similar to the Heisenberg principle: a 'thermodynamic complementarity' where energy and temperature play the role of conjugate variables. In contrast, other authors, such as Kittel, claim that such an idea, as well as the concept of temperature fluctuations, are misleading [2].

The argument is simple: temperature is just a parameter of the canonical ensemble (as well as in any ensemble) which describes the statistics of the system, and, therefore, it is fixed by definition; we share the conclusion of Kittel. In a molecular dynamics simulation at fixed energy, it is common practice to look at the fluctuations of the kinetic energy that can be used to determine the specific heat. Because the mean value of the kinetic energy per particle K is proportional to the temperature, it might be concluded that the fluctuations of K are related to the fluctuations of the temperature. The above argument is wrong: the value of K can fluctuate from sample to sample, but not its mean.

Mandelbrot has shown that the problem of assigning the temperature of the thermal reservoir, to which a system had been in thermal contact, can receive a satisfactory answer within the framework of estimation theory [3]: his analysis helps to understand the confusion on such a topic of the temperature fluctuation. Let us briefly discuss the argument considering an experiment on a system with a given (unknown) β, with $\mathcal{N}$ independent measurements of energy $E_1, E_2, \dots, E_{\mathcal{N}}$ [4]. If we know the Hamiltonian of the system, we have the probability distribution of E:

$$P(E\,|\,\beta) = \frac{\omega(E)}{Z(\beta)} e^{-\beta E},$$

where $\omega(E)$ is the density of states and $Z(\beta)$ is the partition function. From the experimental data we obtain an estimate $\hat{\beta}$ (depending on $E_1, E_2, \dots, E_{\mathcal{N}}$) of the 'true' β, i.e. $\beta = \langle\hat{\beta}\rangle$, and a probability density $F(\hat{\beta}\,|\,\beta)$. The following important Rao–Cramér inequality holds:

$$\overline{\delta\beta^2} = \int (\beta - \hat{\beta})^2 F(\hat{\beta}\,|\,\beta)\, d\hat{\beta} \geqslant \frac{1}{\mathcal{N}\sigma_E^2}, \tag{3.13}$$

where σ_E^2 is the variance of E. The above relation sounds formally similar to equation (3.12) which is obtained by an incorrect use of Einstein's fluctuation formula, but the analogy is misleading: in mathematical statistics the quantity $\overline{\delta\beta^2}$ measures the uncertainty in the determination of the value of β and not the fluctuations of its values.

3.2 Onsager relations

In the previous section we discussed the Einstein statistical theory for small fluctuations at equilibrium; the next natural step is the building of a theory which includes also the dynamical properties. The main contribution to such a problem is due to Onsager; however, in this section we do not present his original approach, and, following the well-known book by de Groot and Mazur [5], we shall use an approach in terms of Langevin equations.

Since $\{\alpha_1, \ldots, \alpha_m\}$ are macroscopic variables it is sensible to adopt a Langevin equation for their time evolution

$$d\alpha_n = F_n(\alpha_1, \ldots, \alpha_m)dt + \sum_j B_{nj}\, dW_j,$$

where the noisy terms give the contribution of the fast part of the dynamics. In general, it is very difficult to find the function $F_n(\alpha_1, \ldots, \alpha_m)$, as well as the matrix $\mathscr{B}$, from the first principles. When the system is close to thermodynamic equilibrium, i.e. the fluctuations $\alpha_i - \alpha_i^*$ are small, we can try to go on with some simplifications. Assuming for simplicity $\alpha_i^* = 0$, the natural candidates are coupled linear Langevin equations of the form

$$d\alpha_n = -\sum_j M_{nj}\alpha_j dt + \sum_j B_{nj}\, dW_j, \tag{3.14}$$

whose associated Fokker–Planck (FP) equation is

$$\frac{\partial P(\alpha_1, \ldots, \alpha_m)}{\partial t} = \sum_{nj} M_{nj} \frac{\partial}{\partial \alpha_n}[\alpha_j P(\alpha_1, \ldots, \alpha_m)] + \frac{1}{2}\sum_{nj} Q_{jn} \frac{\partial^2 P(\alpha_1, \ldots, \alpha_m)}{\partial \alpha_j \partial \alpha_n}, \tag{3.15}$$

where $\mathscr{Q} = \mathscr{B}\mathscr{B}^T$ and ^{T} means transposition. The reasons of such a choice are rather clear: the fact that the quantities $(\alpha_1, \ldots, \alpha_m)$ are small suggests the linear structure of the Langevin equation; in addition the stationary probability distribution of equation (3.14), according to the Einstein theory, is a multivariate Gaussian:

$$P(\alpha_1, \ldots, \alpha_m) = \mathscr{N} \exp\left[\frac{S(\alpha_1, \ldots, \alpha_m)}{k_B}\right] = \mathscr{N} e^{-\frac{1}{2k_B}\sum_{ij} G_{ij}\alpha_i\alpha_j}, \tag{3.16}$$

where $\mathscr{N}$ is a normalization factor. Note that the matrices $\mathscr{M}$, $\mathscr{B}$ and $\mathscr{G}$ cannot be independent; denoting by σ the matrix with elements $\langle \alpha_i \alpha_j \rangle$, i.e. from equation (3.7) $\sigma = k_B \mathscr{G}^{-1}$, it is easy to show the following relation, also known as the Lyapunov equation, which appears in problems of linear stability analysis:

$$\mathscr{Q} = (\mathscr{M}\sigma + \sigma \mathscr{M}^T). \tag{3.17}$$

Indeed, considering the Langevin equation (3.14) and denoting by $\{\alpha_1, \ldots, \alpha_m\}$ and $\{\alpha_1', \ldots, \alpha_m'\}$ the state of the system at time t and $t + dt$, respectively, we have

$$\alpha_i' = \alpha_i - \sum_k M_{ik}\alpha_k dt + \sum_k B_{ik} dW_k, \quad \alpha_j' = \alpha_j - \sum_l M_{jl}\alpha_l dt + \sum_l B_{jl} dW_l.$$

Using the property $\langle dW_k dW_l \rangle = \delta_{kl} dt$, we obtain

$$\langle \alpha_i' \alpha_j' \rangle = \langle \alpha_i \alpha_j \rangle - \sum_l M_{jl} \langle \alpha_l \alpha_i \rangle dt - \sum_k M_{ik} \langle \alpha_k \alpha_j \rangle dt + \sum_k B_{ik} B_{jk} dt + O(dt^{3/2}), \text{ namely}$$

$$\sigma_{ij}' = \sigma_{ij} - \left(\sum_l M_{jl}\sigma_{li} + \sum_k M_{ik}\sigma_{kj} - \sum_k B_{ik} B_{jk}\right)dt + O(dt^{3/2}).$$

In the stationary situation one has $\sigma'_{ij} = \sigma_{ij}$, and using the symmetry property of σ we eventually have equation (3.17).

Let us recall the definitions of the affinities $X_n = \frac{\partial S}{\partial \alpha_n}$, from which we have $dS = \sum_n X_n d\alpha_n$. For instance, in the microcanonical ensemble where S is a function of E, V and $\{n_k\}$, n_k being the number of particles of the kth molecular species, we have $\alpha_1 = E$, $\alpha_2 = V$, and $\alpha_k = n_k$ $(k > 2)$, so that

$$dS = \frac{1}{T}dE + \frac{P}{T}dV + \sum_i \frac{\mu_i}{T}dn_i,$$

and the corresponding conjugated variables are

$$X_1 = \frac{1}{T},\ X_2 = \frac{P}{T},\ X_k = \frac{\mu_k}{T} \text{ for } k > 2.$$

Here μ_i denote the chemical potentials of the different species. Since $\mathscr{G}$ is invertible we can write

$$\alpha_j = -\sum_n [\mathscr{G}^{-1}]_{jn} X_n,$$

and introducing the Onsager matrix

$$\mathscr{L} = \mathscr{M}\mathscr{G}^{-1}, \tag{3.18}$$

we can write equation (3.14) in the form

$$d\alpha_n = \sum_j L_{nj} X_j dt + \sum_j B_{nj}\, dW_j. \tag{3.19}$$

Now we have to determine the constrains on the matrix $\mathscr{L}$ in order to assure the consistency of equation (3.14) with the time reversal symmetry. For this purpose we have to introduce the central concept of detailed balance (DB).

3.2.1 Detailed balance and its relevance

We now discuss the DB condition and its role in statistical mechanics. Only for simplicity we consider the case of Markov chains with discrete states and discrete time: DB holds if

$$P_i^s P_{i\to j} = P_j^s P_{j\to i},$$

where P_i^s denotes the stationary probability to be in the state i and $P_{i\to j}$ the transition probability to jump from state i to state j.

In order to understand the relevance of DB in the mathematical treatment of statistical mechanics, consider a trajectory $\mathscr{T}(N) = (i_1, i_2, \ldots, i_{N-1}, i_N)$ and the time-reversed trajectory obtained inverting the time, i.e. $\mathscr{T}^I(N) = (i_N, i_{N-1}, \ldots, i_2, i_1)$. A simple computation shows that if DB holds, then the two trajectories have the same probability:

$$\text{Prob}(\mathscr{T}(N)) = P_{i_1}^s \prod_{t=1}^{N-1} P_{i_t \to i_{t+1}} = P_{i_N}^s \prod_{j=1}^{N-1} P_{i_{N-j+1} \to i_{N-j}} = \text{Prob}(\mathscr{T}^I(N)).$$

We can say that the DB condition in a stochastic context is analogous to the time reversal property for the Newtonian dynamics.

In the case of a master equation

$$\frac{dP_n}{dt} = \sum_{n' \neq n} W_{n' \to n} P_{n'} - \gamma_n P_n, \tag{3.20}$$

where $W_{n' \to n}$ are the transition rates from the state n' to the state n and $\gamma_n = \sum_{n' \neq n} W_{n \to n'}$ is the escape rate from state n, we have the DB condition if

$$P_n^s W_{n \to n'} = P_{n'}^s W_{n' \to n}; \tag{3.21}$$

in a similar way for a system with continuous states DB corresponds to

$$P^s(\mathbf{y}) W(\mathbf{y} \to \mathbf{x}, t) = P^s(\mathbf{x}) W(\mathbf{x} \to \mathbf{y}, t).$$

Remarkably the validity (or lack) of the DB condition can be checked in a simple way in terms of correlation functions: considering for instance two functions $F(\mathbf{x}(t))$ and $G(\mathbf{x}(t))$, it is easy to show that if DB holds one has

$$\langle F(\mathbf{x}(t)) G(\mathbf{x}(0)) \rangle = \langle F(\mathbf{x}(0)) G(\mathbf{x}(t)) \rangle.$$

Therefore, the DB condition for equation (3.14) is satisfied, i.e. the Langevin equation is consistent with the time reversal property of the original problem, if—for all i and j and couple of times t, t', one has

$$\langle \alpha_i(t) \alpha_j(t') \rangle = \langle \alpha_i(t') \alpha_j(t) \rangle. \tag{3.22}$$

We are now ready to discuss the Onsager relations. Let us start rewriting equation (3.19) in the form

$$\alpha_j(t + dt) = \alpha_j(t) + \sum_k L_{jk} X_k(t) dt + \sum_k B_{jk} dW_k.$$

Then, multiplying by $\alpha_i(t)$ and taking the average, we obtain

$$\langle \alpha_i(t) \alpha_j(t + dt) \rangle = \langle \alpha_i(t) \alpha_j(t) \rangle + \sum_k L_{jk} \langle X_k(t) \alpha_i(t) \rangle dt;$$

in a similar way we have

$$\langle \alpha_i(t + dt) \alpha_j(t) \rangle = \langle \alpha_i(t) \alpha_j(t) \rangle + \sum_k L_{ik} \langle X_k(t) \alpha_j(t) \rangle dt. \tag{3.23}$$

Then, using equation (3.6), we have that equation (3.22) holds if the following Onsager relations are valid:

$$L_{ji} = L_{ij}. \tag{3.24}$$

The relevance of the Onsager relations is rather transparent: when we are not able to derive a dynamical model from the first principles and we are forced to adopt a pragmatic approach with a linear Langevin equation as equation (3.14), if equation (3.22) does not hold, then this implies that the model is not consistent with the time reversal symmetry.

Let us note that, using equations (3.18) and (3.7), the Onsager relations (3.24) correspond to

$$\mathcal{M}\sigma = (\mathcal{M}\sigma)^T, \tag{3.25}$$

which, exploiting equation (3.17), can be also rewritten as

$$\mathcal{M}\mathcal{Q} = (\mathcal{M}\mathcal{Q})^T, \tag{3.26}$$

where now the noise matrix $\mathcal{Q}$ explicitly appears. Indeed, one has

$$\begin{aligned}(\mathcal{M}\mathcal{Q})^T &= \mathcal{Q}^T\mathcal{M}^T = (\sigma^T\mathcal{M}^T + \mathcal{M}\sigma^T)\mathcal{M}^T = (\mathcal{M}\sigma + \mathcal{M}\sigma^T)\mathcal{M}^T \\ &= \mathcal{M}\sigma\mathcal{M}^T + \mathcal{M}\mathcal{M}\sigma = \mathcal{M}(\sigma\mathcal{M}^T + \mathcal{M}\sigma) = \mathcal{M}\mathcal{Q},\end{aligned} \tag{3.27}$$

where in the second equality we have used the definition of $\mathcal{Q}$, equation (3.17), and in the third and in the fourth we have exploited equation (3.25).

3.2.2 The Onsager relations in chemical reactions

It is instructive to discuss the connection between DB and the Onsager relations in the case of a chemical reaction of m chemical reactants $n_1, n_2, \ldots, n_m$, with the condition $n_1 + n_2 + \cdots + n_m = N$. In his paper Onsager discussed the case $N = 3$, but the treatment can be easily extended to the general case. Consider the master equation (3.20) with $P_i = n_i/N$, $P_i^s = n_i^{eq}/N$, and assume that DB (3.21) holds; the entropy $S(n_1, \ldots, n_m)$ of a system with m species of particles $n_1, \ldots, n_m$, with the constraint $\sum_{j=1}^m n_j = N$, can be obtained from the probability $P(n_1, \ldots, n_m)$ given by the multinomial distribution:

$$P(n_1, \ldots, n_m) = \frac{N!}{\prod_j n_j!} \prod_{j=1}^m (P_j^s)^{n_j} = \exp\left[\frac{S(n_1, \ldots, n_m)}{k_B}\right].$$

Using the Stirling approximation one obtains

$$\frac{S(n_1, \ldots, n_m)}{k_B} = -\sum_j n_j \ln\frac{n_j}{NP_j} = -\sum_j n_j \ln\frac{n_j}{n_j^{eq}}.$$

The entropy S takes its maximum value (zero) at $n_j^{eq} = NP_j^s$, and introducing the variables $\alpha_j = n_j - n_j^{eq}$, we can compute S for small α_j:

$$S = \delta S = -\frac{k_B}{2}\sum_j \frac{\alpha_j^2}{n_j^{eq}}.$$

Using the conjugated variables $X_j = \partial S/\partial\alpha_j = -k_B\alpha_j/n_j^{eq}$, we can write the equation for the $\{\alpha_j\}$ (actually for their mean values):

$$\frac{d\alpha_j}{dt} = \sum_n W_{n\to j}\alpha_n = \sum_n L_{jn}X_n = -\sum_n W_{n\to j}n_n^{eq}\frac{X_n}{k_B}, \tag{3.28}$$

which implies

$$L_{jn} = -\frac{1}{k_B} W_{n \to j} n_n^{eq} = -\frac{N}{k_B} W_{n \to j} P_n^s.$$

Therefore, we have that in such a reaction problem DB is equivalent to the Onsager relations.

3.2.3 Onsager relations and ideal gas effusion

We present now, mainly for pedagogical purposes, a discussion of a simple case of thermodiffusion. Following the paper of Patitsas [6] we analyze the effusion of a perfect gas, showing how, in such a system, explicit computations from the elementary kinetic theory allow us to find explicitly the Langevin equation for suitable interesting macroscopic variables, as well as to check the correctness of the Onsager relations.

Consider two identical chambers, each with volume V, containing $2N_0$ atoms of mass m, with total energy $2U_0$. At the initial time the two chambers are isolated; denote by $N_L(N_R)$ the number of atoms in the left (right) chamber and assume that $\Delta N = N_R - N_0 \ll N_0$, and $\Delta U = U_R - U_0 \ll U_0$. At $t = 0$ a small gap of radius r is opened so the atoms can move between the chambers. Using the Sackur–Tetrode formula [7]:

$$\begin{aligned} \frac{S}{k_{\mathrm{B}}} &= (N_0 + \Delta N)\ln\left[\frac{V}{N_0 + \Delta N}\left(\frac{U_0 + \Delta U}{N_0 + \Delta N}\right)^{3/2}\right] \\ &\quad + (N_0 - \Delta N)\ln\left[\frac{V}{N_0 - \Delta N}\left(\frac{U_0 - \Delta U}{N_0 - \Delta N}\right)^{3/2}\right] + \text{constant}, \end{aligned} \tag{3.29}$$

in the limits $\Delta N \ll N_0$ and $\Delta U \ll U_0$, we have

$$\frac{\delta S}{k_{\mathrm{B}}} = -\frac{3N_0}{2U_0^2}\Delta U^2 - \frac{5}{2N_0}\Delta N^2 + \frac{3}{U_0}\Delta N \Delta U. \tag{3.30}$$

Let us introduce the variables $\alpha_1 = \Delta U / U_0$, $\alpha_2 = \Delta N / N_0$, and the corresponding conjugate variables:

$$X_1 = -\frac{3}{2}k_{\mathrm{B}} N_0 \alpha_1 + \frac{3}{2}k_{\mathrm{B}} N_0 \alpha_2,\ X_2 = \frac{3}{2}k_{\mathrm{B}} N_0 \alpha_1 - \frac{5}{2}k_{\mathrm{B}} N_0 \alpha_2.$$

In the following with simple arguments from the kinetic theory we shall find the coefficients of the matrix $\mathscr{L}$, and then verify the Onsager relation.

The flux Φ_N of particles from the left side to the right side is obtained by integrating the velocity distribution over a hemisphere with the z-axis alligned to the gap; in a similar way we have the energy flux Φ_U from the left side to the right side:

$$\Phi_N = \int_{v_z > 0} d^3\mathbf{v}\, f(v_x, v_y, v_z) v_z, \quad \Phi_U = \frac{m}{2}\int_{v_z > 0} d^3\mathbf{v}\, f(v_x, v_y, v_z) |\mathbf{v}\,|^2 v_z,$$

where $f(v_x, v_y, v_z)$ is the Maxwell–Boltzmann distribution at temperature T. With simple computations we have

$$\Phi_N = \frac{\sqrt{N_0 U_0}}{V\sqrt{3\pi m}}, \quad \Phi_U = \frac{4U_0}{3N_0}\Phi_N,$$

and therefore, since the area of the gap is πr^2, we have the equations for the evolution of U and N (actually for their mean values):

$$\frac{dU}{dt} = -\pi r^2\left(\frac{\partial \Phi_U}{\partial U}\Delta U + \frac{\partial \Phi_U}{\partial N}\Delta N\right) = -\frac{1}{2}\pi r^2 \Phi_U(3\alpha_1 - \alpha_2), \tag{3.31}$$

$$\frac{dN}{dt} = -\pi r^2\left(\frac{\partial \Phi_N}{\partial U}\Delta U + \frac{\partial \Phi_N}{\partial N}\Delta N\right) = -\frac{1}{2}\pi r^2 \Phi_N(\alpha_1 + \alpha_2). \tag{3.32}$$

The characteristic timescale is given by $\tau_0 = N_0/(\pi r^2 \Phi_N)$; from (3.31) and (3.32) we have the matrix $\mathcal{M}$

$$\mathcal{M} = \frac{1}{\tau_0}\begin{pmatrix} 2 & -2/3 \\ 1/2 & 1/2 \end{pmatrix}.$$

Since the diagonal matrix element M_{11} is greater than M_{22}, we note that the energy transfer process is inherently faster than the particle transfer. For the Onsager matrix we use equations (3.30) and (3.16) to get $\mathcal{G}$, obtaining

$$\mathcal{L} = \mathcal{M}\mathcal{G}^{-1} = \frac{1}{3k_B N_0 \tau_0}\begin{pmatrix} 8 & 4 \\ 4 & 3 \end{pmatrix};$$

and therefore we have the expected result $L_{12} = L_{21}$. To take into account the fluctuations, we have to find the matrix $\mathcal{Q} = \mathcal{B}\mathcal{B}^T$, which is obtained from equation (3.17).

Let us summarize: in the case of perfect gas effusion using kinetic gas theory, the Onsager symmetry relation can be explicitly verified; in addition the approach to equilibrium is determined by timescales that are explicitly calculated. In the paper by Patitsas the reader can find an interesting analysis of the dynamics of the system approaching equilibrium with different initial conditions, e.g. $\Delta N = 0$, or $\Delta U = 0$ at $t = 0$.

Of course in the more interesting case of interacting gases, it is not possible to give a simple exact treatment; for instance denoting with ℓ the mean-free path of the molecules, the ratio ℓ/r plays a central role (for a discussion of such a topic we refer the reader to the book by Kreuzer [8]).

3.2.4 Variables with different parity under time reversal

A mechanical system described by the canonical variables $(\mathbf{Q}, \mathbf{P})$ evolves according to the Hamilton equations and therefore it is invariant under time reversal, i.e. under the transformations

$$t \to -t, \ \mathbf{Q} \to \mathbf{Q}, \ \mathbf{P} \to -\mathbf{P}.$$

Given a trajectory $\mathscr{T}(T) = \{\mathbf{Q}(t), \mathbf{P}(t); 0 < t < T\}$, the inverse trajectory is $\mathscr{T}^I(T) = \{\mathbf{Q}(T - t), -\mathbf{P}(T - t); 0 < t < T\}$. We define even variables as those that do not change sign under the time reversal (as Q_1, Q_2,...), and odd variables those that do change sign under the time reversal (as P_1, P_2,...). At each variable α_i we associate the parity $\varepsilon_i = 1$ in the even case, and $\varepsilon_i = -1$ for the odd case. Equation (3.22) can be then generalized as follows

$$C_{ij}(t) = \langle \alpha_i(t)\alpha_j(0) \rangle = \varepsilon_i \varepsilon_j C_{ji}(t),$$

and therefore the Onsager relations take the general form:

$$L_{ij} = \varepsilon_i \varepsilon_j L_{ji}.$$

It is not straightforward to determine the value of ε_i;, however, it is not particularly difficult from an analysis of the Langevin equation.

In order to discuss such an aspect we study the electric circuit *RLC*, as in figure 3.1, including a fluctuating voltage due to the system at temperature *T*, and a fluctuating charge due to the contact with a large neutral charge reservoir.

One can write the Langevin equation by equating the rate of gain of charge on the capacitor to the current *i* less a leakage term γq into the reservoir, plus a fluctuating term $\Delta q(t)$ arising from the reservoir; from the conservation of the electric charge *q* and Kirchoff's law, we have the coupled Langevin equations:

$$\frac{d}{dt}q = -\gamma q + i + \Delta q(t),$$

and

$$\frac{d}{dt}i = -\frac{1}{LC}q - \frac{R}{L}i + \frac{\Delta V(t)}{L}.$$

We assume that $\Delta q(t)$ and $\Delta V(t)$ are combinations of white noises η_1 and η_2:

$$\Delta q(t) = B_{11}\eta_1 + B_{12}\eta_2, \quad \frac{\Delta V(t)}{L} = B_{21}\eta_1 + B_{22}\eta_2.$$

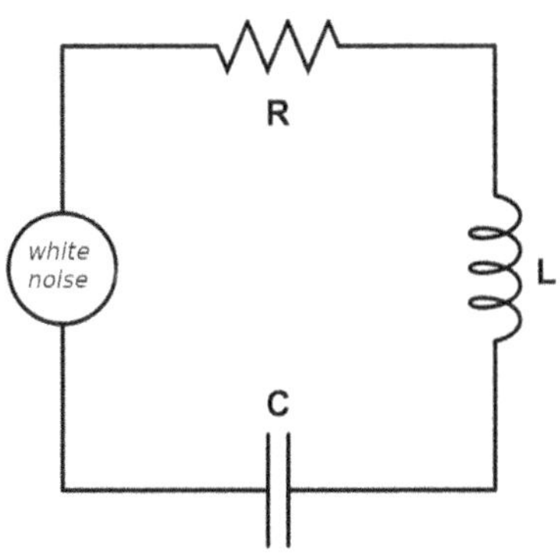

Figure 3.1. *RLC* circuit with white noise.

From the statistical mechanics we have that the stationary distribution is

$$P^s(q, i) = \mathcal{N} \exp\left(-\frac{E}{k_B T}\right) = \mathcal{N} \exp\left(-\frac{Li^2}{2k_B T} - \frac{q^2}{2Ck_B T}\right),$$

and an easy computation shows that, for the off-diagonal entries of the Onsager matrix

$$(\mathcal{M}\sigma)_{12} = -(\mathcal{M}\sigma)_{21}.$$

The origin of the sign is clear: the quantity q is even under time inversion and its ε is $+1$, while the current i is odd and therefore its ε is -1.

Using equation (3.17) we have

$$B_{12} = B_{21} = 0, \;\; B_{11} = \sqrt{2k_B T\gamma c}, \;\; B_{22} = \frac{\sqrt{2k_B TR}}{L},$$

and therefore

$$\langle \Delta V(t) \Delta V(t') \rangle = 2k_B T\delta(t - t').$$

This corresponds to the celebrated Nyquist theorem.

3.2.5 Beyond detailed balance

Let us now briefly discuss systems without DB: such a topic will be treated in detail in chapters 5 as well as in the second part of this book. Consider a random walk on a lattice with N sites and periodic boundary conditions: the walker remains on the same site with probability 1/3, and performs a step to the right or to the left with probabilities $P_{n\to n\pm1} = (1 \pm \varepsilon)/3$ for $n = 1, 2, \ldots, N - 1$, with $P_{N\to1} = (1 + \varepsilon)/3$ and $P_{1\to N} = (1 - \varepsilon)/3$ and $-1 < \varepsilon < 1$. The stationary probabilities $P_j^s = 1/N$ are independent of ε, and DB does not hold for any $\varepsilon \neq 0$. The lack of DB is associated with the presence of a 'current', i.e. a preferential direction of the walker: the variable $v_t = n_t - n_{t-1}$ (which represents a discrete time velocity) takes values 0 and ±1 with probabilities 1/3 and $(1 \pm \varepsilon)/3$, and therefore $\langle v \rangle = 2\varepsilon/3$.

As a continuous version of the above system we can consider the Langevin equation on a ring of length L:

$$dx = vdt + \sqrt{2b}\; dW \;\; \text{with } v = \text{constant}.$$

The stationary probability is still $P^s(x) = 1/L$, but for $v \neq 0$ one has a lack of DB and the presence of a current, as is well clear noting that

$$\langle x(t)|x(0)\rangle = x(0) + vt \bmod L.$$

The above examples are rather simple and therefore it is trivial to identify the current; in general, however, the absence of DB does not allow us immediate evidence of the feature associated with the breaking of the time reversal symmetry. Let us mention the following system, also known as Brownian gyrator:

$$dx_1 = -\alpha x_1 dt + \lambda x_2 dt + \sqrt{2T_1}\; dW_1,$$

$$dx_2 = \mu x_1 dt - \gamma x_2 dt + \sqrt{2T_2}\, dW_1.$$

If $\alpha + \gamma > 0$ and $\alpha\gamma > \mu\lambda$ the real parts of $\mathscr{M}$'s eigenvalues are positive, and one has the convergence of the probability distribution $P(x_1, x_2, t)$ to a stationary distribution. From equation (3.17), with simple computations one finds the covariance matrix σ:

$$\sigma = \begin{pmatrix} \dfrac{T_2\lambda^2 - T_1\mu\lambda + T_1\gamma(\alpha + \gamma)}{(\alpha + \gamma)(\alpha\gamma - \lambda\mu)} & \dfrac{T_2\alpha\lambda + T_1\gamma\mu}{(\alpha + \gamma)(\alpha\gamma - \lambda\mu)} \\ \dfrac{T_2\alpha\lambda + T_1\gamma\mu}{(\alpha + \gamma)(\alpha\gamma - \lambda\mu)} & \dfrac{T_2\mu^2 - T_1\mu\lambda + T_1\alpha(\alpha + \gamma)}{(\alpha + \gamma)(\alpha\gamma - \lambda\mu)} \end{pmatrix}.$$

With a simple computation we can verify that if $T_1 \neq T_2$ the Onsager relations do not hold.

In such a system the absence of DB manifests itself with a 'circular current': introducing the angle

$$\theta(t) = \arctan\left(\frac{x_2(t)}{x_1(t)}\right),$$

one has

$$\langle\theta(t)|\theta(0)\rangle = \theta(0) + \omega t \bmod 2\pi, \text{ with } \omega \neq 0.$$

The current vanishes, $\omega = 0$, if $T_1 = T_2$ or in the trivial case $\lambda = \mu = 0$ which corresponds to decoupling the two variables. Chapter 5 is devoted to a detailed discussion of systems without DB.

3.3 Dynamics of fluctuations

We discuss now the solution of the FP equation (3.15) which is associated with the Langevin equation

$$d\mathbf{x} = -\mathscr{M}\mathbf{x}\, dt + \mathscr{B} d\mathbf{W}. \tag{3.33}$$

Let us start noting that, from the linear structure of the FP equation, given the initial condition $P(\mathbf{x}, 0)$, we can write the solution as

$$P(\mathbf{x}, t) = \int P(\mathbf{y}, 0) W(\mathbf{y} \to \mathbf{x}, t)\, d\mathbf{y}; \tag{3.34}$$

the problem is how to find $W(\mathbf{y} \to \mathbf{x}, t)$.

One can face the problem with two different approaches:

(a) starting from the FP equation which is ruled by a linear operator $\mathscr{L}_{\mathrm{FP}}$

$$\partial_t P = \mathscr{L}_{\mathrm{FP}} P,$$

we can write the spectral decomposition of the operator

$$\mathscr{L}_{\mathrm{FP}} = \sum_n \lambda_n \psi_n(\mathbf{x}) \phi_n(\mathbf{x})$$

in terms of right ($\psi_n(\mathbf{x})$) and left ($\phi_n(\mathbf{x})$) eigenfunctions, where $\{\lambda_n\}$ are the eigenvalues, and then express $W(\mathbf{y} \to \mathbf{x}, t) = \sum_n \psi_n(\mathbf{x})\phi_n(\mathbf{y})\exp\{-\lambda_n t\}$;

(b) noting that equation (3.33) is linear, we have that the stochastic process $\mathbf{x}(t)$ must be Gaussian: this implies that $W(\mathbf{y} \to \mathbf{x}, t)$ is a multivariate Gaussian and so it is enough to find the parameters of such a Gaussian.

In order to present the second approach we discuss the one-dimensional case

$$\partial_t P(x, t) = \partial_x\left(\frac{x}{\tau}P(x, t)\right) + c\partial^2_{xx}P(x, t),$$

corresponding to the Langevin equation

$$dx = -\frac{x}{\tau}dt + \sqrt{2c}\,dW, \tag{3.35}$$

which can be easily solved

$$x(t) = x(0)e^{-\frac{t}{\tau}} + \sqrt{2c}\int_0^t e^{-\frac{t-t'}{\tau}}\,dW(t').$$

One has that $W(y \to x, t)$ is a Gaussian with mean value

$\tilde{x}(t) = \langle x(t)|x(0) = y\rangle$ and variance $\sigma^2(t) = \langle x(t)^2|x(0) = y\rangle - \langle x(t)|x(0) = y\rangle^2$.

From equation (3.35) it is easy to find

$$\tilde{x}(t) = \langle x(t)|x(0) = y\rangle = ye^{-\frac{t}{\tau}},\quad \sigma^2(t) = c\tau\left(1 - e^{-\frac{2t}{\tau}}\right);$$

therefore, we have

$$W(y \to x, t) = \frac{1}{\sqrt{2\pi\sigma^2(t)}}\,e^{-\frac{(x-\tilde{x}(t))^2}{2\sigma^2(t)}}.$$

From equation (3.34), for any $P(x, 0)$ we have

$$P(x, t) \;\to\; P^s(x) = \frac{1}{\sqrt{2\pi c\tau}}\,e^{-\frac{x^2}{2c\tau}},$$

and the relaxation to the $P^s(x)$ occurs with a characteristic time $O(\tau)$.

For the general case we can follow the same approach

$$W(\mathbf{y} \to \mathbf{x}, t) = \sqrt{\frac{|\det \mathscr{Z}(t)|}{(2\pi)^N}}\,\exp\left(-\frac{1}{2}\sum_{ij}(x_i - \tilde{x}_i(t))Z_{ij}(t)(x_j - \tilde{x}_j(t))\right),$$

where

$$\tilde{x}_i(t) = \langle x_i(t)|\mathbf{x}(0) = \mathbf{y}\rangle,$$

and

$$(\mathscr{Z}^{-1}(t))_{ij} = \langle (x_i(t) - \tilde{x}_i(t)(x_j(t) - \tilde{x}_j(t))|\mathbf{x}(0) = \mathbf{y}\rangle.$$

For the above quantities we can use the same approach used in the one-dimensional case. The solution of equation (3.33) is

$$\mathbf{x}(t) - \mathcal{G}(t)\mathbf{x}(0) + \int_0^t \mathcal{G}(t - t')\mathcal{B}\, d\mathbf{W}(t'), \tag{3.36}$$

where

$$\mathcal{G}(t) = e^{-\mathcal{M}t} = \sum_{n=0}^{\infty} \frac{(-\mathcal{M}t)^n}{n!};$$

we find

$$\langle \mathbf{x}(t) | \mathbf{x}(0) = \mathbf{y} \rangle = \mathcal{G}(t)\mathbf{y},$$

and in a similar way we can compute $\mathcal{Z}^{-1}(t)$. Also in this case for any $P(\mathbf{x}, 0)$, we have

$$P(\mathbf{x}, 0) \to P^s(\mathbf{x}) = \sqrt{\frac{\det \mathcal{G}}{(2\pi)^N}} \exp\left(-\frac{1}{2} \sum_{ij} x_i G_{ij} x_i \right),$$

where $\mathcal{G}^{-1} = \mathcal{Z}(\infty)$, and the relaxation time can be easily determined by the eigenvalues of $\mathcal{M}$.

We conclude this section stressing that the above results are valid for any general linear system, i.e. for any linear Langevin equation (3.33), even in the absence of DB. In addition, from equation (3.36) it is easy to show that

$$C_{ij}(t) = \sum_k G_{ik}(t) C_{kj}(0), \tag{3.37}$$

or in compact form $\mathcal{C}(t) = \mathcal{G}(t)\mathcal{C}(0)$. In the determination of the Onsager relations we followed the original presentation, even if it is quite easy to arrive to the same results from equation (3.37). In the limit of small t we have $\mathcal{G}(t) \simeq \mathcal{I} - \mathcal{M}t$; therefore one has

$$\mathcal{C}(t) = \mathcal{C}(0) - \mathcal{M}\mathcal{C}(0)t, \tag{3.38}$$

and reminding that $\mathcal{C}(0) = k_B \mathcal{G}^{-1}$, we have

$$\mathcal{C}(t) = \mathcal{C}(0) + k_B \mathcal{L} t, \tag{3.39}$$

which is nothing but equation (3.23).

3.4 Langevin equation and physics

We conclude this chapter with a simple but important remark on the short time behaviour of the correlation functions: from equation (3.38) for small t one has

$$C_{ii}(t) \simeq C_{ii}(0) - a_i t \text{ with } a_i > 0, \tag{3.40}$$

and this result is valid in a generic (even nonlinear) Langevin equation.

Such a behaviour for small delay is in disagreement with the result for any deterministic system, where instead of equation (3.40) one has

$$C_{ii}(t) \simeq C_{ii}(0) - b_i t^2 \text{ with } b_i > 0.$$

This suggests that in the application of a Langevin equation it is necessary to understand its limits of validity.

Recalling the physics at the base of the Langevin equation for the Brownian particle, it is important to realize that the description in terms of a stochastic differential equation cannot be appropriate for a time delay smaller than the typical collision times between the Brownian particle and the molecules representing the bath. In order to illustrate this point we consider the following solvable system of a harmonic chain of $2N$ particles described by the canonical variables $\{q_n, p_n\}$, $n = \pm 1, \pm 2, \ldots, \pm N$ and a 'heavy' intruder whose variables are (Q, P):

$$H = \frac{P^2}{2M} + \sum_{i=\pm 1,\ldots,\pm N} \frac{p_i^2}{2m} + \frac{k}{2}\sum_{i=-N}^{N+1} (q_i - q_{i-1})^2, \quad Q \equiv q_0. \tag{3.41}$$

In the limit $N \gg 1$ and $M \gg m$ one has a problem somehow similar to the colloidal particle interacting with the molecules of the fluid; in our case the linear structure of the system allows us to show that the momentum P of the intruder evolves according to a linear Langevin equation

$$dP = -\frac{P}{\tau_c}dt + \sqrt{2c}\, dW,$$

where $\tau_c = M/\sqrt{4km}$ and c is determined by the energy of the system [9]. For the correlation function one has:

$$C(t) = \langle P(t)P(0) \rangle = C(0)\exp\left(-\frac{t}{\tau_c}\right). \tag{3.42}$$

The actual correlation function computed with the deterministic dynamic starting with initial conditions at the thermal equilibrium, is in very good agreement with equation (3.42) apart from for a small value of t; this is unavoidable, and it is a consequence of the fact that in deterministic systems one has $C(t)/C(0) = 1 - \frac{t^2}{\tau_D^2} + O(t^3)$, while for a Langevin dynamics $C(t)/C(0) = 1 - \frac{t}{\tau_c} + O(t^2)$. Therefore, there exists a minimal timescale τ_M such that the process can be described in a proper way by the Langevin dynamics only on timescales larger than τ_M. Comparing the behaviour of $C(t)/C(0)$ for the deterministic case and the Langevin equation we have that τ_M must be at least $O(\tau_D^2/\tau_c)$.

References

[1] Uffink J and Van Lith J 1999 Thermodynamic uncertainty relations *Found. Phys.* **29** 655–92

[2] Kittel C 1988 Temperature fluctuation: an oxymoron *Phys. Today* **41** 93

[3] Mandelbrot B B 1989 Temperature fluctuation: a well-defined and unavoidable notion *Phys. Today* **42** 71–3

[4] Falcioni M, Villamaina D, Vulpiani A, Puglisi A and Sarracino A 2011 Estimate of temperature and its uncertainty in small systems *Am. J. Phys.* **79** 777–85

[5] De Groot S R and Mazur P 2013 *Non-Equilibrium Thermodynamics* (North Chelmsford, MA: Courier Corporation)

[6] Patitsas S N 2014 Onsager symmetry relations and ideal gas effusion: a detailed example *Am. J. Phys.* **82** 123–34

[7] Huang K 2008 *Statistical Mechanics* (New York: Wiley)

[8] Kreuzer H J 1981 *Nonequilibrium Thermodynamics and Its Statistical Foundations* (Oxford: Clarendon)

[9] Ford G W, Kac M and Mazur P 1965 Statistical mechanics of assemblies of coupled oscillators *J. Math. Phys.* **6** 504–15

IOP Publishing

Nonequilibrium Statistical Mechanics
Basic concepts, models and applications
Alessandro Sarracino, Andrea Puglisi and Angelo Vulpiani

Chapter 4

Linear response theory and fluctuation–dissipation theorem

> I am of the opinion that the task of the theory consists in constructing a picture of the external world that exists purely internally and must be our guiding star in all thought and experiment.
>
> Ludwig Boltzmann

Surely one of the most relevant, and useful, results of nonequilibrium statistical mechanics is the existence of a precise relation between the spontaneous fluctuations of the system and the response to external perturbations upon physical observables. This result allows for the possibility of studying nonequilibrium features, mainly the response to time-dependent external perturbations, in terms of time-dependent correlations computed for the unperturbed system. The idea dates back to Einstein's work on Brownian motion, and to the Onsager regression hypothesis, according to which the relaxation of a macroscopic perturbation follows the same laws governing the dynamics of fluctuations in equilibrium systems. A very important contribution to the fluctuation–dissipation theorem (FDT) theory for Hamiltonian systems near thermodynamic equilibrium, is due to Kubo. From an equilibrium statistical mechanics study (e.g. via molecular dynamics simulations) one can compute correlation functions and then determine transport coefficients, for instance the electrical conductivity. In this chapter, after a brief historical excursus, we discuss the main results of the theory of FDT, showing that a formulation is possible which, under suitable assumptions, is valid also in nonequilibrium systems. See reference [1] for an accurate history of the FDT, and reference [2] for a general discussion.

4.1 From Einstein and Onsager to Kubo

The first example of FDT is due to Einstein in his study of Brownian motion (see chapter 1). Here we present the result discussing the Langevin equation

doi:10.1088/978-0-7503-6229-0ch4

$$\frac{dv}{dt} = -\frac{v}{\tau} + \sqrt{2c}\,\eta + \frac{F(t)}{m}, \tag{4.1}$$

where m is the mass of the particle, τ a characteristic time, η a white noise with zero mean and unit variance, c the noise amplitude and $F(t)$ an external force. In the case $F = 0$ one has $\langle v \rangle = 0$; a natural problem is to wonder about the effect of switching on the force $F \neq 0$, e.g. $F(t) = F_0\theta(t)$, where $\theta(t)$ is the Heaviside step function: of course the expectation at long times is $\langle v \rangle = F_0\mu$, where μ is called mobility.

More specifically, one can solve equation (4.1):

$$v(t) = v(0)e^{-\frac{t}{\tau}} + \int^t e^{-\frac{t-t'}{\tau}}\left[\sqrt{2c}\,\eta(t') + \frac{F(t')}{m}\right] dt',$$

which yields

$$\langle v(\infty)\rangle = \lim_{t\to\infty}\int^t e^{-\frac{t-t'}{\tau}}\frac{F(t')}{m}\,dt'.$$

If $F(t) = F_0\theta(t)$, one has

$$\langle v(\infty)\rangle = \frac{F_0}{m}\tau.$$

Recalling that $D = \langle v^2\rangle\tau = k_B T\tau/m$ (see chapter 1), we have

$$\mu = \frac{D}{k_B T}. \tag{4.2}$$

The diffusion coefficient D can be expressed in terms of $\langle v(t)v(0)\rangle$: writing $x(t) = x(0) + \int_0^t v(t')\,dt'$, one has

$$\langle (x(t) - x(0))^2\rangle = \int_0^t\int_0^t \langle v(t')v(t'')\rangle\,dt'\,dt'',$$

and for large t we have $\langle (x(t) - x(0))^2\rangle \simeq 2t\int_0^\infty \langle v(t)v(0)\rangle\,dt$, from which

$$D = \int_0^\infty \langle v(t)v(0)\rangle\,dt.$$

The relation (4.2) shows that the mobility, that measures a response to an external perturbation, is determined by D which depends upon the unperturbed autocorrelation, namely $\langle v(t)v(0)\rangle$, and the thermal energetic scale $k_B T$.

Another similar problem is the computation of the electrical conductivity σ which links the electric current $\mathbf{j}$ with the electric field $\mathbf{E}$:

$$\mathbf{j} = \sigma\mathbf{E}.$$

In the (phenomenological) Drude theory one has

$$\sigma = \frac{q^2\rho}{m}\tau, \tag{4.3}$$

where q, ρ and m are the electric charge, the density and the mass of the conducting particles, respectively, while τ is a characteristic time of the unperturbed dynamics. Let us stress the analogy with the Einstein relation (4.2): even here a nonequilibrium property, σ, is expressed in terms of equilibrium features. In the following we will see that it is possible to derive the Drude formula, in the framework of a general approach, and to give a proper interpretation on the time τ.

A fundamental concept, for the link between equilibrium and nonequilibrium[1], that can be traced back to Onsager's 1931 paper, is the regression hypothesis which connects equilibrium fluctuations to the relaxation from a nonequilibrium state: the average regression of fluctuations of a macroscopic variable will obey the same laws as the corresponding macroscopic irreversible process. We do not try to enter into an analysis of the work of Onsager, who often was rather obscure (among the students his course was known as 'sadistical mechanics'); we prefer to discuss the problem in the context of linear Langevin equations which, with good accuracy, describe the dynamics if the system is close to equilibrium. Let us consider the linear system already discussed in chapter 3:

$$d\mathbf{x} = -\mathcal{M}\mathbf{x}\,dt + \mathcal{B}d\mathbf{W}, \tag{4.4}$$

whose solution is

$$\mathbf{x}(t) = \mathcal{G}(t)\mathbf{x}(0) + \int_0^t \mathcal{G}(t-t')\mathcal{B}\,d\mathbf{W}(t') \quad \text{with} \quad \mathcal{G}(t) = e^{-\mathcal{M}t}. \tag{4.5}$$

Let us now introduce the differential response function $R_{ij}(t)$: at time $t = 0$ one performs a small perturbation of the variable x_j, i.e. $x_j(0) \to x_j(0) + \delta x_j(0)$, and we are interested in the mean effect on the variable x_i at time t

$$R_{ij}(t) = \frac{\overline{\delta x_i(t)}}{\delta x_j(0)}.$$

As a consequence of the linear structure of the system, from equation (4.5) we have

$$\overline{\delta \mathbf{x}(t)} = \mathcal{G}(t)\delta\mathbf{x}(0),$$

i.e. $\mathcal{G}(t) = \mathcal{R}(t)$. From equation (4.5) it is easy to show for the correlation function $C_{ij}(t) = \sum_k G_{ik}(t)C_{kj}(0)$, or in compact form $\mathcal{C}(t) = \mathcal{G}(t)\mathcal{C}(0)$. In conclusion, we get the link between the response and the correlation:

$$\mathcal{R}(t) = \mathcal{C}(t)\mathcal{C}^{-1}(0), \tag{4.6}$$

which (as a consequence of what explained in section 3.2.5), is not limited to systems at thermal equilibrium. Let us note that the study of an 'impulsive' perturbation is not a limitation: e.g. in the linear regime from the (differential) linear response one can understand the effect of a generic perturbation. This is true even in nonlinear

[1] The topic of the FDT was first introduced in the context of close-to-equilibrium statistical mechanics. In more recent times a formulation which applies to a much wider framework has been proposed, see section 4.2.

systems; consider a system ruled by the evolution law, which can also include some noisy terms:

$$\frac{d\mathbf{x}}{dt} = \mathbf{Q}(\mathbf{x}),$$

and apply an infinitesimal perturbation, when the system is in its steady state: $\mathbf{Q}(\mathbf{x}) \to \mathbf{Q}(\mathbf{x}) + \delta\mathbf{Q}(t)$ with $\delta\mathbf{Q}(t) = 0$ for $t < 0$. Then one has

$$\overline{\delta x_i(t)} = \sum_j \int_0^t R_{ij}(t - t')\delta Q_j(t')\, dt'.$$

A seminal contribution to the FDT is given by Kubo's formula, which holds in the case of equilibrium dynamics. Let us consider a system described by the Hamiltonian $H_0(\mathbf{X}) - F(t)V(\mathbf{X})$, where $H_0(\mathbf{X})$ is the unperturbed part, $F(t) = 0$ for $t < t_0$, and V is a generic potential, function of the state $\mathbf{X}$, coupled to the external force F. From a perturbative analysis of the Liouville equation, Kubo was able to show the effect of the term $-F(t)V(\mathbf{X})$ on the average of a generic observable A:

$$\overline{\delta A(t)} = \langle A\rangle' - \langle A\rangle = \int_{t_0}^t dt' R(t - t')F(t'),$$

where $R(t)$ is the response function

$$R(t) = \beta\left\langle A(t)\frac{dv}{dt}\bigg|_{t_0}\right\rangle, \tag{4.7}$$

$\langle\ldots\rangle'$ and $\langle\ldots\rangle$ denoting the average performed on the perturbed and unperturbed system, respectively. Let us note that equation (4.7) is nothing but the average $\overline{\delta A(t)}$ in the case $F(t) = \delta(t)$.

As an important application of the above result we mention the computation of the electrical conductivity: just for simplicity we consider the case of an electric field $E(t)$ in the x direction, so we have $F(t) = qE(t)$ and $V = x$. Considering the velocity along the x direction, $A = v_x$, we have

$$\overline{\delta v_x(t)} = q\int^t dt' R(t - t')E(t') \text{ where } R(t) = \beta\langle v_x(t)v_x(0)\rangle,$$

and we can define a characteristic time τ in a rather natural way:

$$\int_0^\infty \langle v_x(t)v_x(0)\rangle\, dt = \frac{k_B T}{m}\tau.$$

A motivation of the previous formula is given by the simple case of an exponential shape $\langle v_x(t)v_x(0)\rangle = \langle v_x^2\rangle e^{-t/\tau}$. Noting that $j_x = q\rho\overline{\delta v_x(\infty)}$ we have the Drude formula (4.3), but now with a non-ambiguous definition of τ.

4.1.1 Response in frequency and Johnson–Nyquist spectrum

Before the general treatment of FDT developed by Onsager and Kubo, some more specific results were obtained in particular cases by Johnson and Nyquist. These are expressed as a relation between response functions and correlations in the frequency representation, that is often the most useful form in experiments.

Let us consider that case where the external perturbation is periodic in time:

$$F(t) = Re[K_0 e^{i\omega t}], \tag{4.8}$$

with frequency ω and amplitude K_0. Because of linearity, the response reads

$$\overline{\delta A(t)} = Re[\chi(\omega) K_0 e^{i\omega t}], \tag{4.9}$$

where $\chi(\omega)$ is the admittance defined as the Fourier transform of the response function $R(t)$:

$$\chi(\omega) = \int_{-\infty}^{\infty} R(t) e^{-i\omega t} dt = \beta \int_{t_0}^{\infty} \langle \dot{V}(t_0) A(t) \rangle e^{-i\omega t} dt. \tag{4.10}$$

To illustrate the general result, let us consider a Langevin equation with a retarded frictional force $\gamma(t)$:

$$\frac{dv(t)}{dt} = -\int_{t_0}^{t} \gamma(t - t') v(t') dt' + \frac{1}{m}\eta(t) + \frac{1}{m}F(t), \qquad t > t_0. \tag{4.11}$$

The usual Langevin equation is obtained in the limit $\gamma(t) = \gamma\delta(t)$. The response of $v(t)$ to a periodic perturbation of the form (4.8) reads

$$\overline{\delta v(t)} = Re[\chi(\omega) K_0 e^{i\omega t}], \tag{4.12}$$

where the admittance is given by

$$\chi(\omega) = \frac{1}{m} \frac{1}{i\omega + \tilde{\gamma}(\omega)}, \tag{4.13}$$

with $\tilde{\gamma}(\omega)$ the Fourier transform of $\gamma(t)$. This result means that, for a periodic perturbation, the admittance is analogous to the mobility. At equilibrium, i.e. for $F \equiv 0$, the autocorrelation of $v(t)$ can be calculated, and its Fourier transform reads

$$\int_0^{\infty} e^{-i\omega t} \langle v(t_0) v(t_0 + t) \rangle dt = \langle v^2 \rangle \frac{1}{i\omega + \tilde{\gamma}(\omega)}, \tag{4.14}$$

and therefore the FDT

$$\chi(\omega) = \frac{1}{m\langle v^2 \rangle} \int_0^{\infty} \langle v(t_0) v(t_0 + t) \rangle e^{-i\omega t} dt \tag{4.15}$$

is recovered. From equation (4.14) one can calculate the power spectrum of $v(t)$ at equilibrium, which yields

$$S_u(\omega) = \int_{-\infty}^{\infty} e^{-i\omega t} \langle v(t_0) v(t_0 + t) \rangle dt = \langle v^2 \rangle \frac{2\tilde{\gamma}(\omega)}{[\tilde{\gamma}(\omega)]^2 + \omega^2}. \tag{4.16}$$

An interesting example of equation (4.11) is represented by the equation governing the charge $Q(t)$ contained in a capacitor of capacitance C, in a simple RC circuit without any externally applied electromotive force:

$$R\frac{dQ}{dt} = -\frac{Q(t)}{C} + \eta(t), \tag{4.17}$$

being $\eta(t)$ a noisy internal electromotive force due to the interactions with a thermostat at temperature T. A series of experiments by Johnson [3], supported by the theory of Nyquist [4], showed the existence of such spontaneous fluctuations of voltage at the edges of the resistor $u = Q/C$, due to the thermal motion of charge carriers. The power spectrum of these voltage fluctuations, i.e. $S_u(\omega)$, was the quantity measured in the experiment. The above equation for the RC circuit can be written in the form

$$\frac{du}{dt} = -\frac{u(t)}{RC} + \eta'(t), \tag{4.18}$$

with $\eta'(t) = \eta/RC$. This is equivalent to equation (4.11) with $\gamma(t) = \frac{1}{RC}\delta(t)$, i.e. $\tilde{\gamma}(\omega) = \frac{1}{RC}$, resulting in

$$S_u(\omega) = \langle u^2 \rangle \frac{2RC}{1 + (RC\omega)^2}. \tag{4.19}$$

The energy of the capacitor is $E = Q^2/2C$, and the equipartition of energy, required at equilibrium, imposes that $\langle E \rangle = \frac{k_B T}{2}$, which yields $\langle u^2 \rangle = \langle Q^2 \rangle / C^2 = 2\langle E \rangle / C = k_B T/C$. This result, placed in equation (4.19), gives the spectrum of the Johnson–Nyquist noise

$$S_u(\omega) = \frac{2Rk_B T}{1 + (\omega RC)^2}. \tag{4.20}$$

At low frequencies, $\omega \ll 1/RC$, the power spectrum becomes constant and does not depend on the capacitance of the circuit:

$$S_u(\omega) \simeq 2Rk_B T. \tag{4.21}$$

For a modern linear response study of chaotic systems in frequency domain we refer the reader to reference [5].

4.1.2 van Kampen's objection

Let us briefly discuss an objection due to van Kampen to the original (perturbative) derivation of the FDT. In the dynamical systems terminology, the argument is the following. Given an impulsive perturbation $\delta\mathbf{x}(0)$ on the state $\mathbf{x}$ of the system at time $t = 0$, the difference between the perturbed trajectory and the unperturbed one is

$$\delta x_i(t) = \sum_j \frac{\partial x_i(t)}{\partial x_j(0)} \delta x_j(0) + O(|\delta \mathbf{x}(0)|^2), \tag{4.22}$$

and averaging over the initial conditions, one has the mean response function:

$$R_{i,j}(t) = \frac{\overline{\delta x_i(t)}}{\delta x_j(0)} = \int \frac{\partial x_i(t)}{\partial x_j(0)} \rho(\mathbf{x}(0)) d\mathbf{x}(0). \tag{4.23}$$

In the presence of chaos the terms $\partial x_i(t)/\partial x_j(0)$ are $O(e^{\lambda t})$, where $\lambda > 0$ is the Lyapunov exponent. Therefore, the linear expansion (4.22) is not accurate for a time larger than $(1/\lambda)\ln(L/|\delta \mathbf{x}(0)|)$, where L is the typical fluctuation of the variable $\mathbf{x}$. Thus, the linear response theory seems to be valid only for extremely small perturbations (or times). According to the previous argument, requiring that the FDT holds up to 1 s when applied to the electrons in a typical conductor, would imply a perturbing electric field smaller than 10^{-20} V m^{-1} [6].

The above conclusion is in clear disagreement with the experience: the success of the linear theory for the computation of transport coefficients (e.g. electrical conductivity) in terms of correlation functions of the unperturbed system, is evident, and its validity has been, directly and indirectly, verified in a huge number of cases. Kubo suggested that the origin of the effectiveness of the FDT theory may reside in the 'constructive role of chaos': '*instability [of the trajectories] instead favours the stability of distribution functions, working as the cause of the mixing*' [7]. Actually, such an intuition is correct: we'll see in the next section that from a technical point of view van Kampen's criticism can be rejected: however, it had the merit of stimulating a deeper understanding of the FDT and its validity range.

4.2 A generalized fluctuation–dissipation theorem even for non-Hamiltonian systems

Parts of this section have been reprinted from [20] copyright (2019), with permission from AIP Publishing.

Let us now present a derivation of the FDT which holds under very general assumptions, even for chaotic systems. Consider a dynamical system $\mathbf{x}(0) \to \mathbf{x}(t) = U^t\mathbf{x}(0)$ with states $\mathbf{x}$ belonging to an N-dimensional space, where U^t describes the time evolution, that may be not deterministic (e.g. stochastic differential equations). Let us assume the existence of an invariant probability distribution $\rho(\mathbf{x})$, for which some 'absolutely continuity'-type conditions are required (see later), and the mixing character of the system (from which its ergodicity follows); no assumption is made on the dimensionality N of the system. Our aim is to study the behaviour of one component of $\mathbf{x}$, say x_i, when the system is subjected to an initial (non-random) perturbation such that $\mathbf{x}(0) \to \mathbf{x}(0) + \Delta x_0$.

An instantaneous kick modifies the density of the system into $\rho'(\mathbf{x})$, related to the invariant distribution by $\rho'(\mathbf{x}) = \rho(\mathbf{x} - \Delta\mathbf{x}_0)$. Let us introduce the probability of transition from $\mathbf{x}_0 = \mathbf{x}(0)$ at time 0 to $\mathbf{x}$ at time t, $w(\mathbf{x}_0, 0 \to \mathbf{x}, t)$ (of course in a deterministic system, with evolution law $\mathbf{x}(t) = U^t x(0)$, one has

$w(\mathbf{x}_0, 0 \to \mathbf{x}, t) = \delta(x - U^t\mathbf{x}_0))$. We can write an expression for the mean value of the variable x_i, computed with the density of the perturbed system:

$$\langle x_i(t)\rangle' = \int\int x_i \rho'(x_0) w(\mathbf{x}_0, 0 \to \mathbf{x}, t)\, d\mathbf{x}\, d\mathbf{x}_0, \tag{4.24}$$

and for the mean value of x_i during the unperturbed evolution one has:

$$\langle x_i(t)\rangle = \int\int x_i \rho(\mathbf{x}_0) w(\mathbf{x}_0, 0 \to \mathbf{x}, t)\, d\mathbf{x}\, d\mathbf{x}_0. \tag{4.25}$$

Defining $\overline{\delta x_i} = \langle x_i\rangle' - \langle x_i\rangle$, if $\rho(\mathbf{x})$ is always different from zero, we have:

$$\begin{aligned}\overline{\delta x_i}\,(t) &= \int\int x_i \frac{\rho(\mathbf{x}_0 - \Delta\mathbf{x}_0) - \rho(\mathbf{x}_0)}{\rho(\mathbf{x}_0)} \rho(\mathbf{x}_0) w(\mathbf{x}_0, 0 \to \mathbf{x}, t)\, d\mathbf{x}\, d\mathbf{x}_0 \\ &= \langle x_i(t)\, F(\mathbf{x}_0, \Delta\mathbf{x}_0)\rangle,\end{aligned} \tag{4.26}$$

where

$$F(\mathbf{x}_0, \Delta\mathbf{x}_0) = \left[\frac{\rho(\mathbf{x}_0 - \Delta\mathbf{x}_0) - \rho(\mathbf{x}_0)}{\rho(\mathbf{x}_0)}\right]. \tag{4.27}$$

Note that the mixing property of the system is required to guarantee the decay to zero of the time-correlation functions and, thus, the switching off of the deviations from equilibrium.

In the case of an infinitesimal perturbation $\delta\mathbf{x}(0) = (\delta x_1(0) \cdots \delta x_N(0))$, if $\rho(\mathbf{x})$ is non-vanishing and differentiable, the function in equation (4.27) can be expanded to first order and one obtains:

$$\begin{aligned}\overline{\delta x_i}\,(t) &= -\sum_j \left\langle x_i(t) \frac{\partial \ln \rho(\mathbf{x})}{\partial x_j}\bigg|_{t=0} \right\rangle \delta x_j(0) \\ &\equiv \sum_j R_{i,j}(t)\delta x_j(0),\end{aligned} \tag{4.28}$$

which gives the linear response

$$R_{i,j}(t) = -\left\langle x_i(t) \frac{\partial \ln \rho(\mathbf{x})}{\partial x_j}\bigg|_{t=0} \right\rangle \tag{4.29}$$

of the variable x_i with respect to a perturbation of x_j. It is easy to repeat the computation for a generic observable $A(\mathbf{x})$, also in the case where $\delta x_j(0)$ depends on $\mathbf{x}(0)$, yielding

$$\overline{\delta A(t)} = -\sum_j \left\langle A(\mathbf{x}(t)) \frac{\partial \ln \rho(\mathbf{x})}{\partial x_j}\bigg|_{t=0} \delta x_j(0) \right\rangle. \tag{4.30}$$

4.2.1 Finite perturbations and the relevance of chaos

In the above derivation of the FDT we never used any approximation on the evolution of $\delta\mathbf{x}(t)$. Starting with the exact expression (4.26) for the response, only a linearization on the initial time perturbed density is needed, and this implies nothing but the smallness of the initial perturbation. In addition it is easy to understand that it is also possible to derive an FDT for finite perturbations: one has

$$\overline{\delta A(t)} = \langle A(\mathbf{x}(t))F(\mathbf{x}(0),\, \Delta\mathbf{x}(0))\rangle, \tag{4.31}$$

where the $F(\mathbf{x}(0), \Delta\mathbf{x}(0))$, given by equation (4.27), depends on the initial perturbation $\Delta\mathbf{x}(0)$ and on the invariant probability distribution.

Let us now discuss the real relevance of van Kampen's criticism. From the evolution of the trajectories difference, one can define the leading Lyapunov exponent λ by considering the absolute values of $\delta\mathbf{x}(t)$: at small $|\delta\mathbf{x}(0)|$ and large enough t one has

$$\langle \ln|\delta\mathbf{x}(t)|\rangle \simeq \ln|\delta\mathbf{x}(0)| + \lambda t. \tag{4.32}$$

On the other hand, in the FDT one deals with averages of quantities with sign, such as $\overline{\delta\mathbf{x}(t)}$. This apparently marginal difference is very important and allows for the possibility to derive the FDT avoiding van Kampen's objection.

Nevertheless the objection that van Kampen raised, has its importance for the accuracy of numerical computations. One can compute $R_{i,j}(t)$ by perturbing the variable x_j at time $t = t_0$ with a small perturbation of amplitude $\delta x_j(0)$ and then evaluating the separation $\delta x_i(t)$ between the two trajectories $\mathbf{x}(t)$ and $\mathbf{x}'(t)$ which are integrated up to a prescribed time $t_1 = t_0 + \Delta t$. At time $t = t_1$ the variable x_j of the reference trajectory is again perturbed with the same $\delta x_j(0)$, and a new sample $\delta\mathbf{x}(t)$ is computed and so forth; repeating $M \gg 1$ times the same procedure, the mean response can be evaluated. In the presence of chaos, the two trajectories $\mathbf{x}(t)$ and $\mathbf{x}'(t)$ typically separate exponentially in time: therefore the mean response is the result of a delicate balance of terms whose absolute values grow exponentially in time. Thus very high statistics (i.e. very large M) is needed in order to properly capture this balance and to compute $R_{i,j}(t)$ for large t: the uncertainty in the computation of $R_{i,j}(t)$ typically increases in time as $e^{\alpha t}/\sqrt{M}$, where $\alpha \simeq \lambda$.

4.3 A series of remarks

This section is devoted to a discussion of the generalized FDT and its use in different contexts. As first we note that the nature of the statistical steady state (either equilibrium, or nonequilibrium) does not play any role for the validity of the FDT. This is evident noting that even in the linear Langevin equation the FDT is surely valid for generic coupling matrix $\mathscr{M}$ and noise matrix $\mathscr{B}$; on the contrary, the system is at equilibrium only for certain $\mathscr{M}$ and $\mathscr{B}$, i.e. as discussed in chapter 3, when the Onsager relations hold.

When the form of $\rho(\mathbf{x})$ is not known, as usually in non-Hamiltonian systems, the relation (4.29) does not give detailed quantitative information. However, even in such a case, it represents a connection between the mean response function $R_{i,j}$ and a suitable correlation function, computed in the non perturbed systems:

$$\left\langle x_i(t) f_j(\mathbf{x}(0)) \right\rangle, \quad \text{with } f_j(\mathbf{x}) = -\frac{\partial \ln \rho}{\partial x_j}. \tag{4.33}$$

In the case of finite perturbations, the FDT (4.31) is nonlinear in the perturbation $\Delta \mathbf{x}_0$, and thus no simple relation analogous to equation (4.30) holds. Nevertheless, equation (4.31) guarantees the existence of a link between the properties of the unperturbed system and the response to finite perturbations.

The existence of such a link has a straightforward consequence for systems with one single characteristic time, e.g. a low-dimensional system such as the Lorenz model: a generic correlation function in principle gives information on the relaxation time of finite size perturbations, even when the invariant measure ρ is not known. In systems with many different characteristic times, such as the fully developed turbulence, one has a more complicated scenario: different correlation functions can show different behaviours, which depend on the observables. In addition, the amplitude of the perturbation can play a major role in determining the response, because different amplitudes may trigger different response mechanisms with different time properties [8].

4.3.1 The Kubo formula

Let us now show how from the generalized FDT one can easily derive the Kubo formula for a perturbation of the form $H_0(\mathbf{x}) \to H_0(\mathbf{x}) - F(t)V(\mathbf{x})$. In such a case the differential linear response function, corresponding to $F(t) - \delta(t)$, for an observable $A(\mathbf{x})$ after a time t is given by

$$R(t) = \beta \left\langle A(\mathbf{x}(t)) \frac{dv(\mathbf{x}(t'))}{dt'} \bigg|_{t'=0} \right\rangle. \tag{4.34}$$

We show that the above formula can be derived by the generalized FDT (4.30). In the canonical ensemble we have $\ln P_s = -\beta H_0 + \text{constant}$; therefore from equation (4.30) one has

$$R(t - t') = \beta \sum_n \left\langle A(\mathbf{q}(t), \mathbf{p}(t)) \left(\frac{\partial H_0}{\partial q_n} \delta q_n + \frac{\partial H_0}{\partial p_n} \delta p_n \right) \bigg|_{t'} \right\rangle, \tag{4.35}$$

and using the Hamilton equations one has that a perturbation $-\delta(t)V(\mathbf{x})$ corresponds to $\delta q_n = \partial V / \partial p_n$ and $\delta p_n = -\partial V / \partial q_n$: therefore, from the previous equation one has

$$R(t) = \beta \sum_n \left\langle A(\mathbf{q}(t), \mathbf{p}(t)) \left(\frac{\partial H_0}{\partial q_n} \frac{\partial V}{\partial p_n} - \frac{\partial H_0}{\partial p_n} \frac{\partial V}{\partial q_n} \right) \bigg|_{t'=0} \right\rangle.$$

Noting that

$$\sum_n \frac{\partial H_0}{\partial q_n}\frac{\partial V}{\partial p_n} - \frac{\partial H_0}{\partial p_n}\frac{\partial V}{\partial q_n} = \sum_n \frac{dp_n}{dt}\frac{\partial V}{\partial p_n} + \frac{dq_n}{dt}\frac{\partial V}{\partial q_n} = \frac{dv}{dt},$$

we have the Kubo formula from the generalized FDT.

4.3.2 Gaussian distributions

When the stationary state is a multivariate Gaussian distribution,

$$\ln \rho(\mathbf{x}) = -\frac{1}{2}\sum_{i,j} \sigma_{ij}^{-1} x_i x_j + \text{const.} \tag{4.36}$$

with $\{\sigma_{ij}\}$ a positive matrix, using equation (4.6) we have the elements of the linear response matrix in terms of the correlation functions:

$$R_{i,j}(t) = \sum_k \sigma_{jk}^{-1} \langle x_i(t) x_k(0) \rangle . \tag{4.37}$$

The above result is nothing but equation (4.6) obtained for linear Langevin equations with a simple direct computation.

It is interesting to note that one can have a Gaussian invariant distribution, even in nonlinear systems: as an important example we can mention the inviscid hydrodynamics, where the Liouville theorem holds, and a quadratic invariant exists [9, 10]. For this class of nontrivial systems, as a consequence of the Gaussian shape of the invariant probability distribution, one has a quite simple relation between the responses and the correlations. In spite of this nice result, the inviscid hydrodynamics is ruled by nonlinear equations and the behaviour of the correlation functions, at variance with the case of linear Langevin equations, cannot be analytically computed.

4.3.3 Apparent violation of the fluctuation–dissipation theorem and marginal distributions

In the FDT, for the responses, one uses the expression of the invariant probability distribution for the whole vector, which describes the system $\mathbf{x} = (x_1, x_2, \ldots, x_N)$. It is natural to wonder whether a partial knowledge of the state, e.g. $\mathbf{y} = (x_1, x_2, \ldots, x_m)$, with $m < N$, can be enough to understand the response features.

It is straightforward to understand that the knowledge of a marginal distribution

$$p_i(x_i) = \int \rho(x_1, x_2, \ldots) \prod_{j \neq i} dx_j \tag{4.38}$$

does not allow for the computation of the auto-response. From equation (4.37), one has that $R_{i,i}(t)$, even in the Gaussian case, is not proportional to $\langle x_i(t) x_i(0) \rangle$:

$$R_{i,i}(t) \neq -\left\langle x_i(t) \frac{\partial \ln p_i(x_i)}{\partial x_i}\bigg|_{t=0} \right\rangle . \tag{4.39}$$

This observation, although trivial, is rather interesting when one wants to use the FDT for understanding the equilibrium (or nonequilibrium) nature of a system. In particular, there are some possible pitfalls.

The Kubo formula can be written as

$$R_{AV}(t) = -\beta \frac{d}{dt}\langle A(t)V(0)\rangle = -\beta \frac{d}{dt} C_{AV}(t);$$

therefore for the quantity $\chi_{AV}(t) = \int_0^t R_{AV}(t')dt'$, which is a kind of susceptibility, one has

$$\chi_{AV}(t) = -\beta[C_{AV}(t) - C_{AV}(0)]. \tag{4.40}$$

Plotting $\chi_{AV}(t)$ versus $C_{AV}(t)$ one has a linear shape whose slope determines the temperature. Some authors propose that a deviation from (4.40) is an indication of a failure of the FDT and a possible mark of being in a nonequilibrium situation. For instance, looking at the quantity

$$\frac{1}{R_{AV}(t)} \frac{dC_{AV}(t)}{dt},$$

it seems that it is possible to introduce a sort of 'effective temperature' [11]. Actually, such an idea can surely be useful in the context of disordered systems with aging features, where clearly equation (4.40) does not hold. On the contrary, a violation of equation (4.40) is not an indication that in the system nontrivial behaviours occur. This can be understood noting that even in a $2d$ linear Langevin equation, at varying the entries of $\mathscr{M}$ and $\mathscr{B}$, the shape of $\chi_{BA}(t)$ versus $C_{BA}(t)$ can show very different behaviours, even rather similar to the results observed in spin glasses. For a more accurate discussion on this topic, see references [12, 13].

As an example, we briefly discuss the motion of an intruder in a granular gas. A granular fluid is made of macroscopic particles subject to external forcing, and therefore is characterized by dissipative interactions (inelastic collisions) and non-equilibrium dynamics (in chapter 8, the physics of granular matter will be discussed in some detail). In order to describe the velocity autocorrelation of the intruder (in one dimension, for simplicity) and its linear response, one can introduce a two-variable model which is—at least formally—in the category of the so-called Brownian gyrators, previously discussed in chapter 3:

$$M\dot{v}(t) = -\Gamma[v(t) - u(t)] + \sqrt{2\Gamma T_{tr}}\mathscr{E}_v(t), \tag{4.41a}$$

$$M'\dot{u}(t) = -\Gamma' u(t) - \Gamma v(t) + \sqrt{2\Gamma' T_b}\mathscr{E}_u(t), \tag{4.41b}$$

where v is the velocity of the intruder with mass M, while u describes a local velocity field (a local average of the velocities of the particles surrounding the intruder) whose dynamics is coupled with that of the tracer, Γ is a viscosity, Γ' and M' are effective parameters to be determined. We denote by T_{tr} the intruder kinetic temperature, and by T_b the kinetic temperature of the granular fluid, playing the role of a nonequilibrium bath, while $\mathscr{E}_v$ and $\mathscr{E}_u$ are delta-correlated noises with zero mean and unitary variance. The dilute limit can be obtained for large values of Γ' and M', which implies small u. Equations (4.41) can describe real systems for not-too-high density; in the elastic limit ($T_{tr} = T_b$), the coupling with u can still be important, but the equilibrium FDT is recovered. Out of equilibrium, one can apply the formula (4.29) to express the response in terms of correlation functions. Since the system is linear, the stationary distribution is a bivariate Gaussian, and from equation (4.29) directly follows

$$R_{vv}(t) = \frac{\overline{\delta v(t)}}{\delta v(0)} = \sigma_{vv}^{-1}\langle v(t)v(0)\rangle + \sigma_{uv}^{-1}\langle v(t)u(0)\rangle, \tag{4.42}$$

where σ_{vv}^{-1} and σ_{uv}^{-1} are the elements of the inverse covariance matrix and can be easily expressed in terms of the model parameters. The main message we want to stress here is the central role played by the correlations between the variable v and the local velocity field u: in general, the knowledge of the statistical properties of v alone, e.g. the measure of its marginalized probability density function, is not enough to reconstruct the response to an external perturbation. Such a difficulty is rather common and it has been stressed in a rather vivid way by Onsager and Machlup in their seminal work on fluctuations and irreversible processes, with the caveat: *how do you know you have taken enough variables, for it to be Markovian?* [14]. The relevance of the use of the proper variables in the applications will be discussed in detail in chapter 6 and, with some examples, in those following it.

4.3.4 Chaotic dissipative systems

One could object that in a chaotic deterministic dissipative system equation (4.29) cannot be applied, because the invariant measure is not smooth at all. Actually, the invariant measure of a chaotic dissipative system is singular with respect to the Lebesgue measure, and its attractor has a multifractal structure [15]. Due to the singular nature of the invariant probability distribution, it is intuitive to understand that the behaviour of a generic perturbation cannot be described by the dynamics of the unperturbed system, and the reason is quite simple: perturbing a state on the attractor one obtains, apart from very peculiar cases, a state out of the attractor. Therefore, a generic response function has two contributions, parallel and perpendicular to the attractor [16]. Each perturbation can be written as the sum of two contributions

$$\delta F(t) = \delta F_{\parallel}(t) + \delta F_{\perp}(t),$$

and the effect of such a perturbation on the response on an observable A attains the form

$$\overline{\delta A(t)} = \int_0^t R_{\parallel}^{(A)}(t - t')\delta F_{\parallel}(t')dt' + \int_0^t R_{\perp}^{(A)}(t - t')\delta F_{\perp}(t')dt'. \tag{4.43}$$

Only for the part $R_{\parallel}^{(A)}$ can one hope to have an FDT, i.e. it can be expressed in terms of a correlation function computed in the unperturbed dynamics on the attractor. In contrast, for the second contribution (from the contracting directions), the response can be obtained only numerically.

On the other hand, in systems with many degrees of freedom at a practical level, there are at least two good reasons which allow us to hope that the singular structure of the invariant measure is not really relevant. First, we note that in systems with many degrees of freedom, for a non-pathological observable the contribution of $R_{\perp}^{(A)}$ should be negligible. In addition, a small amount of noise, that is always present in a physical system, smooths the distribution $\rho(\mathbf{x})$ and the FDT can be derived. We recall that this 'beneficial' noise has the important role of selecting the natural measure, and, in the numerical experiments, it is provided by the round-off errors of the computer [17].

4.3.5 A fluctuation–dissipation theorem in generic noisy systems

The introduction of noise allows for a smooth invariant distribution. On the other hand, often, the explicit shape of P_s remains an open problem: it is therefore desirable to have an alternative way to write the response functions. We mention the generalized FDT derived through path-integral approaches, a method due to Furutsu and Novikov dating back to the 1960s and reformulated later on in the context of statistical field theory. In the following, using an approach not based on path-integral methods, we discuss a unified formulation for the generalized FDT, holding in and out of equilibrium, even in the presence of multiplicative or non-Gaussian noises [18, 19].

In order to explain the idea we can consider a discrete-time Markov process of the form

$$\mathbf{x}(t) = \mathbf{x}(t - \Delta t) + \mathbf{f}(\mathbf{x}(t - \Delta t))\Delta t + \zeta(t), \tag{4.44}$$

where Δt is a fixed time interval and the noise $\{\zeta(t)\}$ is a sequence of random variables distributed according to a given probability distribution $\tilde{P}(\zeta(t)|\mathbf{x}(t - \Delta t))$: of course in the limit $\Delta t \to 0$ and assuming $\zeta(t)$ to be white noise with zero average and variance proportional to Δt, one has a Langevin equation. Assuming that equation (4.44) admits a stationary distribution $P_s(\mathbf{x})$, by using the Markov property one has

$$\frac{\partial P_s(\mathbf{x}(t))}{\partial x_j(t)} = \int d\mathbf{x}(t - \Delta t)\, P_s(\mathbf{x}(t - \Delta t))\frac{\partial}{\partial x_j(t)}\mathscr{P}(\mathbf{x}(t)|\mathbf{x}(t - \Delta t)); \tag{4.45}$$

from equation (4.44) one has that the transition probability $\mathscr{P}(\mathbf{x}(t)|\mathbf{x}(t - \Delta t)$ can be written in terms of the noise distribution:

$$\mathscr{P}(\mathbf{x}(t)|\mathbf{x}(t - \Delta t)) = \tilde{P}(\mathbf{x}(t) - \mathbf{x}(t - \Delta t) - \mathbf{f}(\mathbf{x}(t - \Delta t))\Delta t \mid \mathbf{x}(t - \Delta t)). \tag{4.46}$$

Substituting equations. (4.45) and (4.46) into (4.29) one straightforwardly has

$$R_{ij}(\tau) = -\left\langle x_i(t + \tau)\frac{\partial}{\partial x_j(t)} \ln \tilde{P}(\zeta(t, \mathbf{x}(t), \mathbf{x}(t - \Delta t))|\mathbf{x}(t - \Delta t)) \right\rangle, \tag{4.47}$$

where $\zeta(t, \mathbf{x}(t), \mathbf{x}(t-\Delta t))$ is obtained by inverting equation (4.44). The advantage of equation (4.47) is rather evident: it only requires the knowledge of $\tilde{P}(\zeta(t)|\mathbf{x}(t-\Delta t))$, since P_s does not explicitly appear in equation (4.47).

Let us apply the above result to the case of Langevin equations:

$$\frac{dx_i}{dt} = f_i(\mathbf{x}) + \eta_i,$$

where the $\{\eta_i\}$ are white noises here supposed diagonal for simplicity, satisfying $\langle \eta_i(t)\eta_j(s)\rangle = 2D_j\delta_{ij}\delta(t-s)$; equation (4.47) then becomes

$$R_{ij}(\tau) = -\left\langle x_i(\tau)\frac{\partial}{\partial \dot{x}_j}\ln \tilde{P}(\dot{\mathbf{x}} - \mathbf{f}(\mathbf{x})|\mathbf{x})|_0 \right\rangle, \tag{4.48}$$

and therefore

$$R_{ij}(\tau) = -\frac{1}{2D_j}[\langle x_i(\tau)f^{(j)}(0)\rangle + \langle f^{(i)}(\tau)x_j(0)\rangle]. \tag{4.49}$$

In order to show the advantages of the use of equation (4.49) in chaotic systems, we consider its application to the Lorenz 1963 model, one of the most important and widely studied dynamics showing chaotic behaviour. Consider a stochastic version in which we add a small source of white noise, with evolution equations:

$$\frac{dx}{dt} = \sigma(y-x) + \eta_1, \quad \frac{dy}{dt} = x(\rho - z) - y + \eta_2, \quad \frac{dz}{dt} = xy - \beta z + \eta_3,$$

the $\{\eta_i\}$ are white noises satisfying $\langle \eta_i(t)\eta_j(s)\rangle = 2D\delta_{ij}\delta(t-s)$. For $\sigma = 10$, $\rho = 28$ and $\beta = 8/3$, in the deterministic limit $D = 0$, one has a chaotic behaviour and the attractor has a dimension ≈ 2.05 [18]. For small values of D it is natural to expect that the response functions fairly approximate those obtained from the purely deterministic dynamics: this is well confirmed by numerical simulations [18].

Let us conclude with an important remark on equation (4.49): in gradient systems, i.e. systems such that the vector of deterministic forces can be expressed as the gradient of a potential, the correlation functions are order $O(D)$. In contrast, when the dynamics is chaotic and dissipative, the fluctuations are mainly ruled by the deterministic dynamics. Therefore, the different correlation functions scale, in absolute value, with the size of the attractor, and their sum is order $O(D)$ only by virtue of suitable cancellations: this can produce some practical difficulties in the numerical computations.

References

[1] Darrigol O 2023 A history of the relation between fluctuation and dissipation *Eur. Phys. J.* H **48** 10

[2] Marconi U M B, Puglisi A, Rondoni L and Vulpiani A 2008 Fluctuation-dissipation: response theory in statistical physics *Phys. Rep.* **461** 111–95

[3] Johnson J B 1928 Thermal agitation of electricity in conductors *Phys. Rev.* **32** 97

[4] Nyquist H 1928 Thermal agitation of electric charge in conductors *Phys. Rev.* **32** 110
[5] Topaj D, Kye W-H and Pikovsky A 2001 Transition to coherence in populations of coupled chaotic oscillators: a linear response approach *Phys. Rev. Lett.* **87** 074101
[6] van Kampen N G 1971 The case against linear response theory *Phys. Nor.* **5** 279–84
[7] Kubo R, Toda M and Hashitsume N 2012 *Statistical Physics II: Nonequilibrium Statistical Mechanics* (Berlin: Springer Science)
[8] Boffetta G, Lacorata G, Musacchio S and Vulpiani A 2003 Relaxation of finite perturbations: beyond the fluctuation-response relation *Chaos* **13** 806–11
[9] Kraichnan R H 1959 Classical fluctuation-relaxation theorem *Phys. Rev.* **113** 1181
[10] Kraichnan R H 2000 Deviations from fluctuation-relaxation relations *Physica A: Stat. Mech. Appl.* **279** 30–6
[11] Crisanti A and Ritort F 2003 Violation of the fluctuation-dissipation theorem in glassy systems: basic notions and the numerical evidence *J. Phys. A: Math. Gen.* **36** R181
[12] Cugliandolo L F 2011 The effective temperature *J. Phys. A: Math. Theor.* **44** 483001
[13] Puglisi A, Sarracino A and Vulpiani A 2017 Temperature in and out of equilibrium: a review of concepts, tools and attempts *Phys. Rep.* **709** 1–60
[14] Onsager L and Machlup S 1953 Fluctuations and irreversible processes *Phys. Rev.* **91** 1505
[15] Cencini M, Cecconi F and Vulpiani A 2009 *Chaos: From Simple Models to Complex Systems* **vol 17** (Singapore: World Scientific)
[16] Cessac B 2007 Linear response, susceptibility and resonances in chaotic toy models *Physica* D **225** 13–28
[17] Colangeli M, Rondoni L and Vulpiani A 2012 Fluctuation-dissipation relation for chaotic non-Hamiltonian systems *J. Stat. Mech. Theory Exp.* **2012** L04002
[18] Baldovin M, Caprini L and Vulpiani A 2021 Handy fluctuation-dissipation relation to approach generic noisy systems and chaotic dynamics *Phys. Rev.* E **104** L032101
[19] Caprini L 2021 Generalized fluctuation-dissipation relations holding in non-equilibrium dynamics *J. Stat. Mech. Theory Exp.* **2021** 063202
[20] Sarracino A and Vulpiani A 2019 On the fluctuation-dissipation relation in non-equilibrium and non-Hamiltonian systems *Chaos* **29** 083132

IOP Publishing

Nonequilibrium Statistical Mechanics
Basic concepts, models and applications
Alessandro Sarracino, Andrea Puglisi and Angelo Vulpiani

Chapter 5

Entropy production, fluctuation relations and beyond

> Happy families are all alike; every unhappy family is unhappy in its own way.
>
> *Anna Karenina*, Lev Tolstoj

As we see in other parts of this book, entropy has many faces: related to heat (Clausius heat theorem in standard thermodynamics), being a time-increasing function (H-theorem, see chapter 2), related to fluctuations (Boltzmann–Planck–Einstein theory, see chapter 3), ruling linear response (Kubo FDT, see chapter 4), related to information (Shannon entropy, which will be introduced in chapter 7).

In this chapter we will focus on more general conditions, featuring systems out of equilibrium. These are characterized by the presence of some irreversible process that leads to the breakdown, for the dynamics, of the symmetry under time reversal. Physically, irreversible phenomena are associated with the presence of (matter and/or heat) currents and dissipation, or to relaxation times much longer than observation ones; mathematically, these systems are usually described by stochastic equations that do not satisfy the detailed balance (DB) condition. In this case the Onsager relations derived in chapter 3 do not hold, but still general symmetries can be identified in the fluctuations of some relevant quantities.

In the description of out of equilibrium systems, the uniqueness and universality of concepts such as entropy is lost. The central challenge becomes the quest for other unifying notions and principles. One of the first attempts to develop a theory to describe a weak departure from equilibrium is represented by nonequilibrium thermodynamics [1]. We therefore start by presenting a macroscopic theory of irreversible phenomena valid close to equilibrium, and then we will consider the more general case where a microscopic dynamics violating DB is explicitly introduced. We will focus in particular on the concept of entropy production and we will give general results on the behaviour of its fluctuations.

doi:10.1088/978-0-7503-6229-0ch5

As will be evident from our discussion, nonequilibrium theory goes beyond nonequilibrium thermodynamics. It turns out that considerations about energy and entropy are not enough to describe many phenomena occurring out of equilibrium, and other quantities have to be taken into account. Dynamical activity and kinetic aspects then play a central role, and represent an essential complement to the entropic effects. In particular, we will discuss how the changes in dynamical activity, under external perturbations, affect for instance the fluctuation–dissipation theorem, introducing the coupling with other degrees of freedom in the system and providing a complementary view of this issue with respect to the treatment presented in chapter 4. Other peculiar phenomena will be described in chapters 8 and 9.

5.1 Irreversible thermodynamics

Irreversible thermodynamics is a continuum theory describing macroscopic systems in the presence of irreversible processes [1]. The starting point is the balance equation for the thermodynamic entropy, stating that the total entropy $S(t)$ in a volume changes due to entropy flows from the boundaries (e.g. heat exchanged with the thermostat), or due to some irreversible phenomena occurring inside the volume. This approach is based on an underlying hypothesis of local equilibrium, according to which one can assume the validity of the thermodynamic Gibbs relations connecting the local (within a small volume element) entropy and other thermodynamic quantities (see chapter 3). For the total change of entropy $S(t)$ we can then write

$$\frac{dS}{dt} = \left.\frac{dS}{dt}\right|_{prod} + \left.\frac{dS}{dt}\right|_{ext}, \tag{5.1}$$

where $dS/dt\,|_{prod}$ is the entropy produced internally due to irreversible processes, while $dS/dt\,|_{ext}$ represents the rate of entropy change due to the interaction with the surrounding medium. At equilibrium both terms on the rhs vanish, but in a nonequilibrium state the production term is always positive. In linear nonequilibrium thermodynamics, symmetry arguments contribute to make the above splitting unique; however, more in general, it is not evident that a unique separation can always be achieved.

In the last decade some authors developed a general approach to nonequilibrium systems: the macroscopic fluctuation theory (MFT) [2–4]. MFT can be seen as a next stage beyond the Onsager approach which assumes that the fluctuations follow a Gaussian statistics, for a detailed discussion see [4]. In the following we discuss only the linear case, based on the Onsager ideas.

5.1.1 Thermodynamic forces and fluxes

We start from the local formulation of the conservation laws. In general, the conservation law for a conserved quantity ρ (energy, mass, a component of momentum, etc) in the system can be written as a continuity equation

$$\frac{\partial \rho(t)}{\partial t} = -\nabla \cdot \mathbf{J}_\rho, \tag{5.2}$$

where $\mathbf{J}_\rho$ is the flux associated with ρ. For an open system, where sources (or sinks) for the quantity ρ can be present, the balance equation takes the form

$$\frac{\partial \rho(t)}{\partial t} = -\nabla \cdot \mathbf{J}_\rho + \nu_\rho, \tag{5.3}$$

where ν_ρ represents the production (or absorption) of the quantity ρ in unit time.

From equation (5.1) we have $dS = dS|_{\text{ext}} + dS|_{\text{prod}}$, where for an isolated system, $dS|_{\text{ext}} = 0$ and then $dS = dS|_{\text{prod}} \geqslant 0$, from the second law of thermodynamics, whereas for closed systems exchanging heat Q with a thermostat at temperature T, one has $dS|_{\text{ext}} = \frac{dQ}{T}$ and $dS \geqslant \frac{dQ}{T}$. The change in time of the total entropy S can be written

$$\frac{dS}{dt} = \left.\frac{dS}{dt}\right|_{\text{ext}} + \left.\frac{dS}{dt}\right|_{\text{prod}} = -\int_A \mathbf{J}_s \cdot d\mathbf{A} + \int_V s \; dV, \tag{5.4}$$

where A denotes the contour surface of the system, $\mathbf{J}_s$ is the entropy flow for unit time and unit area, $d\mathbf{A}$ the surface element, s is the entropy production density per unit time and volume, and V the total volume. For open systems in contact with energy and/or particle reservoirs, $\mathbf{J}_s$ is the sum of the heat flow divided by the local temperature, J_q/T, plus the flow of matter from the outside. From the conservation laws and the local equilibrium assumption, the entropy production can be expressed in terms of the different irreversible phenomena in the system. Thus, the structure of the entropy production density is:

$$s = \sum_i \mathbf{J}_i \cdot \mathbf{X}_i, \tag{5.5}$$

where $\mathbf{J}_i$ are fluxes (or currents) of the quantities associated with irreversible phenomena, and $\mathbf{X}_i$ are thermodynamic forces (or affinities), related to gradients of state variables (spatial non-homogeneities) or to external forces, as already encountered in chapter 3. At equilibrium, all the thermodynamic forces vanish and then the entropy production is zero.

As an example, let us consider the case of a metal bar in contact with two thermostats at different temperatures at the edges. A stationary temperature profile sets up along the bar, characterized by a surface entropy flux due to the heat flux across the edges and an internal entropy production due to the temperature gradient in the bulk:

$$s = \mathbf{J}_q \cdot \mathbf{X}_q = -\mathbf{J}_q \cdot \frac{\nabla T}{T^2}, \tag{5.6}$$

where T is the local temperature and $\mathbf{X}_q = -(\nabla T)/T^2$ is the thermodynamic force conjugated to the heat flux.

5.1.2 Onsager coefficients

To close the system of equations for the entropy production, conservation laws and entropy balance equations have to be supplemented by constitutive relations, expressing the fluxes in terms of a linear combination of thermodynamic forces

$$J_i = \sum_j L_{ij} X_j, \tag{5.7}$$

through the Onsager coefficients L_{ij} introduced in chapter 3. We recall that when the thermodynamic forces X_j are spatial gradients of macroscopic fields and J_i are corresponding spatial currents, then L_{ij} are usually called transport coefficients.

The phenomenological relations are constrained by the symmetry properties of the system (according to the Curie symmetry principle[1]). Using these expressions, one obtains

$$s = \sum_{i,j} L_{ij} X_i X_j, \tag{5.8}$$

namely a quadratic form in the thermodynamic forces. Note that at equilibrium both thermodynamic forces and fluxes vanish.

For example, in the case of the system in contact with two thermostats at different temperatures, the phenomenological relation is the Fourier law

$$\mathbf{J}_q = -L_{qq} \frac{\nabla T}{T^2} = -\lambda \nabla T, \tag{5.9}$$

where λ is the heat conductivity, and

$$s = L_{qq} \left(\frac{\nabla T}{T^2} \right)^2. \tag{5.10}$$

Finally, let us note that when a perturbation F_i is applied to a system at equilibrium, namely a system where only conservative forces are present, then the coefficients L_{ij} satisfy the Onsager reciprocal relations introduced in chapter 3. Moreover, the Kubo formula implies an FDT connecting the Onsager (or transport) coefficients with the currents, which is widely known as Green–Kubo relation:

$$L_{ij} = \beta V \int_0^\infty dt \langle j_i(t) j_j(0) \rangle, \tag{5.11}$$

where j_i and j_j are fluctuating current densities (i.e. instantaneous currents per unit of volume) [6–11].

In the following we illustrate the thermoelectric (or Seebeck) effect, as an example among the many phenomena of reciprocal transport that have been studied in the

[1] In its most common form, the Curie symmetry principle claims that when certain causes produce certain effects, the symmetry elements of the causes must be found in their effects. Curie elaborated this principle from his studies on the crystallographic theory of symmetry groups published in a seminal work in 1894, see reference [5] for more details.

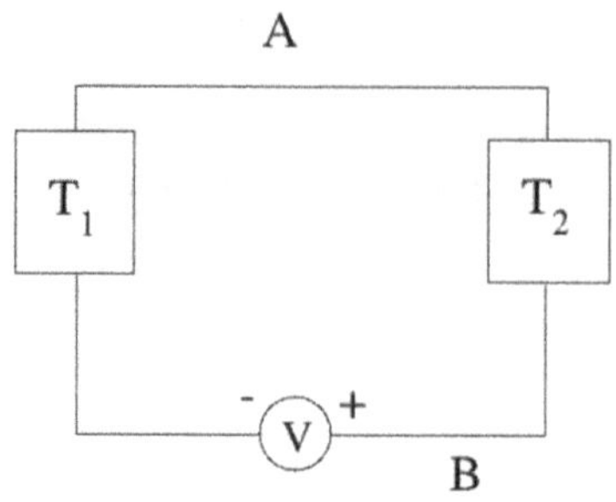

Figure 5.1. Seebeck effect. Two conductors A and B are joined to form a closed circuit and are kept at different temperature T_1 and T_2: an electrical voltage V is generated in the system.

19th and the 20th century and that demonstrate the success of the Onsager reciprocal relations [12]. The effect was observed in 1821 by T J Seebeck, looking at the current that is produced in two different conductors (it can be realized also in semi-conductors with different dopings) joined to form a closed circuit, where the two junctions are kept at different temperatures, see figure 5.1. A small circulating current could be detected thanks to the deflection of a magnetic needle in the proximity. Later H C Ørsted, interpreted the effect as the result of both different electron mobility and different temperatures, that generate a *thermoelectric power*. The idea is that two currents and two thermodynamic forces are present, so that—at the linear order—one has

$$-j_n = L_{nn}\frac{1}{T}\partial_x\mu + L_{nq}\partial_x\frac{1}{T} \tag{5.12}$$

$$j_q = L_{qn}\frac{1}{T}\partial_x\mu + L_{qq}\partial_x\frac{1}{T}, \tag{5.13}$$

where j_n and j_q are electron and heat current densities, respectively, $\mu(x)$ is the electrochemical potential per particle, while $T(x)$ is the temperature, both measured at position x. In order to discuss this effect, we use a setup where the electron current is inhibited by putting a voltmeter in the middle of one of the two conductors, so that the electron current $j_n = 0$ and the thermo-electromotive force is measured as an electric potential difference. The two junctions are set at temperatures T_1 and $T_2 > T_1$. By putting $j_n = 0$ in equations (5.12) one gets a relation between gradients:

$$\partial_x\mu = -T\frac{L_{nq}}{L_{nn}}\partial_x\frac{1}{T}, \tag{5.14}$$

and therefore

$$j_q = -\kappa\partial_x T, \tag{5.15}$$

with

$$\kappa = \frac{\det \mathscr{L}}{T^2 L_{nn}}. \tag{5.16}$$

If the electrical conductivity of one conductor σ is defined as the electric current density ej_n (where e is the electron charge) per unit potential gradient $(1/e)\partial_x\mu$, one has (at isothermal condition) $\sigma = -e^2 j_n/\partial_x\mu$ and therefore

$$L_{nn} = \frac{\sigma T}{e^2}. \tag{5.17}$$

The thermoelectric power ε_{AB} of a thermocouple joining two conductors A and B is defined as the increment of voltage per unit temperature difference (the sign is positive if the increment of voltage drives the current from A to B at the hot junction T_2). In the limit of small $T_2 - T_1$ one can easily see that in each conductor the thermocouple is $\varepsilon_{AB} = -\frac{L_{nq}}{eTL_{nn}}$. This can be inverted to give the expression

$$L_{qn} = L_{nq} = -\frac{T^2\sigma\varepsilon_{AB}}{e}, \tag{5.18}$$

and finally one has

$$L_{qq} = T^3\sigma\varepsilon_{AB}^2 + T^2\kappa. \tag{5.19}$$

5.2 Stochastic thermodynamics

As widely discussed in chapters 1 and 3, fluctuations play a central role. To take into account fluctuations in nonequilibrium systems we have to introduce an appropriate description in terms of stochastic processes. The case of equilibrium has been discussed in chapter 3. Here we consider the more general situations where the presence of energy and/or matter fluxes through the system induce a nonequilibrium dynamics.

5.2.1 Local detailed balance

For simplicity, we start by considering a discrete state system, in contact with a reservoir at inverse temperature β, described by a master equation. The DB condition introduced in chapter 3 implies a symmetry relation between the transition rates

$$\frac{W_{n'\to n}}{W_{n\to n'}} = e^{\beta[E(n')-E(n)]}, \tag{5.20}$$

where $E(n)$ is the system energy in state n. From equation (5.20), there follows the possible structure

$$W_{n\to n'} = k(n, n')e^{-\beta[E(n')-E(n)]/2}, \tag{5.21}$$

where $k(n, n') = k(n', n)$ is a symmetric kinetic factor (or intrinsic rate). Under rather general hypotheses, the DB condition guarantees the convergence of an arbitrary initial density probability towards the equilibrium (Boltzmann–Gibbs) one.

How to generalize such a condition when considering the dynamics of an open system, in contact for instance with several thermal baths at different temperatures or in the presence of some form of dissipation (such as nonlinear friction or inelastic interactions), or driven by non-conservative forces? We face the problem to construct a model that can describe nonequilibrium steady states, based on a physically motivated principle. However, nonequilibrium phenomena are certainly too diversified to be described by a unifying condition. We therefore restrict our initial discussion to open systems weakly coupled to equilibrium baths, which are assumed to be well-separated.

For the sake of simplicity, we first consider the case of a system in contact with a single reservoir at temperature T. Then, according to standard approaches (see for instance the very clear presentation in reference [13, 14]), we introduce nonequilibrium conditions by considering two possible sources of irreversibility: i) the presence of external time-dependent forces (described by a set of parameters $\lambda(t)$) that modify the energy levels of the system, and ii) the coupling with an external agent that exchanges a quantity of energy $\Delta(n' \to n)$ with the system in specific jumps. Then, from the conservation of energy, we can write the heat released to the reservoir in a jump from $n' \to n$ as

$$q(n' \to n) = E(n'; \lambda) - E(n; \lambda) + \Delta(n' \to n), \tag{5.22}$$

where $E(n; \lambda)$ represents the energy of the system in state n with parameters λ. Now we face the problem to define a new transition rate describing the stochastic dynamics of the system in the presence of such nonequilibrium conditions (time-dependent λ and energy exchanges Δ). A natural constraint coming from the generalization of the DB condition is to impose the validity of a local detailed balance (LDB), according to which

$$\frac{W_{n' \to n}}{W_{n \to n'}} = e^{\beta q(n' \to n)} = e^{\beta[E(n';\lambda) - E(n;\lambda) + \Delta(n' \to n)]}. \tag{5.23}$$

Note that the term $E(n; \lambda)$ can be time-dependent in general, but we assume that on the very fast timescale of the interaction between the reservoir and the system λ is constant. It is important to stress that if we interprete the quantity $q(n' \to n)$ as the change of energy due to the heat exchanged with the thermostat at temperature T in the jump $n' \to n$, then the LDB corresponds to the requirement that the log-ratio of the transition rates is proportional to the entropy flux in the transition. The definition of LDB was first introduced by S Katz, J L Lebowitz, and H Spohn in references [15, 16], and a recent interesting discussion on this concept can be found in [17].

From equation (5.23) a possible form for the transition rates is for instance

$$W_{n' \to n} = k(n, n'; \lambda) e^{-\beta[E(n;\lambda) - E(n';\lambda)]/2 + \beta\Delta(n' \to n)}, \tag{5.24}$$

$$W_{n \to n'} = k(n, n'; \lambda) e^{-\beta[E(n';\lambda) - E(n;\lambda)]/2}, \tag{5.25}$$

where $k(n, n'; \lambda)$ can in principle depend on time t and where we have assumed that the energy exchange $\Delta(n' \to n)$ only occurs on one jump direction. The LDB only determines the antisymmetric part of the transition rates, while the symmetric part is encoded in the kinetic factor $k(n, n'; \lambda)$. It is important to stress that, in equilibrium, for instance when λ is constant and $\Delta = 0$, all the static and dynamic quantities in the system do not depend on the explicit form of the kinetic factor. On the other hand, its dependence on the perturbing parameter λ can play an important role in the derivation of FDT out of equilibrium, as we will discuss in the following.

5.2.2 Fluctuating entropy production

The LDB is a requirement that is imposed on the transition rates: it implies that the log-ratio between the transition rate from one state to another and the reversed one is proportional to the entropy flux generated in the transition. It can be introduced as an useful condition to construct a mathematical model based on some information, or some physical intuition, on the thermodynamic properties of the system. More in general, the log-ratio between the probability of a trajectory and its time-reversed one can be introduced to define the degree of irreversibility of the dynamics in the system. Lebowitz and Spohn introduced such a functional in reference [18].

5.2.2.1 Fokker–Planck equation

Before discussing in detail the connection between LDB and irreversibility, we first introduce the notion of entropy production in the context of the Fokker–Planck equation. We start rewriting it as a continuity equation for the probability distribution of the system with variables $\mathbf{z}$

$$\frac{\partial P(\mathbf{z}, t)}{\partial t} = -\nabla \cdot \mathbf{J}(\mathbf{z}, t), \tag{5.26}$$

where $\mathbf{J}$ is the current probability with components

$$J_i(\mathbf{z}, t) = [A_i(\mathbf{z}, t) - \partial_{z_i} B_i(\mathbf{z}, t)]P(\mathbf{z}, t), \tag{5.27}$$

$A_i(\mathbf{z}, t)$ being the drift term and $B_i(\mathbf{z}, t)$ the diffusive term. This can be decomposed as

$$\mathbf{J}(\mathbf{z}, t) = \mathbf{J}^{irr}(\mathbf{z}, t) + \mathbf{J}^{rev}(\mathbf{z}, t), \tag{5.28}$$

according to the definition of time reversal

$$J_i^{irr}(\mathbf{z}, t) = \frac{1}{2}[A_i(\mathbf{z}, t) + \varepsilon_i A_i(\varepsilon\mathbf{z}, t)] - \partial_{z_i} B_i(\mathbf{z}, t)P(\mathbf{z}, t), \tag{5.29}$$

$$J_i^{rev}(\mathbf{z}, t) = \frac{1}{2}[A_i(\mathbf{z}, t) - \varepsilon_i A_i(\varepsilon\mathbf{z}, t)], \tag{5.30}$$

where $\varepsilon_i = \pm 1$ denotes the parity of the variable z_i. Note that $\mathbf{J}^{irr} = 0$ is a necessary condition for DB to hold [19].

Then we consider the Gibbs entropy introduced in chapter 2, associated with the probability distribution $P(\mathbf{z}, t)$

$$S_G(t) = -\int d\mathbf{z}P(\mathbf{z},\, t)\log P(\mathbf{z},\, t). \tag{5.31}$$

Taking the time derivative, inserting the equation (5.26) and integrating by parts twice, one obtains

$$\begin{aligned}\frac{dS_G(t)}{dt} &= \int d\mathbf{z}P(\mathbf{z},\, t)\nabla\cdot\left(\frac{\mathbf{J}^{rev}(\mathbf{z},\, t)}{P(\mathbf{z},\, t)}\right) - \int d\mathbf{z}\mathbf{J}^{irr}(\mathbf{z},\, t)\cdot\left(\frac{\nabla P(\mathbf{z},\, t)}{P(z,\, t)}\right)\\ &= -\int d\mathbf{z}\mathbf{J}^{irr}(\mathbf{z},\, t)\cdot\left(\frac{\nabla P(\mathbf{z},\, t)}{P(z,\, t)}\right),\end{aligned} \tag{5.32}$$

where we have exploited the fact that the quantity $\mathbf{J}^{rev}/P$ is divergenceless.

To illustrate the meaning of this result, let us consider for simplicity the case of an inertial particle of mass m, in one dimension with position x and velocity v, in contact with a thermal bath at temperature T with viscosity γ, and in the presence of a potential $U(x)$. Then $\mathbf{z} = (x,\, v)$ and one has

$$J_x^{rev} = vP(x,\, v), \qquad J_v^{rev} = -U'(x)P(x,\, v), \tag{5.33}$$

$$J_x^{irr} = 0, \qquad J_v^{irr} = -\frac{\gamma}{m}vP(x,\, v) - \frac{\gamma T}{m^2}\partial_v P(x,\, v). \tag{5.34}$$

Substituting the last equality into equation (5.32) one has

$$\frac{dS_G(t)}{dt} = \frac{m^2}{\gamma T}\int d\mathbf{z}\frac{[J_v^{irr}(z,\, t)]^2}{P(\mathbf{z},\, t)} + \frac{m}{T}\int d\mathbf{z}J_v^{irr}v. \tag{5.35}$$

The first term is identified with rate of entropy production $\dot{S}_{prod} \geqslant 0$, where the equality holds only in equilibrium, namely when $J_v^{irr} = 0$ and the DB condition is satisfied. The second term is associated with the entropy exchanged with the environment, and in this case, replacing equation (5.34) takes the form $\dot{S}_{env} = (\gamma/T)(T/m - \langle v^2\rangle) = \langle\dot{Q}\rangle/T$, where Q is the heat exchanged with the reservoir. Therefore, we get the relation

$$\dot{S}_{prod} = \frac{dS_G(t)}{dt} - \frac{1}{T}\langle\dot{Q}\rangle \geqslant 0, \tag{5.36}$$

analogously to equation (5.1). The above discussion can be generalized to cases of many interacting particles and arbitrary dimension.

5.2.2.2 Discrete systems

We now discuss more in detail the relation between LDB and entropy production. We start by considering a system with discrete states governed by a master equation. We denote by $\Omega_0^t = \{n(\tau)\}$ a trajectory in the state space of the system under the protocol $\lambda(\tau)$ in the time interval $\tau \in [0,\, t]$, where jumps occur at times τ_i, with $i \in [1,\, k]$, with initial state $n_0 = n(0)$ and final state $n_k = n(t)$. The probability of the occurrence of the trajectory Ω_0^t (forward in time) is

$$P_\lambda(\Omega_0^t) = P_{n_0}^s e^{-\int_0^{t_1} d\tau \gamma_{n_0}(\tau)} W_{n_0 \to n_1} e^{-\int_{t_1}^{t_2} d\tau \gamma_{n_1}(\tau)} W_{n_1 \to n_2} \dots W_{n_{k-1} \to n_k} e^{-\int_{t_k}^{t} d\tau \gamma_{n_k}(\tau)}, \quad (5.37)$$

where $P_{n_i}^s$ is the stationary probability to be in the state n_i and $\gamma_{n_i}(t) = \sum_{n_j \neq n_i} W_{n_i \to n_j}$ are the escape rates. We then define the time-reversed trajectory $\overline{\Omega_0^t} = \{n(t - \tau)\}$ under the reversed protocol $\bar{\lambda}(\tau) = \lambda(t - \tau)$, namely the trajectory where the same states are visited in reversed order. We assume that the states are symmetric under such a transformation. Then the probability of the occurrence of the trajectory $\overline{\Omega_0^t}$ (backward in time) is

$$P_{\bar{\lambda}}(\overline{\Omega_0^t}) = P_{n_k}^s e^{-\int_{t_k}^{t} d\tau \gamma_{n_k}(\tau)} W_{n_k \to n_{k-1}} e^{-\int_{t_{k-1}}^{t_k} d\tau \gamma_{n_{k-1}}(\tau)} \dots W_{n_1 \to n_0} e^{-\int_0^{t_1} d\tau \gamma_{n_0}(\tau)}. \quad (5.38)$$

We now can define the functional $\Sigma(\Omega_0^t)$ as the ratio between the two probabilities

$$\Sigma(\Omega_0^t) \equiv \ln \frac{P(\Omega_0^t)}{P_{\bar{\lambda}}(\overline{\Omega_0^t})} = \ln \frac{P_{n_0}^S W_{n_0 \to n_1} \dots W_{n_{k-1} \to n_k}}{P_{n_k}^S W_{n_k \to n_{k-1}} \dots W_{n_1 \to n_0}} = \ln \prod_{i=0}^{k-1} \frac{W_{n_i \to n_{i+1}}}{W_{n_{i+1} \to n_i}} + \ln \frac{P_{n_0}^s}{P_{n_k}^s}, \quad (5.39)$$

where we have used the fact that the exponential factors cancel out.

To make the link with the macroscopic entropy production introduced in the previous section, we observe that, using the LDB condition (5.23), we have

$$\Sigma(\Omega_0^t) = \beta \sum_{i=0}^{k-1} q(n_i \to n_{i+1}) + \ln \frac{P_{n_0}^s}{P_{n_k}^s} = \beta q(\Omega_0^t) + \ln \frac{P_{n_0}^s}{P_{n_k}^s}, \quad (5.40)$$

where by $q(\Omega_0^t)$ we denote the total (stochastic) heat released by the system to the reservoir during the trajectory Ω_0^t.

Now, let us introduce a stochastic entropy associated with the system state, using the Shannon entropy that will be discussed in more detail in chapter 7,

$$S_n^{\rm sys} = -k_{\rm B} \ln P_n^s. \quad (5.41)$$

Then, we can include the boundary terms related to the probabilities $P_{n_0}^s$ and $P_{n_k}^s$ to get

$$\Sigma(\Omega_0^t) = \frac{S_{n_k}^{\rm sys} - S_{n_0}^{\rm sys}}{k_{\rm B}} + \beta q(\Omega_0^t) = \frac{S^{\rm tot}(\Omega_0^t)}{k_{\rm B}}, \quad (5.42)$$

or

$$\frac{P(\Omega_0^t)}{P_{\bar{\lambda}}(\overline{\Omega_0^t})} = e^{S^{\rm tot}(\Omega_0^t)/k_{\rm B}}, \quad (5.43)$$

where $S^{\rm tot}(\Omega_0^t) = S_{n_k}^{\rm sys} - S_{n_0}^{\rm sys} + q(\Omega_0^t)/T$ is the total entropy production along the trajectory, taking into account both the heat exchanged with the reservoir and the internal changes of the system. This represents an explicit relation between the entropy production and irreversibility, defined as an asymmetry between the probability of a trajectory and its time-reversed. In equilibrium, when the DB condition holds, $P(\Omega_0^t) = P_{\bar{\lambda}}(\overline{\Omega_0^t})$ for all trajectories Ω_0^t, and then $S^{\rm tot}(\Omega_0^t) = 0$ at the level of single trajectory.

5.2.2.3 Coupled Langevin equations

The above discussion can be repeated also for systems with continuous degrees of freedom, with time-continuous trajectories described by Langevin equations, where nonequilibrium conditions are introduced through non-conservative forces or through the coupling with thermal baths at different temperatures. In particular, in cases which are relevant for nonequilibrium spatially extended systems, the functional (5.39) can take a clear thermodynamic meaning, since it can be expressed in terms of macroscopic currents and forces, in strict analogy with the form of entropy production (5.5) discussed in section 5.1.1.

Here we specialize to a continuous space and time Markov process whose state is described by a d-dimensional vector $\mathbf{z}$, with components z_i ($i = 1, \ldots, d$). The time reversal transformation on the vector $\mathbf{z}$ is defined as an operator which changes $z_i \to \bar{\alpha}_i \equiv \varepsilon_i \alpha_i$ with $\varepsilon_i \in \{+1, -1\}$, according to the parity of the component i. As shorthand notations we will use $\bar{\mathbf{z}}$ or $\varepsilon \mathbf{z}$ to indicate the vector made of time-reversed components $\{\varepsilon_i z_i\}$. We assume that the dynamics of each component of the vector $\mathbf{z}$ is described by a Langevin equation:

$$\frac{dz_i(t)}{dt} = F_i[\mathbf{z}(t); \lambda(t)] + \xi_i(t), \tag{5.44}$$

where F_i is a drift which depends on the protocol $\lambda(t)$, and ξ_i a Gaussian process with zero mean and $\langle \xi_i(t) \xi_j(t') \rangle = 2\delta_{ij} \delta(t - t') D_{ii}$.

If we consider a trajectory Ω_0^t in the time interval $[0, t]$ as the sequence $\{\mathbf{z}(\tau)\}$ with $\tau \in [0, t]$, then we can define its time-reversed $\overline{\Omega_0^t}$ as the sequence $\{\bar{\mathbf{z}}(\tau)\} = \{\varepsilon \mathbf{z}(t - \tau)\}$ with $\tau \in [0, t]$. Now, let us first neglect the noise terms in equation (5.44) and consider a solution $\mathbf{z}(\tau)$. The time-reversed trajectory then satisfies the following differential equation:

$$\frac{d\bar{z}_i(\tau)}{d\tau} = -\varepsilon_i \frac{dz_i(t - \tau)}{dt} = -\varepsilon_i F_i[\mathbf{z}(t - \tau); \lambda(t - \tau)] = -\varepsilon_i F_i[\varepsilon \bar{\mathbf{z}}(\tau); \bar{\lambda}(\tau)]. \tag{5.45}$$

We notice two particular cases for $F_i(\mathbf{z}; \lambda)$:

$$F_i(\varepsilon \mathbf{z}; \lambda) = -\varepsilon_i F_i(\mathbf{z}; \lambda) \tag{5.46}$$

$$F_i(\varepsilon \mathbf{z}) = \varepsilon_i F_i(\mathbf{z}; \lambda). \tag{5.47}$$

In the first case $\bar{z}_i(\tau)$ satisfies exactly the forward equation (5.44). In the second case it satisfies the same equation where the drift sign is changed. Following these two limit cases, we can in general decompose the ith component of the drift as

$$F_i(\mathbf{z}; \lambda) = F_{i,\mathrm{rev}}(\mathbf{z}; \lambda) + F_{i,\mathrm{irr}}(\mathbf{z}; \lambda), \tag{5.48}$$

with

$$F_{i,\mathrm{rev}}(\mathbf{z}; \lambda) = \frac{1}{2}[F_i(\mathbf{z}; \lambda) - \varepsilon_i F_i(\varepsilon \mathbf{z}; \lambda)] = -\varepsilon_i F_{i,\mathrm{rev}}(\varepsilon \mathbf{z}; \lambda) \tag{5.49}$$

$$F_{i,\mathrm{irr}}(\mathbf{z};\lambda)=\frac{1}{2}[F_i(\mathbf{z};\lambda)+\varepsilon_i F_i(\varepsilon\mathbf{z};\lambda)]=\varepsilon_i F_{i,\mathrm{irr}}(\varepsilon\mathbf{z};\lambda), \tag{5.50}$$

where $F_{i,\mathrm{rev}}$ and $F_{i,\mathrm{irr}}$ represent the reversible and irreversible contributions to the drift, respectively, as in the splitting of the probability currents discussed before.

Now, assuming that the probability density of the system (5.44) converges to a unique stationary state in the long time limit, we can write the Lebowitz and Spohn functional for the entropy production. Following the Onsager–Machlup prescription [20, 21] (see also reference [22] for a more general treatment), in the case of an additive Gaussian noise, the probability of a trajectory Ω_0^t is

$$P_\lambda(\Omega_0^t)=P^s[\mathbf{z}(0)]\exp[\mathscr{A}(\Omega_0^t)], \qquad P_{\bar{\lambda}}(\overline{\Omega_0^t})=P^s[\bar{\mathbf{z}}(0)]\exp[\mathscr{A}(\overline{\Omega_0^t})], \tag{5.51}$$

where $P^s(\mathbf{z})$ is the probability of state $\mathbf{z}$ in the steady state, and the *action* $\mathscr{A}$ is given by

$$\mathscr{A}(\Omega_0^t)=-\int_0^t d\tau\sum_i{}'\frac{1}{4D_{ii}}\left[\frac{dz_i(\tau)}{d\tau}-F_i[\mathbf{z}(\tau);\lambda(\tau)]\right]^2. \tag{5.52}$$

We have used the notation $\sum_i'$ to indicate that the sum runs only on those indexes such that $D_{ii}\neq 0$ and the Stratonovich prescription for the integration is assumed. Using the form of the action for the time-reversed trajectory

$$\begin{aligned}\mathscr{A}(\overline{\Omega_0^t})&=-\int_0^t d\tau\sum_i{}'\frac{1}{4D_{ii}}\left[\frac{d\bar{z}_i(\tau)}{d\tau}-F_i[\bar{\mathbf{z}}(\tau);\bar{\lambda}(\tau)]\right]^2\\&=-\int_0^t d\tau\sum_i{}'\frac{1}{4D_{ii}}\left[\frac{dz_i(\tau)}{d\tau}+F_i[\mathbf{z}(\tau);\lambda(\tau)]\right]^2,\end{aligned} \tag{5.53}$$

and recalling that $\int_0^t d\tau f(t-\tau)=\int_0^t d\tau f(\tau)$, we get the entropy production along a path Ω_0^t [23]:

$$\Sigma(\Omega_0^t)=\ln\frac{P(\Omega_0^t)}{P_{\bar{\lambda}}(\overline{\Omega_0^t})}=\mathscr{A}(\Omega_0^t)-\mathscr{A}(\overline{\Omega_0^t})+\ln\frac{P^s[\mathbf{z}(0)]}{P^s[\bar{\mathbf{z}}(t)]}=\int_0^t d\tau\ \zeta(\tau)+\ln\frac{P^s[\mathbf{z}(0)]}{P^s[\bar{\mathbf{z}}(t)]}, \tag{5.54}$$

where, using definitions (5.49) and (5.50), we can rewrite the entropy production rate as

$$\zeta(\tau)=\sum_i{}'\frac{1}{D_{ii}}\underbrace{F_{i,irr}[\mathbf{z}(\tau);\lambda]}_{\text{force}}\circ\underbrace{(\dot{z}_i(\tau)-F_{i,\mathrm{rev}}[\mathbf{z}(\tau);\lambda])}_{\text{current}}, \tag{5.55}$$

with $\circ$ used to remind ourselves about the Stratonovich integration needed. This formula expresses the entropy production as the product of forces (the irreversible part of the drift) multiplied by currents (time derivative of the variables). Moreover, according to the phenomenological relations (5.7), the irreversible fluxes should vanish when the driving forces are switched off. This identification of forces and fluxes is consistent with such requirements: indeed, when $\langle F_{i,\mathrm{irr}}[\mathbf{z}(\tau);\lambda]\rangle=0$ we also have $\langle\dot{z}_i(\tau)-F_{i,\mathrm{rev}}[\mathbf{z}(\tau);\lambda]\rangle=0$, as can be seen by averaging equation (5.44) over the noise. This is in analogy with the macroscopic expressions obtained from irreversible thermodynamics.

In the case of a single thermostat at temperature T, we have the very simple interpretation as in the case discussed previously:

$$\Sigma(\Omega_0^t) = S^{\text{tot}}(\Omega_0^t)/k_{\text{B}}, \tag{5.56}$$

where we have introduced the stochastic heat for continuous systems according to Sekimoto [24]

$$\int_0^t d\tau \ \zeta(\tau) = \frac{q(t)}{T} = \int_0^t \sum_i F_i[\mathbf{z}(\tau);\ \lambda(\tau)] \circ dz_i(\tau), \tag{5.57}$$

and the total entropy production $S^{\text{tot}}(\Omega_0^t) = S^{\text{sys}}(\mathbf{z}(t)) - S^{\text{sys}}(\mathbf{z}(0)) + q(t)/T$.

As an example, we consider again an inertial particle with position x and velocity v, in contact with a thermal bath with temperature T and viscosity γ subjected to an external force $F(x)$

$$\begin{aligned} \dot{x} &= v(t) \\ \dot{v} &= -\gamma v + F(x) + \xi(t), \end{aligned} \tag{5.58}$$

with $\langle \xi(t)\xi(t') \rangle = 2\gamma T\delta(t - t')$. Applying (5.49) and (5.50) with the identification $z_1 = x$ with $\varepsilon_1 = 1$, $z_2 = v$ with $\varepsilon_2 = -1$, and $D_{22} = \gamma T$, we have

$$\begin{aligned} F_1^{\text{rev}} &= v, \quad F_1^{\text{irr}} = 0 \\ F_2^{\text{rev}} &= F(x), \quad F_2^{\text{irr}} = -\gamma v, \end{aligned} \tag{5.59}$$

and, using equation (5.55), we get

$$\zeta(\tau) = -\frac{1}{T}\dot{v}(\tau)v(\tau) + \frac{1}{T}F[x(\tau)]v(\tau). \tag{5.60}$$

Integrating in time, the first term on the rhs represents the kinetic energy difference between initial and final state dissipated in the environment, while the second term gives the dissipated power as heat due to the external force, in agreement with physical intuition.

5.2.2.4 Linear processes

In the linear case with constant protocol $F[\mathbf{z};\ \lambda] \equiv -\mathcal{M}\mathbf{z}$:

$$F_i[\mathbf{z}] = -\sum_j M_{ij} z_j = -\sum_{j \in \text{rev}(i)} M_{ij} z_j - \sum_{j \in \text{irr}(i)} M_{ij} z_j, \tag{5.61}$$

where the shorthand notation $j \in \text{rev}(i)$ means that the index j runs on the set of indices such that $\varepsilon_j = -\varepsilon_i$, while $j \in \text{irr}(i)$ stands for the set of indices such that $\varepsilon_j = \varepsilon_i$. In this case, from equation (5.55), one has

$$\zeta = \sum_i{}' \frac{1}{D_{ii}} \underbrace{\left(\sum_{j \in \text{irr}(i)} M_{ij} z_j \right)}_{\text{force}} \circ \underbrace{\left(\dot{z}_i - \sum_{j \in \text{rev}(i)} M_{ij} z_j \right)}_{\text{current}} \tag{5.62}$$

$$= \sum_i' \frac{1}{D_{ii}} \left(\sum_{j \in \mathrm{irr}(i)} M_{ij} z_j \dot{z}_i - \sum_{\substack{j \in \mathrm{irr}(i) \\ l \in \mathrm{rev}(i)}} M_{ij} M_{il} z_j z_l \right). \tag{5.63}$$

At equilibrium $\langle \zeta \rangle = 0$, as expected. Formula (5.62) is analogous to the macroscopic thermodynamic result (5.5): it expresses the entropy production as the product of a force by a current.

In the next section we discuss an explicit example where irreversible currents are present.

5.2.3 Langevin equations with irreversible currents

Steady states with irreversible currents in a confined space (e.g. without periodic boundary conditions) can be built in more than one dimension, in order to allow for the existence of rotating currents. A particularly simple class of those systems are the so-called Brownian gyrators, already introduced in chapter 3. The simplest one is described by an overdamped Langevin dynamics of a particle moving in a two-dimensional space in the presence of an external potential $U(x, y)$ and two different thermal baths. The particle instantaneous position is defined by coordinates $x(t)$ and $y(t)$, respectively. The time evolution of $x(t)$ and $y(t)$ is described by the following equations:

$$\frac{d}{dt}x(t) = -\frac{\partial}{\partial x}U(x, y) + \xi_x(t),$$
$$\frac{d}{dt}y(t) = -\frac{\partial}{\partial y}U(x, y) + \xi_y(t).$$

Here $\xi_{x,y}(t)$ is anisotropic stochastic noise, with zero mean and correlation functions

$$\left\langle \xi_\alpha(t)\xi_\beta(t') \right\rangle = 2T_\alpha \delta_{\alpha,\beta}\delta(t - t'), \quad (\alpha, \beta = x, y)$$

where T_x and T_y are two 'temperatures', different in general, and $U(x, y)$ has the following skewed parabolic form:

$$U(x, y) = \frac{1}{2}x^2 + \frac{1}{2}y^2 + uxy;$$

the shape of the potential is controlled by the parameter u. To have a confining potential the constraint $|u| < 1$ is necessary. This follows from the requirement that both eigenvalues of the matrix which determines the above quadratic form, $\lambda_{1,2} = 1 \pm u$, must be positive. In the stationary regime, the probability distribution function $P(x, y)$ of the particle position obeys the stationary Fokker–Planck equation:

$$\frac{\partial}{\partial x}\left[T_x \frac{\partial P(x, y)}{\partial x} + P(x, y)\frac{\partial U(x, y)}{\partial x}\right] + \frac{\partial}{\partial y}\left[T_y \frac{\partial P(x, y)}{\partial y} + P(x, y)\frac{\partial U(x, y)}{\partial y}\right] = 0.$$

In the trivial isothermal case, $T_x = T_y = T$, the steady state is simply the equilibrium Gibbs distribution $P_{\text{iso}}(x, y) \propto \exp\{-\frac{1}{T}U(x, y)\}$. In the generic anisotropic case with arbitrary T_x and T_y, the solution of the stationary equation reads [25]:

$$P(x, y) = Z^{-1}\exp\left\{-\frac{1}{2}\gamma_1 x^2 - \frac{1}{2}\gamma_2 y^2 - u\gamma_3 xy\right\},$$

where

$$\gamma_1 = \frac{T_x + \frac{1}{2}u^2(T_x - T_y)}{T_x T_y(1 + u^2\Delta^2)}, \qquad \gamma_2 = \frac{T_y + \frac{1}{2}u^2(T_y - T_x)}{T_x T_y(1 + u^2\Delta^2)},$$
$$\gamma_3 = \frac{T_x + T_y}{2T_x T_y(1 + u^2\Delta^2)}, \qquad \Delta = \frac{(T_y - T_x)}{2\sqrt{T_y T_x}}.$$

The normalization Z is

$$Z = \iint_{-\infty}^{+\infty} dx\, dy \exp\left\{-\frac{1}{2}\gamma_1 x^2 - \frac{1}{2}\gamma_2 y^2 - u\gamma_3 xy\right\} = 2\pi\sqrt{\frac{T_x T_y(1 + u^2\Delta^2)}{1 - u^2}}, \tag{5.64}$$

and of course Z is well defined only for $|u| < 1$. In the stationary case the probability currents $\mathbf{J} = (J_x, J_y)$ (which are fully irreversible) are:

$$J_x = [(1 - T_x\gamma_1)x + u(1 - T_x\gamma_3)y]P(x, y),$$
$$J_y = [(1 - T_y\gamma_2)y + u(1 - T_y\gamma_3)x]P(x, y).$$

In the isothermal case, $T_x = T_y = T$, one has $\gamma_1 = \gamma_2 = \gamma_3 = 1/T$, so that $\mathbf{J} \equiv 0$. In the out-of-equilibrium case $T_x \neq T_y$ the above currents field can be characterized in terms of its rotor field:

$$\text{Rot}(x, y) \equiv \boldsymbol{\nabla} \times \mathbf{J}(x, y) = \frac{\partial}{\partial x}J_y - \frac{\partial}{\partial y}J_x,$$

which is a rather complicated function of two variables x and y, but it has a simple expression at the origin $x = y = 0$:

$$\text{Rot}(0) = u(T_x - T_y)\gamma_3 Z^{-1} = \frac{u}{4\pi}\frac{T_x^2 - T_y^2}{T_x^2 T_y^2}\sqrt{\frac{T_x T_y(1 - u^2)}{(1 + u^2\Delta^2)}}.$$

Note that this quantity changes sign when going from $T_y > T_x$ to $T_y < T_x$.

The non-zero particle's current rotor implies a non-zero mean particle's rotation velocity, which can be characterized by its angular velocity $\omega(t) = \frac{1}{r^2}(\mathbf{v} \times \mathbf{r})$ where $(\mathbf{v} \times \mathbf{r})$ is the vector product directed between the particle velocity v and the position vector r along the z-axis. Thus, the mean rotation velocity $\langle\omega\rangle$ in the limit of an infinite observation time can be defined as $\langle\omega\rangle = \lim_{\mathcal{T}\to\infty}\frac{1}{\mathcal{T}}\int_0^{\mathcal{T}} dt\omega(t)$. If an ensemble average replaces the time-average, one gets

$$\langle \omega \rangle = \int d^2\mathbf{r}\frac{1}{r^2}(\mathbf{J} \times \mathbf{r}) = \int_0^{2\pi} d\phi \int_0^{\infty} dr(J_x \sin\phi - J_y \cos\phi),$$

and simple calculations (see reference [25] for details) give

$$\langle \omega \rangle = u\Delta\sqrt{\frac{1-u^2}{1+u^2\Delta^2}}.$$

The Brownian gyrator with harmonic potential can be generalized to a wider category of coupled linear Langevin equations [26]:

$$\dot{x} = -\alpha x + \lambda y + \xi_x(t),$$
$$\dot{y} = -\gamma x + \mu y + \xi_y(t),$$

where the case discussed before corresponds to $\alpha = \gamma = 1$ and $\lambda = \mu = -u$.

The entropy production for this model can be obtained along the general lines described in the previous section. Assuming both variables even under time reversal, applying the definitions in equations (5.49), (5.50), and (5.55), we obtain

$$\begin{aligned}\int_0^t d\tau\zeta(\tau) = &\frac{1}{T_x}\int_0^t d\tau[\lambda y(\tau)\dot{x}(\tau) - \alpha x(\tau)\dot{x}(\tau)]\\ &+\frac{1}{T_y}\int_0^t d\tau[\mu x(\tau)\dot{y}(\tau) - \gamma y(\tau)\dot{y}(\tau)],\end{aligned} \tag{5.65}$$

that, for large times, becomes

$$\int_0^t d\tau\ \zeta(\tau) \simeq \left[\frac{\lambda}{T_x} - \frac{\mu}{T_y}\right]\int_0^t d\tau\ y(\tau)\dot{x}(\tau). \tag{5.66}$$

The (steady) average entropy production rate of this process can be then recast into

$$\left\langle \dot{S}_{\text{prod}} \right\rangle = \langle \zeta(\tau) \rangle = \left[\frac{\lambda}{T_x} - \frac{\mu}{T_y}\right]\langle y\dot{x} \rangle = \frac{(T_y\lambda - T_x\mu)^2}{T_xT_y(\alpha+\gamma)}.$$

This is identically zero in equilibrium, namely for potential condition ($\mu = \lambda$) and isothermal baths ($T_x = T_y$), as expected.

5.3 Fluctuation relations

We now focus on the study of the fluctuations of entropy production. These satisfy very general symmetries that represent some of the few general results available, today, for nonequilibrium systems [27, 28]. These symmetry relations descend from the LDB condition, which is assumed for the transition rates.

5.3.1 Integral and detailed fluctuation theorem

From equation (5.43) we have

$$P_{\bar{\lambda}}(\overline{\Omega_0^t}) = e^{-S^{\text{tot}}/k_{\text{B}}}P_{\lambda}(\Omega_0^t). \tag{5.67}$$

It is possible to sum over all trajectories on both sides. On the left-hand side we get 1 by normalization, on the right-hand side we obtain the average of $e^{-S^{\rm tot}/k_B}$ weighted by the probability generated by the forward protocol $\lambda(t)$, immediately giving

$$1 = \langle e^{-S^{\rm tot}/k_B} \rangle. \tag{5.68}$$

This represents the so-called integral fluctuation theorem (IFT) [29]. From this result, using the Jensen inequality

$$\log\langle e^{-x} \rangle \geqslant \log [e^{-\langle x \rangle}] = -\langle x \rangle, \tag{5.69}$$

there follows

$$\langle S^{\rm tot} \rangle \geqslant 0, \tag{5.70}$$

which expresses the second law of thermodynamics. In equilibrium, the equality holds, while out of equilibrium, since $e^{-S^{\rm tot}/k_B} < 1$ for $S^{\rm tot} > 0$, positive values of $S^{\rm tot}$ must be compensated by the occurrence of trajectories with $S^{\rm tot} < 0$, which represents (unlikely but observable) 'violations' of the second law for stochastic systems.

Assuming an involution property for the entropy production, namely the condition

$$S^{\rm tot}(\overline{\Omega_0^t}) = -S^{\rm tot}(\Omega_0^t), \tag{5.71}$$

one can show a stronger relation. Indeed, denoting by $d\Omega_0^t$ a measure in the space of the possible trajectories, with $\int d\Omega_0^t = \int d\overline{\Omega_0^t}$, we have

$$\begin{aligned}
\text{Prob}[S^{\rm tot}(\Omega_0^t) = x] &= \int d\Omega_0^t \ \delta[S^{\rm tot}(\Omega_0^t) - x] P_\lambda(\Omega_0^t) \\
&= \int d\Omega_0^t \ \delta[S^{\rm tot}(\Omega_0^t) - x] P_{\bar\lambda}(\overline{\Omega_0^t}) e^{S^{\rm tot}(\Omega_0^t)} \\
&= e^x \int d\overline{\Omega_0^t} \ \delta[-S^{\rm tot}(\overline{\Omega_0^t}) - x] P_{\bar\lambda}(\overline{\Omega_0^t}) \\
&= e^x \text{Prob}[S^{\rm tot}(\overline{\Omega_0^t}) = -x].
\end{aligned} \tag{5.72}$$

Then one gets

$$\frac{\text{Prob}[S^{\rm tot}(\Omega_0^t) = x]}{\text{Prob}[S^{\rm tot}(\overline{\Omega_0^t}) = -x]} = e^x, \tag{5.73}$$

which is known as the detailed fluctuation theorem (DFT) [29]. The involution property always holds for stationary systems, and if the initial and final states are equilibrium states.

This kind of relation has been first numerically observed in a fluid under shear by Evans and coworkers in [27]. Then, in the context of dynamical systems, it was proved by Gallavotti and Cohen [28]. Later, Kurchan [30] and Lebowitz and Sphon [18] derived the same relation for stochastic systems. Maes rigorously showed the mathematical relation with the LDB condition in [31].

5.3.2 Jarzynski and Crooks relations

From the above relations many other interesting relations can be derived [29]. We consider an experiment where one prepares the system (described by discrete variables) in an initial equilibrium state with parameter $\lambda(0)$, described by the equilibrium distribution

$$P_{n_0}^{eq}(\lambda(0)) = e^{[F(0) - E(n_0;\ \lambda(0))]/k_B T}, \tag{5.74}$$

where

$$F(0) = -k_B T \log \sum_n e^{-E(n;\lambda(0))/k_B T} \tag{5.75}$$

is the free energy of the equilibrium state.

Then the system evolves in time according to a generic nonequilibrium dynamics, eventually relaxing to a final (different) equilibrium state corresponding to the parameter $\lambda(t)$

$$P_{n_k}^{eq}(\lambda(t)) = e^{[F(t) - E(n_k;\ \lambda(t))]/k_B T}. \tag{5.76}$$

The change in the system entropy is

$$\Delta S^{\mathrm{sys}} = k_B \log \frac{P_{n_0}^{eq}(\lambda(0))}{P_{n_k}^{eq}(\lambda(t))} = \frac{1}{T}[E(n_k;\ \lambda(t)) - E(n_0;\ \lambda(0)) - \Delta F], \tag{5.77}$$

where $\Delta F = F(t) - F(0)$. The total entropy production is

$$S^{\mathrm{tot}} = \frac{q}{T} + \Delta S^{\mathrm{sys}} = \frac{W - \Delta F}{T}, \tag{5.78}$$

where we have used the first law in the form $W = q + E(n_k;\ \lambda(t)) - E(n_0;\ \lambda(0))$. Substituting into the IFT, one obtains

$$\langle e^{-W/k_B T} \rangle = e^{-\Delta F/k_B T}. \tag{5.79}$$

This result is known as the Jarzynski relation [32]. Following similar steps, and using the DFT, if initial and final states are at equilibrium, one gets the more general result known as the Crooks relation [33]

$$\frac{\mathrm{Prob}[W = x]}{\mathrm{Prob}[W = -x]} = e^{(x - \Delta F)/k_B T}. \tag{5.80}$$

Analogous relations can also be derived for continuous variables, in more general conditions [29].

5.4 Thermodynamic uncertainty relations

As discussed above, entropy production is one of the central quantities to characterize the nonequilibrium behaviour of a system. In this section we will show that it also enters general relations, known as thermodynamic uncertainty relations (TURs), that constrain the ratio between the average value of a generic current and its fluctuations. These kinds of relations can be exploited to obtain estimations of entropy production in systems where only partial information on the relevant degrees of freedom is available, for instance due to experimental limitations. Indeed, as we know from irreversible thermodynamics (see above section 5.1.1) the entropy production rate (or its density s, see equation (5.4)) can be expressed as

$$\dot{S}_{\text{prod}} = \sum_{\alpha=1}^{\nu} X_\alpha J_\alpha, \tag{5.81}$$

where X_α is an affinity or thermodynamic force, while J_α is the associated average current. A rigorous and general treatment of this decomposition principle can be found in the Schnakenberg network theory [34], which decomposes a nonequilibrium Markov process (with discrete state space) into fundamental cycles, each crossed by its own current:

$$\dot{S}_{\text{prod}} = \sum_{\alpha=1}^{\nu} X(\vec{C}_\alpha) J(\vec{C}_\alpha), \tag{5.82}$$

where $\vec{C}_\alpha$ are the ν fundamental cycles in the graph of the process, $X(\vec{C}_\alpha)$ is the affinity or thermodynamic force that acts directly in cycle α, representing the total asymmetry in the transition rates of the edges of that cycle, while $J(\vec{C}_\alpha)$ is the average net current in that cycle. We note that equations (5.81) and (5.82) require a lot of information to retrieve a valid estimate of s, making its direct measurement a hard task in experiments. TURs represent an interesting alternative to obtaining information on entropy production rate. These relations express a bound between the entropy production rate and the first two cumulants of the fluctuations of any kind of current measured in the system.

In particular, one of the most common TURs discussed in the recent literature is a lower bound for the integrated (in a time t) entropy production rate S_{prod}, in the form of a precision rate for the fluctuations of *any* nonequilibrium current integrated for the same time t, J_t, in the system (in the following we take $k_B = 1$ for the Boltzmann constant):

$$S_{\text{prod}} \geqslant 2\frac{\langle J_t \rangle^2}{\text{Var}(J_t)}, \tag{5.83}$$

where $\text{Var}(J_t)$ denotes the variance of the fluctuating current J_t. We note that in a steady state, for large time t, one has $S_{\text{prod}} = \dot{S}_{\text{prod}} t$, $\langle J_t \rangle = Jt$ and $\text{Var}(J_t) \sim 2D_J t$, where D_J is the diffusivity associated with the current rate (whose average we denote by J), leading to an expression of equation (5.83) for rates:

$$\dot{S}_{\text{prod}} \geqslant \frac{J^2}{D_J}. \tag{5.84}$$

This relation is general and involves any observable current. However, it only provides a bound. Equation (5.83) was first derived for some simple models in [35] and then generalized to Markov processes in steady states in [36]. See also [37] for a review with perspectives.

As discussed in section 5.1.1, in the close-to-equilibrium limit the linearity between currents and affinities makes the entropy production rate a bilinear form $\dot{S}_{\rm prod} = \sum_{\beta,\gamma} L_{\beta\gamma} X_\gamma X_\beta$. The Einstein relation then implies that the diffusion coefficient for the fluctuations of the time-integral of the current J_α is $D_\alpha = L_{\alpha\alpha}$, and therefore the TUR for the α current reads

$$\frac{\dot{S}_{\rm prod}}{J_\alpha^2/D_\alpha} = \frac{L_{\alpha\alpha}\sum\limits_{\beta,\gamma} L_{\beta\gamma} X_\beta X_\gamma}{\sum\limits_{\beta,\gamma} L_{\alpha\beta} L_{\beta\gamma} X_\beta X_\gamma} \geqslant \left(1 + \frac{\sum\limits_{\beta,\gamma\neq\alpha} G_{\beta\gamma} X_\beta X_\gamma}{J_\alpha^2}\right) \geqslant 1, \qquad (5.85)$$

where in the last passage we have used the fact that $G_{\beta\gamma} = (L_{\alpha\alpha} L_{\beta\gamma} - L_{\alpha\beta} L_{\alpha\gamma})$ is a positive semi-definite matrix (as a consequence of the fact that the Onsager matrix $\mathscr{L}$ is also positive semi-definite). Thus, in the equilibrium limit the equality is obtained if $X_\beta = 0$ for each $\beta \neq \alpha$.

TURs have been applied to inference problems, in particular for molecular motors and enzymatic networks [38]. The connection with large deviations theory has been discussed in [36, 39–41]. It is also interesting to note the link with information theory and with the generalized Cramér–Rao inequality [42]. This approach is based on the evaluation of the Fisher information for the Onsager–Machlup measure of the path where an external perturbation is applied to the dynamics, and requires particular care in the case of underdamped systems, see [43, 44]. Further generalizations have been applied to obtain a lower bound for diffusivity of a tracer particle under the action of nonlinear friction and non-equilibrium baths with multiple timescales and multiple temperatures [45, 46]. An alternative direct derivation of the TUR is also discussed in reference [47].

It is clear that the bound provided by the TUR is particularly useful if it is tight. This condition is in general not satisfied when the distribution of the current fluctuations is non-Gaussian or when the choice of the current is sub-optimal, i.e. it does not contain enough information to retrive the total entropy production. In these cases, tighter bounds can be obtained if more information is known rather than just the fluctuations of a current. This has been illustrated for instance in reference [48], where it has been shown that

$$\eta_J + \chi^2_{J_t, Z} \leqslant 1, \qquad (5.86)$$

$$\eta_J = \frac{2\langle J_t\rangle^2}{\mathrm{Var}(J_t) S_{\rm prod}}, \qquad (5.87)$$

$$\chi_{J_t, Z} = \mathrm{Cov}(J_t, Z)/\sqrt{\mathrm{Var}(J_t)\mathrm{Var}(Z)}, \qquad (5.88)$$

where the usual TUR states that $\eta_J \leqslant 1$, while the improving term $\chi_{J_t,Z}$ is the Pearson correlation coefficient between J_t and any variable Z satisfying $-1 \leqslant \chi_{J_t,Z} \leqslant 1$, and $\mathrm{Cov}(a, b)$ is the covariance between variables a and b.

5.5 Fluctuation–dissipation theorem: alternative approaches

We now present an alternative treatment of the FDT introduced in chapter 4. Such an approach allows us to put in evidence its relation with the entropy production, the LDB and with the fluctuation relations. Moreover, we will discuss in detail the role played by other quantities such as dynamical activity and kinetic factors. Readers interested in more details on this approach are referred to references [49, 50].

5.5.1 Fluctuation–dissipation theorem and dynamical activity

We are interested in the study of the behaviour of the system in response to an external perturbation $h(t)$, for Hamiltonian systems. Such a perturbation modifies the original Hamiltonian in such a way that $H \rightarrow H_h(x) = H(x) - V(x)h(t)$, where x represents a state variable (even under time reversal, e.g. the position) and $V(x)$ a generic function. Therefore, given a trajectory $\Omega_0^t = \{x(\tau)\}$, we introduce the probability density

$$\mathscr{P}_h(\Omega_0^t) \equiv \frac{P_h(\Omega_0^t)}{P(\Omega_0^t)} = e^{-\mathscr{A}_h(\Omega_0^t)}, \tag{5.89}$$

which gives the ratio between the likelihood of a trajectory when a perturbation h is applied and the unperturbed one. Decomposing the action $\mathscr{A}_h$ in two contributions,

$$\mathscr{A}_h = \frac{1}{2}(\mathscr{T}_h - \mathscr{S}_h), \tag{5.90}$$

where

$$\mathscr{T}_h = \overline{\mathscr{A}}_h + \mathscr{A}_h, \qquad \mathscr{S}_h = \overline{\mathscr{A}}_h - \mathscr{A}_h, \tag{5.91}$$

with $\overline{\mathscr{A}}_h = \mathscr{A}_h(\overline{\Omega_0^t})$, we can rewrite

$$\mathscr{P}_h(\Omega_0^t) = e^{-\mathscr{T}_h(\Omega_0^t)/2} e^{\mathscr{S}_h(\Omega_0^t)/2}. \tag{5.92}$$

Equations (5.91) can be also rewritten as

$$\begin{aligned} \mathscr{S}_h &= \log\left(\frac{P(\overline{\Omega_0^t})}{P(\Omega_0^t)} \Big/ \frac{P^h(\overline{\Omega_0^t})}{P^h(\Omega_0^t)}\right) = \Sigma_h - \Sigma, \\ \mathscr{T}_h &= \log\frac{P(\overline{\Omega_0^t})P(\Omega_0^t)}{P^h(\overline{\Omega_0^t})P^h(\Omega_0^t)} = D_h - D, \end{aligned} \tag{5.93}$$

where Σ is the entropy production defined in (5.42) and $D = -\log[P(\Omega_0^t)P(\overline{\Omega_0^t})]$ is the dynamical activity. Indeed, according to the LDB condition, the antisymmetric part $\mathscr{S}_h$ represents the excess in physical entropy flux due to the external perturbation, namely the further contribution to the entropy production induced by the perturbation. The time symmetric term $\mathscr{T}_h$ is the excess in the dynamical activity and is related to the escape rates. Notice that, considering two trajectories having the same entropy flux, the one with the lowest excess in activity turns out to be the most probable.

For a given observable A, the linear response function can be then written as

$$\begin{aligned} R_{AV}(t, t') &\equiv \left.\frac{\delta\langle A(t)\rangle_h}{\delta h(t')}\right|_{h=0} = \frac{\delta\langle A(t)e^{-\mathcal{T}_h/2+\mathcal{S}_h/2}\rangle}{\delta h(t')} \\ &= \frac{1}{2}\left\langle A(t)\left.\frac{\delta \mathcal{S}_h}{\delta h(t')}\right|_{h=0}\right\rangle - \frac{1}{2}\left\langle A(t)\left.\frac{\delta \mathcal{T}_h}{\delta h(t')}\right|_{h=0}\right\rangle. \end{aligned} \tag{5.94}$$

To be more specific, let us consider a continuous-time homogeneous Markov process with transition rates $W(y|x)$ from state x to state y and persistence probability of remaining in the state x for a time interval Δt, $\pi(x; \Delta t)$. As a perturbed process, we consider the one evolving in the presence of the external field $h(t)$ coupled to the observable $V(x)$. The persistence probabilities get changed into $\pi_h(x; \Delta t)$. Assuming the LDB condition for the perturbed transition rates $W[y|x; h(t)]$, we can write the general expression as

$$W[y|x; h(t)] = W(y|x)K[x, y; h(t)]e^{\beta h(t)[V(y)-V(x)]/2}, \tag{5.95}$$

where the $W(y|x)$ are the unperturbed transition rates and $K[x, y; h(t)]$ is a generic function symmetric in its arguments, such that $W[y|x; h(t)]$ is a transition rate, namely positive and normalizable, with the constraint $K(x, y; 0) = 1$. This quantity represents a kinetic factor that can in principle depend on the applied perturbation. Then let us write explicitly the probabilities of a trajectory over a time interval $[0, t]$ with jumps at times t_i ($i = 1, \cdots n$) and of its time reversal in the presence and in the absence of perturbation, as done in equations (5.37) and (5.38) (but without the assumption of an initial stationary state)

$$\begin{aligned} P(\Omega_0^t) &= P(x_0)\pi(x_0; t_1)W(x_1|x_0)\pi(x_1; t_2 - t_1)W(x_2|x_1)\dots \\ &\quad\times W(x_n|x_{n-1})\pi(x_n; t - t_n) \\ P(\overline{\Omega_0^t}) &= P'(x_n)\pi(x_n; t - t_n)W(x_{n-1}|x_n)\pi(x_{n-1}; t_n - t_{n-1})W(x_{n-2}|x_{n-1})\dots \\ &\quad\times W(x_0|x_1)\pi(x_0; t_1) \\ P^h(\Omega_0^t) &= P[x_0; h(0)]\pi_h(x_0; t_1)W[x_1|x_0; h(t_1)]\pi_h(x_1; t_2 - t_1)W[x_2|x_1; h(t_2)]\dots \\ &\quad\times W[x_n|x_{n-1}; h(t_n)]\pi_h(x_n; t - t_n) \\ P^h(\overline{\Omega_0^t}) &= P'[x_n; h(t)]\pi_h(x_n; t - t_n)W[x_{n-1}|x_n; h(t_n)]\pi_h(x_{n-1}; t_n - t_{n-1})\dots \\ &\quad\times W[x_0|x_1; h(t_1)]\pi_h(x_0; t_1). \end{aligned} \tag{5.96}$$

Note that the initial probability P' for the time reversal trajectories can be in principle arbitrarily chosen. Now we can give an explicit meaning to the entropy and dynamical activity excesses appearing in equaions (5.93). Let us start by computing the entropic contribution. Enforcing the LDB condition one obtains

$$\begin{aligned} \frac{P^h(\Omega_0^t)}{P^h(\overline{\Omega_0^t})} &= \frac{P[x_0; h(0)]}{P'[x_n; h(t)]}\exp\left\{\beta\int_0^t d\tau\ h(\tau)\dot{V}(\tau)\right\} \\ &\quad\times\frac{W(x_1|x_0)W(x_2|x_1)\ \dots\ W(x_n|x_{n-1})}{W(x_{n-1}|x_n)W(x_{n-2}|x_{n-1})\ \dots\ W(x_0|x_1)}, \end{aligned} \tag{5.97}$$

and

$$\frac{P(\Omega_0^t)}{P(\overline{\Omega_0^t})} = \frac{P(x_0)}{P'(x_n)} \frac{W(x_1|x_0)W(x_2|x_1)\ \dots\ W(x_n|x_{n-1})}{W(x_{n-1}|x_n)W(x_{n-2}|x_{n-1})\ \dots\ W(x_0|x_1)}. \tag{5.98}$$

Hence,

$$\mathscr{S}_h = \log\left(\frac{P^h(\Omega_0^t)}{P^h(\overline{\Omega_0^t})}\right) - \log\left(\frac{P(\Omega_0^t)}{P(\overline{\Omega_0^t})}\right) = \beta\int_0^t d\tau\ \ h(\tau)\dot{V}(\tau) + \Delta\mathscr{S}_h, \tag{5.99}$$

where

$$\Delta\mathscr{S}_h = \log\left[\frac{P[x_0;\ h(0)]}{P(x_0)}\frac{P'(x_n)}{P'[x_n;\ h(t)]}\right], \tag{5.100}$$

and

$$\left\langle A(t)\frac{\delta\mathscr{S}_h}{\delta h(t')}\bigg|_{h=0}\right\rangle = \beta\langle A(t)\dot{V}(t')\rangle + \left\langle A(t)\frac{\delta\Delta\mathscr{S}_h}{\delta h(t')}\bigg|_{h=0}\right\rangle. \tag{5.101}$$

The term related to dynamical activity can be obtained in the same way

$$\begin{aligned}\mathscr{T}_h =& -\log\frac{P^h(\overline{\Omega_0^t})P^h(\Omega_0^t)}{P(\overline{\Omega_0^t})P(\Omega_0^t)}\\ =& -2\log\{\pi_h(x_0;\ t_1)W[x_1|x_0;\ h(t_1)]\pi_h(x_1;\ t_2-t_1)W[x_2|x_1;\ h(t_2)]\dots\\ &\times W[x_n|x_{n-1};\ h(t_n)]\pi_h(x_n;\ t-t_n)\}\\ &+2\log\{\pi(x_0;\ t_1)W(x_1|x_0)\pi(x_1;\ t_2-t_1)W(x_2|x_1)\dots\\ &\times W(x_n|x_{n-1})\pi(x_n;\ t-t_n)\}\\ &+\Delta\mathscr{T}_h,\end{aligned} \tag{5.102}$$

where

$$\Delta\mathscr{T}_h = -\log\frac{P[x_0;\ h(0)]}{P(x_0)}\frac{P'[x_n;\ h(t)]}{P'(x_n)}. \tag{5.103}$$

Using the definition of persistence probability

$$\pi_h(x;\ t_2-t_1) = e^{-\sum_{y\neq x}\int_{t_1}^{t_2} d\tau\ \ W[y|x;h(\tau)]}, \tag{5.104}$$

together with the LDB condition (5.95), one obtains

$$\mathscr{T}_h = 2\int_0^t d\tau\left\{\sum_{y\neq x} W(y|x)[K[x,\ y;\ h(\tau)]e^{\beta h(\tau)[V(y)-V(x)]/2} - 1]\right\} + \Delta\mathscr{T}_h, \tag{5.105}$$

and then

$$\left\langle A(t)\frac{\delta\mathscr{T}_h}{\delta h(t')}\bigg|_{h=0}\right\rangle = \beta\langle A(t)B(t')\rangle + \langle A(t)C(t')\rangle + \left\langle A(t)\frac{\delta\Delta\mathscr{T}_h}{\delta h(t')}\bigg|_{h=0}\right\rangle, \tag{5.106}$$

where

$$B(t) \equiv \sum_{y \neq x} W(y|x)[V(y) - V(x(t))], \tag{5.107}$$

and

$$C(t) \equiv \sum_{y \neq x} W(y|x) \frac{\partial K(x, y; h)}{\partial h}\bigg|_{h=0}. \tag{5.108}$$

This last quantity only appears when kinetic factors explicitly depend on the perturbation h and represents their contribution to the response function. However, in most cases, such a dependence is not considered, and for the linear response function one eventually finds the expression

$$R_{AV}(t, t') = \frac{\beta}{2}[\langle A(t)\dot{V}(t')\rangle - \langle A(t)B(t')\rangle] + \left\langle A(t) \frac{\delta \Delta_h}{\delta h(t')}\bigg|_{h=0} \right\rangle, \tag{5.109}$$

where

$$\Delta_h \equiv \frac{1}{2}(\Delta \mathcal{S}_h - \Delta \mathcal{T}_h) = \log \frac{P[x_0; h(0)]}{P(x_0)}. \tag{5.110}$$

If $P[x_0; h(0)] = P(x_0)$, namely if the perturbation modifies the propagator of the process and not the initial condition, then $\Delta_h = 0$ and one gets

$$R_{AV}(t, t') = \frac{\beta}{2}[\langle A(t)\dot{V}(t')\rangle - \langle A(t)B(t')\rangle]. \tag{5.111}$$

This formula expresses the response function in terms of unperturbed correlators: the first term is the usual equilibrium contribution, related to the entropy production induced by the perturbation (see Kubo relation in chapter 4), whereas the second term involves correlations with other degrees of freedom in the system, through the quantity B, and is related to the dynamical activity change due to the perturbation. This quantity describes the evolution in time of the perturbing potential $V(x)$, as it can be checked by choosing the observable $A \equiv 1$ in equation (5.111), so that

$$\langle \dot{V}(t)\rangle = \langle B(t)\rangle. \tag{5.112}$$

The equilibrium form of the FDT can be immediately recovered from equation (5.111), observing that, exploiting time reversal and time translation invariance, we have

$$\langle A(t)B(t')\rangle = \langle B(t)A(t')\rangle = \frac{\partial}{\partial t}\langle V(t)A(t')\rangle = \frac{\partial}{\partial t}\langle A(t)V(t')\rangle = -\langle A(t)\dot{V}(t')\rangle, \tag{5.113}$$

where in the second equality we have used the definition (5.107). This yields

$$R_{AV}(t, t') = \beta\langle A(t)\dot{V}(t')\rangle. \tag{5.114}$$

This result also shows that, in equilibrium conditions, the contributions from entropy production and from dynamical activity are equivalent.

On the other hand, if the perturbation is applied only at the initial instant, shifting the initial distribution, i.e. $P[x_0; h(0)] = P[x_0 - h(0)]$ and $\partial P[(x_0 - h(0)]/\partial h = -\partial P[x_0 - h(0)]/\partial x_0$, the first two terms in equation (5.109) vanish identically because $h(t) = 0$ for $t > 0$, and the last term gives the general formula derived in chapter 4 in terms of the probability distribution

$$R_{AV}(t, t') = \left\langle A(t)\frac{\delta\Delta}{\delta h(t')}\bigg|_{h=0}\right\rangle = -\left\langle A(t)\frac{\partial \log P(x, t')}{\partial x}\right\rangle. \tag{5.115}$$

The above derivations can be worked out along the same lines for systems described by Langevin equations. In this case, one can show that the role of the quantity B coming from the dynamical activity contribution is played by the drift in the Langevin equation. Indeed, for the perturbed action (5.52) in the presence of the field $h(t)$ one has

$$\mathscr{A}_h(\Omega_0^t) = -\int_0^t d\tau \sum\nolimits_i' \frac{1}{4D_{ii}}\left[\frac{dx_i(\tau)}{d\tau} - F_i[x(\tau); \lambda(\tau)] + h(\tau)\right]^2 \tag{5.116}$$

and the functional derivative with the respect to $h(t')$ appearing in the definition of the response function yields two contributions analogous to those obtained in the previous derivation.

We explicitly illustrate this case with the simple example of a two-variable Brownian gyrator previously introduced. If we apply a perturbation on the variable x, which modifies the Hamiltonian as $H \to H - h(t)x$, and we study the response of the variable x itself, the FDT then reads

$$R_{xx}(t, t') = \frac{1}{2T_x}[\langle x(t)\dot{x}(t')\rangle + \alpha\langle x(t)x(t')\rangle - \lambda\langle x(t)y(t')\rangle]. \tag{5.117}$$

In the stationary state, this result is equivalent to the one obtained with the approach discussed in chapter 4, namely with equation (5.115), which gives

$$R_{xx}(t - t') = \sigma_{11}^{-1}\langle x(t)\dot{x}(t')\rangle + \sigma_{12}^{-1}\langle x(t)y(t')\rangle, \tag{5.118}$$

where we have exploited the fact that the stationary distribution is a bivariate Gaussian and where σ^{-1} represents the inverse covariance matrix.

Finally, let us stress that the above relations can be generalized to systems with discrete variables (such as Ising model) [51, 52] and to nonlinear orders [53–55], providing relations between nonlinear response functions and multi-point correlators. For instance, in the context of Ising models and spin glasses, the non-equilibrium FDT in the form (5.109) has been exploited to develop field-free algorithms for the efficient numerical measurement of the response function from unperturbed correlators [51, 56–58]. We report here the explicit formula for an Ising system in contact with a bath at temperature T. We denote the spin configurations

$[s] = \{s_i\}$, with $s_i = \pm 1$ and $i \in [1, N]$. For the spin variable at site i, s_i, the linear response to the magnetic field applied on site j is

$$R_{ij}(t, t') \equiv \left.\frac{\delta\langle s_i(t)\rangle}{\delta h_j(t')}\right|_{h=0} = \frac{1}{2T}\left[\frac{\partial}{\partial t'}\langle s_i(t)s_j(t')\rangle - \langle s_i(t)B_j(t')\rangle\right], \tag{5.119}$$

with $B_j([s]) = -\sum_{[s']}(s_j - s_j')W([s']|[s])$, where $\sum_{[s']}$ denotes a sum over all possible configurations $[s']$ and $W([s']|[s])$ is the (unperturbed) transition rate from the configuration $[s]$ to the configuration $[s']$.

5.5.2 The fluctuation relation close to equilibrium and its connection to the fluctuation–dissipation theorem

We stress again that the fluctuation relation expresses a property of the fluctuations of entropy production in any physical regime, including far from equilibrium steady states. The same remark applies to the generalized FDTs that we have discussed in chapter 4 and in the previous section. More importantly, as illustrated above, the FDT, when used to predict the linear response of an arbitrarily far from equilibrium stationary state, is not directly related to the entropy production only and therefore it is not expected to be connected to the fluctuation relation. Nevertheless, when the FDT is applied to small perturbations around an *equilibrium* steady state, it can be immediately connected to the fluctuation relation, as briefly explained below [59].

We assume that a system close to equilibrium, in a stationary state, has a small non-zero average current $\langle j\rangle$ which can be understood as the $t \to \infty$ response to a small external perturbation $F_0\Theta(t)$, applied since time 0, in an equilibrium system. For this reason it has to satisfy the Kubo relation (see chapter 4) that is the near equilibrium FDT. In this particular case it takes the form

$$\langle j\rangle = F_0\int_0^\infty dt' R_{JF}(t') = \frac{F_0}{k_B T}\int_0^\infty dt'\langle j(t')\dot{A}(0)\rangle_0, \tag{5.120}$$

with $A(t)$ the observable conjugated to the perturbation F. It is interesting to study the fluctuations, e.g. the variance, of the τ-averaged current J_τ, defined as

$$J_\tau = \frac{1}{\tau}\int_0^\tau dt' j(t'). \tag{5.121}$$

At equilibrium (when $\langle j\rangle = 0$), and for large τ, it is always possible to write

$$\tau^2\langle J_\tau^2\rangle_0 = \left\langle\left[\int_0^\tau dt' j(t')\right]^2\right\rangle_0 \approx 2\tau\int_0^\tau dt'\langle j(t')j(0)\rangle_0. \tag{5.122}$$

In the framework of linear response theory, where $O(F_0^2)$ and higher order terms are neglected, the variance of a current is not changed too much by the perturbation, that is $\langle J_\tau^2\rangle - \langle J_\tau\rangle^2 \approx \langle J_\tau^2\rangle_0$. Taking the Kubo FDT and equation (5.122) we get, for large τ:

$$\langle J_\tau^2\rangle - \langle J_\tau\rangle^2 \approx \frac{2}{\tau}\int_0^\tau dt'\langle j(t')j(0)\rangle_0 = 2\frac{k_\mathrm{B}T\langle J_\tau\rangle}{\tau F_0}. \tag{5.123}$$

At equilibrium, and close to it (at first order in the perturbation), one can disregard large deviations of J_τ and focus on the central part of its statistics, that is a Gaussian probability density function, which—according to the last result—reads at large τ

$$p(J_\tau) \propto \exp\left[-\tau\frac{(J_\tau - \mu F_0)^2}{4\mu k_\mathrm{B}T}\right], \tag{5.124}$$

where we have defined the generalized mobility $\mu = \langle J_\tau\rangle / F_0$.

Remarkably, equation (5.124)—which is a mere consequence of the FDT (and the limit $\tau \to \infty$) appears to be identical to the fluctuation relation

$$p(J_\tau)/p(-J_\tau) = \exp\left(\tau\frac{\dot{S}_\tau}{k_\mathrm{B}}\right), \tag{5.125}$$

where the role of 'fluctuating entropy production rate' $\dot{S}_\tau$ (averaged over the time τ) is played by

$$\dot{S}_\tau = \frac{J_\tau \delta F_0}{T}. \tag{5.126}$$

5.5.3 Harada–Sasa relation

The extra terms appearing in the generalized FDT represent the effect of non-equilibrium conditions and indeed can be connected with the energy (or heat) dissipation in the system. To illustrate such a result, we consider the simplest case of a one-dimensional overdamped Langevin equation

$$\gamma\frac{dx}{dt} = F[x(t), t] + \xi(t), \tag{5.127}$$

where γ is the fluid friction, F an external force and ξ a white noise with zero average and variance $2\gamma T$. We define the velocity fluctuations as

$$C(t) = \langle[\dot{x}(t) - v_s][\dot{x}(0) - v_s]\rangle, \tag{5.128}$$

where v_s is the stationary velocity, while the energy dissipation, or heat flow to the bath, according to the prescription of stochastic thermodynamics discussed previously [13, 24], is

$$J(t)dt = [\gamma\dot{x}(t) - \xi(t)] \circ dx(t); \tag{5.129}$$

here we are using the Stratonovich convention for integration, by denoting it with the $\circ$ in the product. Recalling the definition of the response of the particle velocity to an external field $h(t)$ which modifies the energy as $H \to H_h = H(x) - xh(t)$, in the stationary state

$$R_{vx}(t-s) = \left.\frac{\delta\langle \dot{x}(t)\rangle_h}{\delta h(s)}\right|_{h=0}, \tag{5.130}$$

one can obtain the following result

$$\langle J\rangle = \gamma\left\{v_s^2 + \int_{-\infty}^{+\infty}[\tilde{C}(\omega) - 2T\tilde{R}'(\omega)]\frac{d\omega}{2\pi}\right\}, \tag{5.131}$$

where the tilde denotes the Fourier transform and the prime indicates the real part. This result is known as Harada–Sasa relation [60, 61].

We report here a simple derivation, along the lines of reference [61]. Since for the Langevin equation (5.127) in the presence of the perturbation $h(t)$, the probability of a trajectory $\Omega_{t_0}^{t_1} = \{x(t)\}$ in the time interval $[t_0, t_1]$ can be written as

$$P_h(\Omega_{t_0}^{t_1}) \propto \exp\left\{-\frac{1}{4\gamma T}\int_{t_0}^{t_1}[\gamma\dot{x}(t) - F[x(t)] - h(t)]^2 dt\right\}, \tag{5.132}$$

from the definition of response (5.130) one has

$$\begin{aligned}
R_{vx}(t-s) &= \left.\frac{\delta}{\delta h(s)}\right|_{h=0}\int d\Omega_{t_0}^{t_1} P_h(\Omega_{t_0}^{t_1})\dot{x}(t)\\
&= \int d\Omega_{t_0}^{t_1}\left.\frac{\delta P_h(\Omega_{t_0}^{t_1})}{\delta h(s)}\right|_{h=0}\dot{x}(t)\\
&= \frac{1}{2\gamma T}\langle \dot{x}(t)\circ\{\gamma\dot{x}(s) - F[x(s)]\}\rangle|_{h=0}\\
&= \frac{1}{2T}[v_s^2 + C(t-s)] - \frac{1}{2\gamma T}\langle\dot{x}(t)\circ F[x(s)]\rangle|_{h=0}.
\end{aligned} \tag{5.133}$$

Adding this expression with that obtained by exchanging t and s, one gets

$$\begin{aligned}
R_{vx}(t-s) + R_{vx}(s-t) = &\frac{1}{T}[v_s^2 + C(t-s)]\\
&- \frac{1}{2\gamma T}\langle\dot{x}(t)\circ F[x(s)] + \dot{x}(s)\circ F[x(t)]\rangle|_{h=0}.
\end{aligned} \tag{5.134}$$

Taking then the limit $t = s$, one observes from equation (5.129) that the last term on the right-hand side is the average energy dissipation $\langle J\rangle$, and therefore one finally obtains the Harada–Sasa relation in the form (5.131).

Generalizations of the Harada–Sasa relation to the case of bath fluctuations non-local in time can be found in reference [62], while the extension to systems described by a master equation is considered in reference [63].

5.5.4 Fluctuation–dissipation theorem from local detailed balance

We conclude this chapter with a discussion on a further relation between the FDT and the fluctuation relation. In particular, we show how to derive the FDT in the form (5.109) directly from the LDB. A first derivation was presented in reference [64] for a stationary dynamics. Here we present the general derivation valid for transient out of equilibrium evolution [54].

For simplicity, the derivation is presented for a one-dimensional system described by a continous variable $x(t)$, however, it can be obtained analogously also for discrete variables. The system is initially in equilibrium with a reservoir at inverse temperature β. We call experimental protocol the assigned time dependence of the external field $h(\tau)$, coupled with the potential $V(x)$, in some time interval $\tau \in [0, t]$. Then, from the LDB condition in the form (5.54) follows that the probability $P[\Omega_0^t|x_0; h(t)]$ of a trajectory Ω_0^t, taking place under the protocol $h(t)$ and conditioned to the initial value $x_0 = x(0)$, is related to the probability of the reverse trajectory $\overline{\Omega_0^t}$ under the reverse protocol $\bar{h}(t)$ and conditioned to $x_t = x(t)$, by

$$P[\Omega_0^t|x_0; h(t)]\exp\left\{-\beta\int_0^t d\tau\ h(\tau)\dot{V}(\tau)\right\} = P[\overline{\Omega_0^t}|x_t; \bar{h}(t)]\exp\{\beta[H(x_0) - H(x_t)]\}. \quad (5.135)$$

Multiplying both sides by $A(x_t)P_I(x_0)$, where $P_I(x)$ is an arbitrary probability distribution and $A(x)$ is a generic observable function of the state variable x, and summing over all trajectories in the interval $[0, t]$, one finds

$$\begin{aligned}&\int d\Omega_0^t A(x_t)P[\Omega_0^t|x_0; h(t)]\ \exp\left\{-\beta\int_0^t d\tau\ h(\tau)\dot{V}(\tau)\right\}P_I(x_0)\\ &= Z_0\int d\Omega_0^t P_I(x_0)e^{\beta H(x_0)}P[\overline{\Omega_0^t}|x_t; \bar{h}(t)]A(x_t)P_0(x_t).\end{aligned} \quad (5.136)$$

Here we have introduced P_0 and Z_0, denoting the equilibrium distribution and the corresponding partition function, respectively, in the absence of the external field. Hence, the above result can be rewritten more compactly as

$$\left\langle A(x_t)\exp\left\{-\beta\int_0^t d\tau\ h(\tau)\dot{V}(\tau)\right\}\right\rangle_{I\to\beta,[h(t)]} = Z_0\langle P_I(x_t)e^{\beta H(x_t)}A(x_0)\rangle_{\beta,0\to\beta,[\bar{h}(t)]} \quad (5.137)$$

where $\langle\cdot\rangle_{I\to\beta,[h(t)]}$ stands for the average in the process starting with the initial condition P_I, thereafter in contact with the thermal resorvoir β and evolving with the protocol $[h(t)]$, while $\langle\cdot\rangle_{\beta,0\to\beta,[\bar{h}(t)]}$ stands for the process in contact with the thermal resorvoir β, starting with the unperturbed equilibrium distribution P_0 and evolving with the reverse protocol $[\bar{h}(t)]$.

The next step is to expand both sides in powers of $[h(t)]$ about $h(t) = 0$ and to compare terms of the same order. At zero order one gets the identity

$$Z_0\langle P_I(x_t)e^{\beta H(x_t)}A(x_0)\rangle_{\beta,0} = \langle A(x_t)\rangle_{I\to\beta,0} \quad (5.138)$$

where $\langle\cdot\rangle_{I\to\beta,0}$ stands for the average in the off-equilibrium process starting with P_I and evolving in contact with the thermal reservoir β in the absence of the external field. Indeed the factor $Z_0P_I(x_t)e^{\beta H(x_t)}$ in the left-hand side has the effect of replacing the equilibrium initial condition with P_I. At the first order one has

$$
\begin{aligned}
&Z_0 \frac{\delta \langle P_I(x_t) e^{\beta H(x_t)} A(x_0) \rangle_{\beta,0\to\beta,[\bar{h}(t)]}}{\delta \bar{h}(t')}\Bigg|_{\bar{h}=0} = \\
&-\beta \frac{\partial}{\partial t'} \langle A(x_t) V(t') \rangle_{I\to\beta,0} + \frac{\delta \langle A(x_t) \rangle_{I\to\beta,[h(t)]}}{\delta h(t')}\Bigg|_{h=0}
\end{aligned}
\tag{5.139}
$$

Recalling the definition of the response function, equation (5.139) can be rewritten as

$$
R_{AV}(t, t') = \beta \frac{\partial}{\partial t'} \langle A(t) V(t') \rangle_{I\to\beta,0} + Z_0 \frac{\delta \langle P_I(x_t) e^{\beta H(x_t)} A(x_0) \rangle_{\beta,0\to\beta,[\bar{h}(t)]}}{\delta \bar{h}(t')}\Bigg|_{\bar{h}=0} . \tag{5.140}
$$

Using equation (5.116) to treat the derivative with respect to $\bar{h}(t')$, the second term in the right-hand side can be rewritten as

$$
\begin{aligned}
&Z_0 \frac{\delta \langle P_I(x_t) e^{\beta H(x_t)} A(x_0) \rangle_{\beta,0\to\beta,[\bar{h}(t)]}}{\delta \bar{h}(t')}\Bigg|_{\bar{h}=0} = \\
&Z_0 \frac{\beta}{2} \frac{\partial}{\partial t'} \langle P_I(x_t) e^{\beta H(x_t)} V(t') A(x_0) \rangle_0 - Z_0 \frac{\beta}{2} \langle P_I(x_t) e^{\beta H(x_t)} B(t') A(x_0) \rangle_0 .
\end{aligned}
\tag{5.141}
$$

Furthermore, since after setting to zero the external field the averages become equilibrium averages, using the Onsager relations for time correlation functions, we get

$$
\begin{aligned}
&Z_0 \frac{\delta \langle P_I(x_t) e^{\beta H(x_t)} A(x_0) \rangle_{\beta,0\to\beta,[\bar{h}(t)]}}{\delta \bar{h}(\bar{t}')}\Bigg|_{\bar{h}=0} = \\
&- Z_0 \frac{\beta}{2} \frac{\partial}{\partial t'} \langle A(t) V(t') P_I(x_0) e^{\beta H(x_0)} \rangle_0 - Z_0 \frac{\beta}{2} \langle A(t) B(t') P_I(x_0) e^{\beta H(x_0)} \rangle_0 .
\end{aligned}
\tag{5.142}
$$

Next, using the identity (5.138), equation (5.142) can be rewritten as

$$
Z_0 \frac{\delta \langle P_I(x_t) e^{\beta H(x_t)} A(x_0) \rangle_{\beta,0\to\beta,[\bar{h}(t)]}}{\delta \bar{h}(\bar{t}')}\Bigg|_{\bar{h}=0} = -\frac{\beta}{2} \frac{\partial}{\partial t'} \langle A(t) V(t') \rangle_{I\to\beta,0} - \frac{\beta}{2} \langle A(t) B(t') \rangle_{I\to\beta,0} \tag{5.143}
$$

and inserting it into equation (5.140), eventually one finds

$$
R_{AV}(t, t') = \frac{\beta}{2} \left[\frac{\partial}{\partial t'} \langle A(t) V(t') \rangle_{I\to\beta,0} - \langle A(t) B(t') \rangle_{I\to\beta,0} \right], \tag{5.144}
$$

thus recovering equation (5.109), valid in generic transient regimes.

References

[1] De Groot S R and Mazur P 2013 *Non-Equilibrium Thermodynamics* (North Chelmsford, MA: Courier Corporation)

[2] Bertini L, De Sole A, Gabrielli D, Jona-Lasinio G and Landim C 2002 Macroscopic fluctuation theory for stationary non-equilibrium states *J. Stat. Phys.* **107** 635–75

[3] Derrida B, Lebowitz J L and Speer E R 2007 Entropy of open lattice systems *J. Stat. Phys.* **126** 1083–108

[4] Bertini L, De Sole A, Gabrielli D, Jona-Lasinio G and Landim C 2015 Macroscopic fluctuation theory *Rev. Mod. Phys.* **87** 593–636

[5] Castellani E and Ismael J 2016 Which Curie's principle? *Phil. Sci.* **83** 1002–13

[6] Green M S 1951 Brownian motion in a gas of noninteracting molecules *J. Chem. Phys.* **19** 1036–46

[7] Kubo R 1957 Statistical-mechanical theory of irreversible processes. I. General theory and simple applications to magnetic and conduction problems *J. Phys. Soc. Japan* **12** 570–86

[8] Mori H 1958 Statistical-mechanical theory of transport in fluids *Phys. Rev.* **112** 1829

[9] Green M S 1960 Comment on a paper of Mori on time-correlation expressions for transport properties *Phys. Rev.* **119** 829

[10] Kubo R, Toda M and Hashitsume N 2012 *Statistical Physics II: Nonequilibrium Statistical Mechanics* (Berlin: Springer Science)

[11] Brandner K, Saito K and Seifert U 2015 Thermodynamics of micro- and nano-systems driven by periodic temperature variations *Phys. Rev.* X **5** 031019

[12] Livi R and Politi P 2017 *Nonequilibrium Statistical Physics: A Modern Perspective* (Cambridge: Cambridge University Press)

[13] Peliti L and Pigolotti S 2021 *Stochastic Thermodynamics: An Introduction* (Princeton, N J: Princeton University Press)

[14] Peliti L 2024 *Statistical Mechanics in a Nutshell* (Princeton, NJ: Princeton University Press)

[15] Katz S, Lebowitz J L and Spohn H 1983 Phase transitions in stationary nonequilibrium states of model lattice systems *Phys. Rev.* **B 28** 1655

[16] Katz S, Lebowitz J L and Spohn H 1984 Nonequilibrium steady states of stochastic lattice gas models of fast ionic conductors *J. Stat. Phys.* **34** 497–537

[17] Maes C 2021 Local detailed balance *Sci. Post Phys. Lect. Notes* **32** 1–17

[18] Lebowitz J L and Spohn H 1999 A Gallavotti-Cohen-type symmetry in the large deviation functional for stochastic dynamics *J. Stat. Phys.* **95** 333–65

[19] Risken H 1996 *Fokker-Planck equation* (Berlin: Springer)

[20] Onsager L and Machlup S 1953 Fluctuations and irreversible processes *Phys. Rev.* **91** 1505

[21] Machlup S and Onsager L 1953 Fluctuations and irreversible processes II. Systems with kinetic energy *Phys. Rev.* **91** 1512

[22] Spinney R E and Ford I J 2012 Entropy production in full phase space for continuous stochastic dynamics *Phys. Rev.* E **85** 051113

[23] Puglisi A and Villamaina D 2009 Irreversible effects of memory *Europhys. Lett.* **88** 30004

[24] Sekimoto K 2010 *Stochastic Energetics* (Berlin: Springer)

[25] Dotsenko V, Maciołek A, Vasilyev O and Oshanin G 2013 Two-temperature Langevin dynamics in a parabolic potential *Phys. Rev.* E **87** 062130

[26] Crisanti A, Puglisi A and Villamaina D 2012 Nonequilibrium and information: the role of cross correlations *Phys. Rev.* E **85** 061127

[27] Evans D J, Cohen E G D and Morriss G P 1993 Probability of second law violations in shearing steady states *Phys. Rev. Lett.* **71** 2401

[28] Gallavotti G and Cohen E G D 1995 Dynamical ensembles in stationary states *J. Stat. Phys.* **80** 931–70

[29] Seifert U 2012 Stochastic thermodynamics, fluctuation theorems and molecular machines *Rep. Prog. Phys.* **75** 126001

[30] Kurchan J 1998 Fluctuation theorem for stochastic dynamics *J. Phys. A: Math. Gen.* **31** 3719
[31] Maes C 1999 The fluctuation theorem as a Gibbs property *J. Stat. Phys.* **95** 367–92
[32] Jarzynski C 1997 Equilibrium free-energy differences from nonequilibrium measurements: a master-equation approach *Phys. Rev.* E **56** 5018
[33] Crooks G E 1999 Entropy production fluctuation theorem and the nonequilibrium work relation for free energy differences *Phys. Rev.* E **60** 2721
[34] Schnakenberg J 1976 Network theory of microscopic and macroscopic behavior of master equation systems *Rev. Mod. Phys.* **48** 571
[35] Barato A C and Seifert U 2015 Thermodynamic uncertainty relation for biomolecular processes *Phys. Rev. Lett.* **114** 158101
[36] Gingrich T R, Horowitz J M, Perunov N and England J L 2016 Dissipation bounds all steady-state current fluctuations *Phys. Rev. Lett.* **116** 120601
[37] Horowitz J M and Gingrich T R 2020 Thermodynamic uncertainty relations constrain non-equilibrium fluctuations *Nat. Phys.* **16** 15–20
[38] Seifert U 2019 From stochastic thermodynamics to thermodynamic inference *Annu. Rev. Condens. Matter Phys.* **10** 171–92
[39] Gingrich T R, Rotskoff G M and Horowitz J M 2017 Inferring dissipation from current fluctuations *J. Phys. A: Math. Theor.* **50** 184004
[40] Dechant A and Sasa S-i 2018 Current fluctuations and transport efficiency for general Langevin systems *J. Stat. Mech. Theory Exp.* **2018** 063209
[41] Koyuk T and Seifert U 2020 Thermodynamic uncertainty relation for time-dependent driving *Phys. Rev. Lett.* **125** 260604
[42] Liu K, Gong Z and Ueda M 2020 Thermodynamic uncertainty relation for arbitrary initial states *Phys. Rev. Lett.* **125** 140602
[43] Hasegawa Y and Van Vu T 2019 Uncertainty relations in stochastic processes: an information inequality approach *Phys. Rev.* E **99** 062126
[44] Lee J S, Park J-M and Park H 2021 Universal form of thermodynamic uncertainty relation for Langevin dynamics *Phys. Rev.* E **104** L052102
[45] Plati A, Puglisi A and Sarracino A 2023 Thermodynamic bounds for diffusion in non-equilibrium systems with multiple timescales *Phys. Rev.* E **107** 044132
[46] Plati A, Puglisi A and Sarracino A 2024 Thermodynamic uncertainty relations in the presence of non-linear friction and memory *J. Phys. A: Math. Theor.* **57** 155001
[47] Dieball C and Godec A 2023 Direct route to thermodynamic uncertainty relations and their saturation *Phys. Rev. Lett.* **130** 087101
[48] Dechant A and Sasa S-i 2021 Improving thermodynamic bounds using correlations *Phys. Rev.* X **11** 041061
[49] Maes C 2017 *Non-Dissipative Effects in Nonequilibrium Systems* (Berlin: Springer)
[50] Maes C 2020 Frenesy: time-symmetric dynamical activity in nonequilibria *Phys. Rep.* **850** 1–33
[51] Lippiello E, Corberi F and Zannetti M 2005 Off-equilibrium generalization of the fluctuation dissipation theorem for Ising spins and measurement of the linear response function *Phys. Rev.* E **71** 036104
[52] Corberi F, Lippiello E and Zannetti M 2007 Fluctuation dissipation relations far from equilibrium *J. Stat. Mech.: Theory Exp.* **2007** P07002
[53] Lippiello E, Corberi F, Sarracino A and Zannetti M 2008 Nonlinear susceptibilities and the measurement of a cooperative length *Phys. Rev.* B **77** 212201

[54] Lippiello E, Corberi F, Sarracino A and Zannetti M 2008 Nonlinear response and fluctuation-dissipation relations *Phys. Rev.* E **78** 041120

[55] Basu U, Krüger M, Lazarescu A and Maes C 2015 Frenetic aspects of second order response *Phys. Chem. Chem. Phys.* **17** 6653–66

[56] Chatelain C 2003 A far-from-equilibrium fluctuation–dissipation relation for an Ising–Glauber-like model *J. Phys. A: Math. Gen.* **36** 10739

[57] Ricci-Tersenghi F 2003 Measuring the fluctuation-dissipation ratio in glassy systems with no perturbing field *Phys. Rev.* E **68** 065104

[58] Corberi F, Lippiello E, Sarracino A and Zannetti M 2010 Fluctuation-dissipation relations and field-free algorithms for the computation of response functions *Phys. Rev.* E **81** 011124

[59] Derrida B and Brunet E 2005 *Le mouvement Brownien et le théoreme de fluctuation-dissipation Einstein Aujourd'hui* (London: EDP Sciences)

[60] Harada T and Sasa S-i 2005 Equality connecting energy dissipation with a violation of the fluctuation-response relation *Phys. Rev. Lett.* **95** 130602

[61] Harada T and Sasa S-i 2006 Energy dissipation and violation of the fluctuation-response relation in nonequilibrium Langevin systems *Phys. Rev.* E **73** 026131

[62] Deutsch J M and Narayan O 2006 Energy dissipation and fluctuation response for particles in fluids *Phys. Rev.* E **74** 026112

[63] Lippiello E, Baiesi M and Sarracino A 2014 Nonequilibrium fluctuation-dissipation theorem and heat production *Phys. Rev. Lett.* **112** 140602

[64] Semerjian G, Cugliandolo L F and Montanari A 2004 On the stochastic dynamics of disordered spin models *J. Stat. Phys.* **115** 493–530

Part II

Models and applications

Chapter 6

Model building in systems with multiple scales

> To develop the skill of correct thinking is in the first place to learn what you have to disregard. In order to go on, you have to know what to leave out: this is the essence of effective thinking.
>
> Kurt Gödel

> It is better to be roughly right than precisely wrong.
>
> John Maynard Keynes

> The history of the trials and errors of science is largely, perhaps entirely, the history of different and successive applications of the notion of the negligible[a].
>
> Simone Weil

6.1 Introduction

Parts of this section have been reprinted with permission from [49] copyright Cambridge University Press & Assessment 2008.

Both in science and engineering, there is a large class of interesting problems characterized by the presence of more than one significant scale, namely systems with degrees of freedom showing a very different evolution in time and space. Important examples include protein folding and climate. Indeed, the vibration timescale of covalent bonds is very small (order 10^{-12} s), whereas the folding time for proteins may be much larger, of order of seconds; in the climate dynamics the characteristic times of the involved processes vary from days for the atmosphere, to the order of 10^3–10^4 years for the deep ocean and ice shields. In these cases the system has a multiscale character (in the literature one also finds the term multiple scale). Therefore, one has to treat the 'slow dynamics' using effective equations:

[a] 'L'histoire des tâtonnements de la science est en grande partie, peut-être toute entière, l'histoire des applications différentes et successives de la notion de négligeable.'

doi:10.1088/978-0-7503-6229-0ch6

indeed, even modern supercomputers cannot simulate all the relevant scales involved. Moreover, effective equations allow one to highlight general features and important ingredients which can remain hidden in the detailed description.

Unfortunately, only in a few cases is it possible to derive effective equations with a systematic approach: important examples are dilute gases, harmonic chains and the Markovian limit of Hamiltonian dynamics. However, there is a series of clever practical approaches for the study of multiscale problems that do not rely on rigorous derivations, e.g. the averaging method in celestial mechanics, the Langevin equation for colloids, the homogenization for partial differential equations, the Born–Oppenheimer 'approximation' and the Car–Parrinello method, for a general discussion see reference [1].

The origin of the study of multiscale problems goes back to Newton, with his explanation of the precession of the equinoxes. Let us briefly discuss the basic idea of the averaging method in mechanics [2]: consider the Hamilton equations written in terms of the action-angle variables:

$$\frac{\mathrm{d}\phi}{\mathrm{d}t} = \frac{1}{\varepsilon}[\omega(I) + f(\phi, I)], \quad \frac{\mathrm{d}I}{\mathrm{d}t} = g(\phi, I), \tag{6.1}$$

where the functions f and g are 2π-periodic in ϕ. In the limit $\varepsilon \ll 1$, ϕ and I are the fast and slow variables whose timescales are $O(\varepsilon)$ and $O(1)$, respectively. According to this approach, one introduces a smoothed action J, which describes the 'slow motion' of I, as obtained by the averaging of the fast variable oscillations. The dynamics of J is governed by the force acting on I averaged over the fast variable ϕ

$$\frac{dJ}{dt} = G(J) = \frac{1}{2\pi}\int_0^{2\pi} g(\phi, J)d\phi, \tag{6.2}$$

and the evolution of J gives the leading order behaviour of I.

6.2 Many levels of description: molecular dynamics, Brownian motion and beyond

Parts of this section have been reprinted with permission from [49] copyright Cambridge University Press & Assessment 2008.

The building of appropriate equations to describe the macroscopic world starting from the fundamental level, is surely one of the most important multiscale problems: the ambition is a rigorous derivation of the hydrodynamics equations from the microscopic laws of motion, that are—for classical mechanics—the deterministic Newton's equations [3]. Here we do not discuss in detail this noble and difficult topic (a sketch of a possible derivation of hydrodynamics is summarized in section 8.4); our more modest aim is to remind the basic facts that allows us to derive macroscopic equations from the microscopic dynamics.

Let us start from the fact that particle configurations at the molecular scale are quickly randomized by collisions. This means that in a short time a local equilibrium is reached, for the few macroscopic quantities which evolve more slowly (mass density, temperature, momentum density, pressure and so on). This behaviour relies on the local conservation of mass, momentum and energy, which implies a large-scale separation between the microscopic characteristic time $\tau_{micro} = \ell/v_t$ (ℓ is a characteristic microscopic length, such as the mean free path, and v_t is the typical molecule velocity), and the macroscopic time $\tau_{\mathrm{macro}} = L/U$ (L is a macroscopic length, such as the size of the box, and U is the characteristic velocity at hydrodynamic level).

We can briefly sketch the steps from the microscopic level to the macroscopic one by the following scheme:

- Microscopic level, Γ-space description (Hamilton equations, Liouville equation).
- Microscopic level, μ-space description (Boltzmann equation).
- Mesoscopic level, μ-space description at large scale for colloidal particles (Fokker–Planck equation).
- Macroscopic level, hydrodynamic description (Navier–Stokes equations, diffusion equation, Fourier law).

The crossing between different levels of description is obtained through a coarse-graining and/or a projection procedure, which implies a 'loss of information', where rather delicate mathematical singular limits can be involved [4].

6.2.1 The microscopic level: Hamiltonian dynamics

Let us now briefly review the steps (via coarse-graining procedures) from the molecular dynamics up to the Brownian motion, stressing mainly their conceptual aspects [5]. Let us consider a system of colloids in a liquid. At the microscopic level the system is described by the canonical coordinates $(\mathbf{Q}_i, \mathbf{P}_i)$ of colloidal particles and $(\mathbf{q}_n, \mathbf{p}_n)$ of solvent molecules. In the absence of external potentials, the Hamiltonian reads

$$H = \sum_i \frac{\mathbf{P}_i^2}{2M} + \sum_n \frac{\mathbf{p}_n^2}{2\,m} + \sum_{i,j} V^{ss}(\mathbf{q}_j - \mathbf{q}_i) + \sum_{n,l} V^{cc}(\mathbf{Q}_n - \mathbf{Q}_l) + \sum_{n,i} V^{sc}(\mathbf{Q}_n - \mathbf{q}_i), \tag{6.3}$$

where m is the mass of a solvent molecule, M is the mass of a colloid, V^{ss}, V^{sc} and V^{cc} are the interaction potentials between solvent molecules, solvent and colloids, and colloidal particles, respectively. The evolution of the system is governed by the Hamilton equations for $(\mathbf{Q}_i, \mathbf{P}_i)$ and $(\mathbf{q}_n, \mathbf{p}_n)$; for the colloidal particles we have

$$\frac{\mathrm{d}\mathbf{Q}_i}{\mathrm{d}t} = \frac{\partial H}{\partial \mathbf{P}_i} = \frac{\mathbf{P}_i}{M}, \quad \frac{\mathrm{d}\mathbf{P}_i}{\mathrm{d}t} = -\frac{\partial H}{\partial \mathbf{Q}_i} = \mathbf{F}_i^{cc} + \mathbf{F}_i^{sc}, \tag{6.4}$$

where $\mathbf{F}_i^{cc}$ is the force given by the interaction among the colloidal particles while $\mathbf{F}_i^{sc}$ also depends on the particles of the solvent. This implies that the equations of motion for $(\mathbf{Q}_i, \mathbf{P}_i)$ must include memory terms originated by the interactions with the solvent. In order to go on and write down an effective equation for the colloidal particles we note that, since in comparison with the solvent molecules the colloidal particles have a much larger mass (we assume $M \gg m$), they have a much slower evolution. As a consequence of the timescale separation between the two subsystems, and the huge number of the solvent particles, we can expect that the fast solvent dynamics can be consistently decoupled from the slow colloid dynamics, by approximating its effects on the big suspended particles by means of an effective force. The latter may be decomposed into a systematic part, of viscous type, and a truly stochastic fluctuating part. Basically, the aim is to rationalize, starting from the first principles, the intuition of Einstein and Langevin for the Brownian motion.

Just for simplicity, instead of discussing the general case, we consider a very dilute colloidal suspension: in such a limit the mutual influence among the colloidal particles is negligible, and $\mathbf{F}_i^{cc}$ is only due to possible external potentials. Therefore, we have the well-known Langevin equation, for the independent evolution of each colloidal particle

$$\frac{\mathrm{d}\mathbf{Q}}{\mathrm{d}t} = \mathbf{V} = \frac{\mathbf{P}}{M}, \quad \frac{\mathrm{d}\mathbf{P}}{\mathrm{d}t} = \mathbf{F}(Q) - \gamma\mathbf{V} + \sqrt{2k_{\mathrm{B}}T\gamma}\;\boldsymbol{\eta}, \tag{6.5}$$

where we simplified the notation writing $\mathbf{F}$ instead of $\mathbf{F}^{cc}$, γ is the friction coefficient, and for the random Gaussian vector $\boldsymbol{\eta}$, one has $\langle \eta^k(t)\rangle = 0$ and $\langle \eta^k(t)\eta^l(t')\rangle = \delta^{kl}\delta(t - t')$. The fundamental problem here, at least at thermal equilibrium, is to find γ. This point will be discussed in the next section.

6.2.2 Mesoscopic level: the Kramers and Smoluchowski equations

For simplicity, we consider the one-dimensional case:

$$\frac{dQ}{dt} = V = \frac{P}{M}, \quad \frac{dP}{dt} = F(Q) - \gamma V + \sqrt{2k_{\mathrm{B}}T\gamma}\eta, \tag{6.6}$$

where η is white noise. The evolution of the probability distribution $\rho_c(P, Q, t)$ is ruled by the corresponding Fokker–Planck equation, which is usually indicated as the Kramers equation:

$$\frac{\partial}{\partial t}\rho_c = -\left(V\frac{\partial}{\partial Q} + F(Q)\frac{\partial}{\partial P}\right)\rho_c + \gamma k_{\mathrm{B}}T\frac{\partial}{\partial P}\left(\frac{\partial}{\partial P} + \frac{P}{Mk_{\mathrm{B}}T}\right)\rho_c. \tag{6.7}$$

Let us now discuss the case of large γ: we expect that the velocity variables, for $t \gg 1/\gamma$, are described by a statistical equilibrium so that energy equipartition holds, and one has just to describe the features of Q. It is possible to show that, for $t \gg 1/\gamma$, the probability distribution of Q is ruled by the so-called Smoluchowski equation:

$$\frac{\partial}{\partial t}\rho(Q, t) = -\gamma\frac{\partial}{\partial Q}(F(Q)\rho(Q, t)) + \frac{k_{\mathrm{B}}T}{\gamma}\frac{\partial^2}{\partial Q^2}\rho(Q, t), \tag{6.8}$$

and the corresponding Langevin equation for the slow variable is:

$$\frac{dQ}{dt} = \frac{1}{\gamma}F(Q) + \sqrt{\frac{2k_B T}{\gamma}}\eta. \tag{6.9}$$

Let us give an intuitive argument to justify the above result. Consider equations (6.6): if γ is large the timescale of the velocity variable is much shorter than that of the spatial variable. If one is interested only in the slower Q-evolution, one can treat V as a relaxed variable and one can put formally $dP/dt = 0$ in the second equation of (6.6). Therefore, one can express V in terms of Q and η: namely $V = \frac{1}{\gamma}F(Q) + \sqrt{\frac{2k_B T}{\gamma}}\,\eta$, and we recover the Langevin equation (6.9).

The above simple argument gives the correct result; however, a consistent perturbative derivation of (6.8) from (6.7) has been obtained only about 50 years ago. Actually, also in well-known textbooks one can find inconsistent derivations, based on a perturbative scheme, see e.g. Section VII.7 of reference [6]. The key point is the singular nature of the limit $1/\gamma \to 0$ which does not allow for a straightforward application of perturbation theory, see reference [7] for a clear detailed discussion.

6.2.3 Beyond the Kramers and Smoluchowski equations

The Kramers equation describes the colloidal system at a level which is less accurate than the microscopic one ruled by the complete Hamiltonian. The Smoluchowski equation represents an even coarser level of description, i.e. a mesoscopic level restricted to the evolution of the variable Q, the position of the colloidal particle, which changes along timescales much larger than the typical time M/γ.

Beyond the Kramers and Smoluchowski equations there is another level of description, which rules the evolution of $\rho(\mathbf{Q}, t)$ at very long times and large spatial scales. Let us discuss this point in the $1d$ case where $F(Q) = -\partial U(Q)/\partial Q$ is a periodic function of period L. Denote by $\{Q_n\}$ the minima of U: for small values of the diffusion coefficient $D = k_B T/\gamma$, it easy to realize that $Q(t)$ performs a kind of random walk jumping at random times from one minimum Q_k to one of the two nearest neighbour minima $Q_{k\pm1}$. The interval Δt between two successive jumping times is a stochastic process whose mean value, in the limit of small D, is given by the celebrated Kramers formula [8]:

$$\langle \Delta t \rangle \simeq \tau_0\, e^{\frac{\Delta U}{D}}, \tag{6.10}$$

where ΔU is the difference of potential between the maximum and the minimum and τ_0 is a characteristic time which depends on the second derivatives of U computed at the minimum and the maximum. In chapter 7 we will discuss the relevance of the exit time problem and the Kramers formula.

At large times, i.e. $t \gg \langle \Delta t \rangle$, the dynamics of the field $\tilde{\rho}(Q, t)$, computed from a local average over a region of size much larger than L, is described by the Fick equation:

$$\frac{\partial}{\partial t}\,\tilde{p}(Q,\,t) = D^E\frac{\partial^2}{\partial Q^2}\,\tilde{p}(Q,\,t), \tag{6.11}$$

where D^E is a macroscopic diffusion coefficient. In the limit of small D, one has

$$D^E \sim \frac{L^2}{\langle \Delta t \rangle} \simeq \frac{L^2}{\tau_0} e^{-\frac{\Delta U}{D}}. \tag{6.12}$$

In other words, for time much larger than $\langle \Delta t \rangle$, and spatial scales much larger that L, the variable Q is described by the Langevin equation

$$\frac{dQ}{dt} = \sqrt{2D^E}\,\eta. \tag{6.13}$$

We can conclude the above discussion with the following summary for the evolution of the colloidal particle:

(a) for short times, order τ_{micro}, the correct description of the system is given by the full Hamilton equations (the colloidal particle bounces with the molecules of the surrounding fluid);

(b) for t smaller than M/γ the proper level is given by the Langevin equation (6.6) for Q and P (single collisions are washed out and only an average effect remains on the decay of velocity correlation);

(c) for large value of t, one can assume that P is thermalized and the Q follows equation (6.9);

(d) at time much larger than $\langle \Delta t \rangle$ and large-scale resolution one can safely use equation (6.13).

6.2.4 A parenthesis on the diffusion at large scale in stirred fluids

In the previous section we discussed the general aspects of the modelling problem, i.e. what are the effective equations for the slow dynamics of multiscale systems. However, some important details are still open, for instance how to determine γ in terms of the parameters of the system under investigation. Before discussing—in the next section—some existing methods to solve this problem, we briefly discuss another important issue, which is for some aspects, rather similar: the asymptotic behaviour of the passive scalar in incompressible velocity fields [9, 10]. Taking into account also the molecular diffusion, the time evolution of a test particle is ruled by the Langevin equation

$$\frac{d\mathbf{x}}{dt} = \mathbf{u}(\mathbf{x},\,t) + \sqrt{2D_0}\eta, \tag{6.14}$$

where $\mathbf{u}(\mathbf{x},\,t)$ is the Eulerian velocity field at the position $\mathbf{x}$ and time t, and D_0 is the molecular diffusion. The density of tracers $\theta(\mathbf{x},\,t)$ evolves according to the transport equation (which is nothing but the Fokker–Planck equation):

$$\frac{\partial}{\partial t}\theta + \nabla \cdot (\mathbf{u}\theta) = D_0\Delta\theta, \tag{6.15}$$

where Δ denotes the Laplacian operator. Such a topic has a great practical and theoretical interest in many disciplines such as geophysics, chemical engineering and astrophysics [11].

For the sake of notation simplicity, we assume that $\langle \mathbf{u} \rangle = 0$, and denote by U the typical speed of the field $\mathbf{u}(\mathbf{x}, t)$ and by L its typical length. At time much larger than the characteristic time $T = U/L$, it is reasonable to expect that the field $\Theta(x, t)$, obtained by locally averaging $\theta(\mathbf{x}, t)$ over a volume of linear dimension much larger than L, evolves according to the Fick equation[1]

$$\frac{\partial}{\partial t}\Theta(\mathbf{x}, t) = \sum_{i,j} D_{ij}^{E} \frac{\partial^2}{\partial x_i \partial x_j}\Theta(\mathbf{x}, t); \tag{6.16}$$

this is equivalent to saying that the tracer is well described by

$$dx_n = \sum_j (\sqrt{2D^E})_{nj} dW_j.$$

The validity of the above scenario has been proved under very general assumptions on the velocity field: the effects of the advecting field $\mathbf{u}$ on the asymptotic properties of the Lagrangian behaviour, are taken into account by the coefficients D_{ij}^{E}, which—in the context of transport in fluids—are called eddy diffusivity coefficients [9, 10]. Typically, the elements D_{ij}^{E} are much larger than D_0, and depend both on the shape of $\mathbf{u}$ and on D_0: often such a dependence is rather non-trivial. For instance, in a shear flow along x for small D_0 one has $D_{11}^{E} \sim D_0^{-1}$, while in $2d$ convective cells $D^E \sim \sqrt{D_0}$; in both cases $D^E \gg D_0$. It is interesting, for instance, in geophysics, the fact that one can have large values of D^E on $\mathbf{u}$ also for laminar field: such a feature is associated with the so-called Lagrangian chaos [11].

There is a very powerful multiscale method which can be applied in a given generic incompressible velocity field $\mathbf{u}(\mathbf{x}, t)$ and allows us to derive Fick's equation from the transport equation and compute D_{ij}^{E} [9, 10]. Let us present just the practical recipe: given a velocity field $\mathbf{u}(\mathbf{x}, t)$, with $\mathbf{x} \in \mathbb{R}^2$ or $\mathbb{R}^3$, spatially periodic with an 'elementary' cell $\mathcal{C}$ of size L, one introduces an auxilary vector $\mathbf{w}$ evolving with the equation

$$\frac{\partial}{\partial t}\mathbf{w} + (\mathbf{u} \cdot \nabla)\mathbf{w} - D_0 \Delta \mathbf{w} = -\mathbf{u}. \tag{6.17}$$

From the solution of the above equation in $\mathcal{C}$, as a final result we can compute D_{ij}^{E}:

$$D_{ij}^{E} = D_0 \delta_{ij} - \frac{1}{2}[\langle u_i w_j \rangle + \langle u_j w_i \rangle], \tag{6.18}$$

where the average is only on the cell $\mathcal{C}$ and, in the case of non-stationary velocity field, on time (for details see reference [9]). Only in a few special cases, e.g. shear flows, is it possible to solve analytically the problem. Therefore, it is necessary to use

[1] Such an idea dates back to Maxwell.

suitable perturbative methods or numerical computations; however, we have the remarkable result that it is enough to find the solution of the field $\mathbf{w}$ just on $\mathscr{C}$.

Instead, to enter into the mathematical details of the multiscale method, we discuss the diffusion equation in one spatial dimension:

$$\frac{\partial}{\partial t}\theta = \frac{\partial}{\partial x}\left(D(x)\frac{\partial}{\partial x}\theta\right), \tag{6.19}$$

where $D(x)$ is a periodic function with period L. The aim is to write an effective (Fick's) equation valid at long time and large scale (much larger than L):

$$\partial_t \Theta = D^E \partial^2_{xx}\Theta,$$

and to find D^E in terms of $D(x)$. Let us present a heuristic argument to determine D^E: the Langevin equation associated with (6.19) is

$$\frac{dx}{dt} = u(x) + \sqrt{2D(x)}\,\eta,\ u(x) = \frac{\partial}{\partial x}D(x), \tag{6.20}$$

where the Ito formulation has been used: the stochastic process $x(t)$ spends a time interval $\Delta t(x) \sim (\Delta x)^2/2D(x)$ in a segment of length Δx centred in x. Let us now follow $x(t)$ up to the time t_N such that N jumps (among Δx-segments) occur. Elementary considerations give

$$\langle x(t_N) - x(0)\rangle = 0,\ \langle (x(t_N) - x(0))^2\rangle = N\Delta x^2, \tag{6.21}$$

and for $N \gg 1$

$$t_N = \frac{\Delta x^2}{2}\sum_{j=1}^{N}\frac{1}{D(x(t_j))} \simeq \frac{N\Delta x^2}{2}\left\langle\frac{1}{D(x(t_j))}\right\rangle = \frac{N\Delta x^2}{2L}\int_0^L \frac{dx}{D(x)}, \tag{6.22}$$

where we used the fact that the process $x(t)$, if considered in the interval $(0, L)$ with periodic boundary conditions, is ergodic and its stationary probability distribution is constant. Noting that $\langle (x(t_N) - x(0))^2\rangle \simeq 2D^E t_N$, from equations (6.21) and (6.22) one obtains the eddy diffusion coefficient

$$D^E = \left\langle\frac{1}{D(x(t_j))}\right\rangle^{-1} = \left(\frac{1}{L}\int_0^L \frac{dx}{D(x)}\right)^{-1}. \tag{6.23}$$

The above results can be obtained in a rigorous way with a multiscale analysis [12, 13].

6.3 Coarse-graining: from the micro-world to the meso-level

The purpose of this section is to discuss some among the historically relevant approaches for the derivation of a Langevin equation starting from the microscopic Hamiltonian dynamics. We begin with the Smoluchovski approach based upon kinetic theory, which is able to establish one part of the Langevin equation (the average 'drift') and exploits the Einstein relation to derive the missing part (noise): therefore, it can be used only at thermal equilibrium. We then describe the van

Kampen size expansion, which is more general because it does not require the use of the Einstein relation, i.e. it can be used also out of equilibrium. Actually, such an approach has been used to derive Langevin equations for tracers in granular gases and for massive ratchet particles surrounded by out-of-equilibrium baths [14–16] (see chapter 8). Later we discuss the derivation of a Langevin equation for a massive tracer in a solid, i.e. interacting with the bath through an elastic chain/network. Finally, we consider the generalization of the latter—so-called projection method—which is able to establish equilibrium-consistent generalized Langevin equations (GLEs), i.e. stochastic equations with memory.

6.3.1 A big tracer in a gas, assuming thermodynamic equilibrium

We briefly review, here, Smoluchovski's approach to derive the dynamics of a large and heavy impurity colliding with many lighter particles [17]. We consider N hard disks of radius r and mass m plus an impurity consisting in another disk of radius R and mass M. All particles are in a square box of side L, with periodic boundary conditions. Crucial parameters are the number density $\rho = N/L^2$ and the volume fraction $\psi = N\pi r^2/L^2$. In the following, we shall always consider very dilute systems ($\psi \ll 1$) so that the properties of the system will be those of a rarefied gas. This system simplifies in two limits: for $M/m \to 0$, it approaches the Lorentz gas model [18], where the impurity is much faster than the (now heavier) gas particles which can be treated as immobile obstacles. For $R/r \to \infty$, the Rayleigh flight model is recovered if the N small disks are very dilute so that most of the collisions involve the impurity [19, 20]. Here, being interested in Brownian motion, we consider situations in which $R \gg r$ and $M \gg m$.

We focus on the collisions of the colloidal disk with mass M, radius R and precollisional velocity $\boldsymbol{V}$ with the gas particles which are characterized by m, r, $\mathbf{v}$, respectively. The implulse transferred to the large particle in a collision is

$$\boldsymbol{M}\Delta\boldsymbol{V} = \boldsymbol{M}(\boldsymbol{V}' - \boldsymbol{V}) = \frac{2mM}{m+M}\boldsymbol{g},$$

$\boldsymbol{g} = \mathbf{v} - \boldsymbol{V}$ being the precollisional relative velocity. The rate of such collisions can be obtained by considering the equivalent problem of a tracer, at rest, with radius $r + R$, and hit by a flux of point-like particles moving at relative velocity $\boldsymbol{g}$: the number of point-like particles hitting the unit surface per unit time for a given orientation $\boldsymbol{e}$ corresponds to the particles contained in the collisional cylinder of base $(R + r)d\theta$ and height $\rho|\boldsymbol{g}\cdot\boldsymbol{e}|\Theta(-\boldsymbol{e}\cdot\boldsymbol{g})\delta t$. The unitary step function $\Theta(s)$ selects the condition, $\boldsymbol{g}\cdot\boldsymbol{e} < 0$, to have a collision, see figure 6.1. Accordingly, the mean force in the normal direction $\boldsymbol{e} = \{\cos(\theta), \sin(\theta)\}$ selected by the θ-angle that $\boldsymbol{e}$ forms with vector $\boldsymbol{V}$ (taken as x-axis direction) is given by

$$\left\langle M\frac{\delta V_n}{\delta t}\right\rangle = \frac{2mM}{m+M}\rho(R+r)\int_0^{2\pi} d\theta \int d\mathbf{v}P(\mathbf{v})\Theta(-\boldsymbol{e}\cdot\boldsymbol{g})|\boldsymbol{g}\cdot\boldsymbol{e}|(\boldsymbol{g}\cdot\boldsymbol{e})\boldsymbol{e}, \qquad (6.24)$$

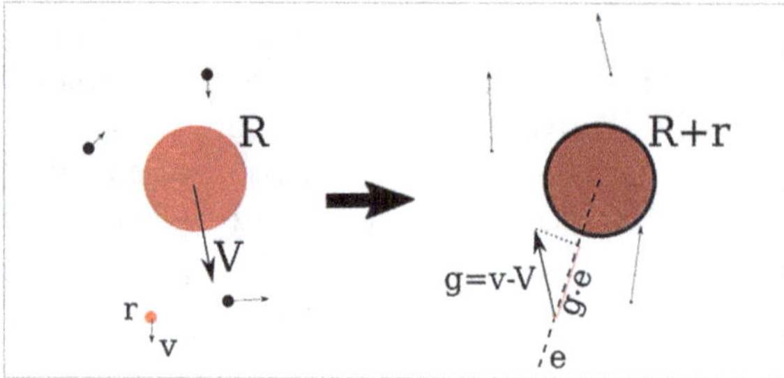

Figure 6.1. An example of the collision kinematics. On the left we show a configuration of large and small particles in the laboratory system. On the right we show the quantities in the frame where the large particle is at rest, with effective radii.

where the average is taken over the equilibrium distribution of the gas velocities $P(\mathbf{v}) = m/(2\pi T)\exp[-m|\mathbf{v}|^2/(2T)]$ (in this section for simplicity we use $k_B = 1$). It is convenient to make the change of variable $(v_x, v_y) \to (g_x, g_y)$ and then perform the approximation $P(\mathbf{v}) \simeq P(\boldsymbol{g})[1 - (m/T)\boldsymbol{g}\cdot\boldsymbol{V}]$, which holds in the limit $M \gg m$. Then the integral (6.24) becomes

$$\left\langle M\frac{\delta V_n}{\delta t}\right\rangle \simeq -\frac{4m^2M}{T(m+M)}\rho(R+r)\int_0^{2\pi} d\theta \int d\boldsymbol{g}P(\boldsymbol{g})\Theta(-\boldsymbol{e}\cdot\boldsymbol{g})(\boldsymbol{g}\cdot\boldsymbol{e})^2(\boldsymbol{g}\cdot\boldsymbol{V})\,\boldsymbol{e},$$

which, recast in the form $M\dot{V} = -\gamma MV$, allows us to extract the friction coefficient γ. In polar coordinates, $g_x = g\cos(\alpha)$, $g_y = g\sin(\alpha)$, the integral simplifies (α being the angle between $\boldsymbol{g}$ and $\boldsymbol{V}$, whose direction coincides with x-axis). If ϕ is the angle between $\boldsymbol{e}$ and $\boldsymbol{g}$, we get

$$\int_0^{2\pi} d\theta \int_0^{\infty} dgP(g)g^4 \int_0^{2\pi} d\alpha|\cos(\phi)|\cos(\phi)\Theta[-\cos(\phi)]V\cos(\alpha)\cos(\theta),$$

where the angles α, θ, ϕ are related by $\theta - \alpha = \pi - \phi$. The integration over g yields $3/(2\pi)\sqrt{\pi/2}(m/T)^{-3/2}$, and that one on the angles gives the value $4\pi/3$. Taking into account finite size and mass corrections, the friction coefficient finally reads

$$\gamma = 2\sqrt{2\pi}\,\frac{\rho R\sqrt{mT}}{M}\left(\frac{1+r/R}{1+m/M}\right) \simeq 2\sqrt{2\pi}\,\frac{\rho R\sqrt{mT}}{M}. \tag{6.25}$$

Now, thanks to the Green–Kubo relation, linking the auto-correlation function to the diffusion coefficient D

$$D = \int_0^{\infty}\langle V(t)V(0)\rangle dt = \frac{\langle V^2\rangle}{\gamma} = \frac{T}{M\gamma}, \tag{6.26}$$

it is immediate to derive the diffusion constant of the colloidal particle

$$D_c = \frac{1}{2\sqrt{2\pi}}\frac{1}{\rho R}\sqrt{\frac{T}{m}}\left(\frac{1+m/M}{1+r/R}\right) \simeq \frac{1}{2\sqrt{2\pi}}\,\frac{1}{\rho R}\sqrt{\frac{T}{m}}. \tag{6.27}$$

Notice that D_c is proportional to $\sqrt{T}$ and not to T as in liquids. This comes from the fact that in liquids the friction is temperature independent, while in rarefied gases it is proportional to the square root of the temperature. Summing up, we have derived—thanks to the large timescale separation between the impurity and gas particles motions, the following Langevin equation holding for each velocity component V of the colloid

$$\frac{dV}{dt} = -\gamma V + \sqrt{2\gamma \frac{T}{M}}\eta, \tag{6.28}$$

where $\eta(t)$ is a white noise Gaussian process. Of course, equation (6.28) describes the effective dynamics of the impurity for times much larger than the average collision time with the gas particles. The friction constant γ sets the decay of the velocity–velocity correlation function

$$C_V(t) = \langle V(t)V(0)\rangle = \langle V^2\rangle e^{-\gamma t} = \frac{T}{M}e^{-\gamma t}, \tag{6.29}$$

where brackets indicate time or ensemble averages.

Another interesting quantity is the self-diffusion coefficient D_g of a tagged gas particle, i.e. a particle not larger but of the same mass and radius of the solvent particles. This is an important transport coefficient ruling the density field evolution. In the dilute limit, the particle self-diffusion coefficient takes the form [21]:

$$D_g = \lim_{t\to\infty}\frac{1}{2t}\langle[x(t) - x(0)]^2\rangle = \frac{1}{4\sqrt{\pi}}\frac{1}{\rho r}\sqrt{\frac{T}{m}}, \tag{6.30}$$

where x indicates the x-component of the position of a tagged gas particle. At high volume fractions, corrections to the formula must be considered. Although the expression is formally similar to equation (6.27), only the prefactor changes, the Langevin description (6.28) does not apply in this case due to the absence of timescale separation. Moreover, an important warning is in order when considering a gas in two dimensions: the gas velocity auto-correlation function develops small, slowly decaying tails that may lead to an ill-defined diffusion coefficient [22, 23]. However, when $\psi \to 0$ this problem becomes relevant only for enormously long times [24]. Therefore, disregarding this issue is usually justified from the practical point of view.

6.3.2 A tracer in a gas without assuming equilibrium

The same system analysed in the previous subsection can be studied through a different, more general, approach, where both terms (dissipation and fluctuations) are derived consistently. Such an approach has been introduced by van Kampen; here we follow the discussion in reference [25]. The basic idea is to expand the master equation characterizing the Markovian evolution for the quantity $a(t)$

$$\frac{\partial P(a, t)}{\partial t} = \int \{W(a|a')P(a', t) - W(a'|a)P(a, t)\}da',$$

through the so-called Kramers–Moyal expansion

$$\frac{\partial P(a, t)}{\partial t} = \sum_{n=1}^{\infty}\frac{1}{n!}\left(-\frac{\partial}{\partial a}\right)^n \alpha_n(a)P(a, t),$$

where the coefficients represent moments of the changing rate of a

$$\alpha_n(a) = \int (a' - a)^n W(a'|a)da'.$$

One could think that the Fokker–Planck equation may be obtained by just cutting the expansion at the order $n = 2$, but this is not justified as the above expansion is not done as an expansion in powers of some small parameter. The idea of van Kampen was exactly to transform it into such a power expansion and justify the cutting. This can be done by first rewriting the transition probability $W(a|a')$ as a function of the length of the jump, $\Delta a = a - a'$, and its initial value a',

$$W(a|a') = W(a'; \Delta a). \tag{6.31}$$

At this point it becomes useful to assume that a is an extensive quantity, i.e that an intensive quantity X exists such that

$$a = \Omega X,$$

where Ω is a large parameter (e.g. the size of the system, the mass of the tracer etc). The crucial point is that the jump, Δa, is a function of the extensive quantity, whereas the dependence on a' in equation (6.31) may be expressed in terms of the intensive quantity X. More precisely, writing

$$W(a'; \Delta a) = \Phi(X'; \Delta a) = \Phi\left(\frac{a'}{\Omega}; \Delta a\right),$$

one has that the function Φ no longer involves Ω implicitly, in contrast with $W(a'; \Delta a)$. Now it is possible to expand it in reciprocal powers of Ω. The moments α_n, when expressed as functions of X, read

$$\int (\Delta a)^n \Phi(X; \Delta a) d\Delta a = \alpha_n(X),$$

and no longer contain Ω.

The application of this general procedure to the case of the massive tracer in the gas can be done, without loss of generality, in one dimension: a heavy particle with mass M is subjected to collisions with light gas molecules with mass m and velocity distribution $f(v)$. A collision of a molecule with velocity v changes the velocity V of the heavy particle into

$$V' = V + \frac{2m}{M + m}(v - V) = V + \Delta V.$$

One thus obtains a master equation for the velocity distribution $P(V, t)$, with transition rates

$$W(V'|V) = \nu\left(\frac{M + m}{2m}\right)^2 |V' - V| f\left(\frac{M + m}{2m}V' - \frac{M - m}{2m}V\right), \tag{6.32}$$

where ν is the number of molecules per unit length. The individual jumps in V are due to collisions with the gas molecules, and are therefore appropriately measured in terms of the momentum MV, while the probability distribution of these jumps depends on the velocity V itself. Therefore, we can identify

$$a = \frac{M+m}{m}V, \quad X = V, \quad \Omega = \frac{M+m}{m}.$$

The transition rate takes the form

$$W(a;\, \Delta a) = \frac{1}{4}\nu|\Delta a|f\left(X + \frac{1}{2}\Delta a\right) = \Phi(X;\, \Delta a),$$

entirely independent of Ω. Hence we can apply the results of the previous sections. The zeroth order approximation is the linear Fokker–Planck equation

$$\frac{\partial P(a,\, t)}{\partial t} = -\frac{\partial}{\partial a}\alpha_1(a)P(a,\, t) + \frac{1}{2}\frac{\partial^2}{\partial a^2}\alpha_2(a)P(a,\, t),$$

with

$$\alpha_1 = \frac{1}{4}\nu\int_{-\infty}^{+\infty}(\Delta a)|\Delta a|f\left(\frac{1}{2}\Delta a\right)d\Delta a,$$

$$\alpha_2 = \frac{1}{4}\nu\int_{-\infty}^{+\infty}(\Delta a)^2|\Delta a|f\left(\frac{1}{2}\Delta a\right)d\Delta a;$$

if f is the Maxwell distribution, we have

$$\alpha_1 = -\frac{m}{2k_B T} \quad \alpha_2 = 4\nu\left(\frac{2k_B T}{m}\right)^{\frac{1}{2}}.$$

After transforming back to the usual variables $V = \Omega^{-\frac{1}{2}}x$, $t = \Omega\tau$, the Fokker–Planck equation becomes

$$\frac{\partial P(V,\, t)}{\partial t} = \nu\frac{4m}{M+m}\left(\frac{2k_B T}{\pi m}\right)^{\frac{1}{2}}\frac{\partial}{\partial V}\left\{VP + \frac{k_B T}{M+m}\frac{\partial^2 P}{\partial V^2}\right\}.$$

This is the well-known Rayleigh equation, with $M+m$ replacing M. Applications of this formalism—which can be used in out-of-equilibrium systems—will be discussed in chapters 8 and 9.

6.3.3 A tracer in a solid

Parts of this section have been reprinted with permission from [49] copyright Cambridge University Press & Assessment 2008.

An interesting example is the derivation of a Langevin equation for a system interacting with a heat bath made of harmonic oscillators. This constitutes a simplified model for many classical and quantum systems. For simplicity we assume that the system has one physical dimension with a coordinate x and its conjugate momentum p. The bath in contrast has coordinates $\{q_j\}$ and their conjugate momenta $\{p_j\}$. The oscillator masses are put to 1 without loss of generality. The Hamiltonian of the system H_s is

$$H_s = \frac{p^2}{2m} + U(x),$$

while the Hamiltonian of the heat bath H_B is that of several harmonic oscillators coupled to the system in the following way

$$H_B = \sum_j \left[\frac{p_j^2}{2} + \frac{1}{2}\omega_j^2 \left(q_j - \frac{\gamma_j}{\omega_j^2} x \right)^2 \right],$$

which includes three terms: the first is the usual harmonic oscillator with its frequency ω_j; the second represents the bilinear oscillator–system coupling, $(\Sigma_j \gamma_j q_j)x$ where γ_j is a coupling constant; and the third contains x and can be considered as part of the potential $U(x)$. The bilinear coupling simplifies the derivation. The motion equations are derived from the total Hamiltonian $H_S + H_B$, reading:

$$\frac{dx}{dt} = \frac{p}{m}, \quad \frac{dp}{dt} = -U'(x) + \sum_j \gamma_j \left(q_j - \frac{\gamma_j}{\omega_j^2} x \right),$$

$$\frac{dq_j}{dt} = p_j, \quad \frac{dp_j}{dt} = -\omega_j^2 q_j + \gamma_j x.$$

Once the time dependence of $x(t)$ is known, the motion of the bath oscillators can be solved as functions of $x(t)$ and of their initial values,

$$q_j(t) = q_j(0)\cos \omega_j t + p_j(0)\frac{\sin \omega_j t}{\omega_j} + \gamma_j \int_0^t dsx(s)\frac{\sin \omega_j (t-s)}{\omega_j}.$$

An integration by parts leads to:

$$q_j(t) - \frac{\gamma_j}{\omega_j^2}x(t) = \left(q_j(0) - \frac{\gamma_j}{\omega_j^2}x(0) \right)\cos \omega_j t + p_j(0)\frac{\sin \omega_j t}{\omega_j}$$
$$- \gamma_j \int_0^t ds \frac{p(s)}{m} \frac{\cos \omega_j (t-s)}{\omega_j^2}.$$

By inserting this into the equation for dp/dt, the following Langevin equation is obtained

$$\frac{dp(t)}{dt} = -U'(x(t)) - \int_0^t dsK(s)\frac{p(t-s)}{m} + F_p(t),$$

with a memory kernel $K(t)$

$$K(t) = \sum_j \frac{\gamma_j^2}{\omega_j^2}\cos \omega_j t,$$

and a 'noise' $F_p(t)$ given by

$$F_p(t) = \sum_j \gamma_j p_j(0)\frac{\sin \omega_j t}{\omega_j} + \sum_j \gamma_j \left(q_j(0) - \frac{\gamma_j}{\omega_j^2}x(0) \right)\cos \omega_j t.$$

In the case of a continuous spectrum, by replacing the sum over j with an integral weigthed by the density of states $g(\omega)$, i.e. $\int d\omega g(\omega)$, assuming that γ is a function of ω, leads us to transforming the memory function $K(t)$ into a Fourier integral,

$$K(t) = \int_0^\infty d\omega g(\omega) \frac{\gamma(\omega)^2}{\omega^2} \cos \omega t.$$

With the choice $g(\omega) \propto \omega^2$ and constant γ, then $K(t)$ is proportional to $\delta(t)$ and a Markovian Langevin equation is finally obtained.

In this example the 'noise' $F_p(t)$ is a deterministic function that depends on the initial positions and momenta of the oscillators; in the limit of a large number of independent degrees of freedom for the bath, the noise becomes the sum of a large number of independent terms, and we can exploit the central limit theorem to get simple noise properties. For instance, in the case of an ensemble of numerical simulations where the oscillator initial conditions are at thermal equilibrium (with respect to a frozen or constrained system coordinate $x(0)$),

$$f_{\text{eq}}\,(p, q) \propto \exp\,(-H_B/k_{\text{B}}T),$$

then one has

$$\left\langle q_j(0) - \frac{\gamma_j}{\omega_j^2} x(0) \right\rangle = 0, \quad \left\langle p_j(0) \right\rangle = 0.$$

The noise is a linear combination of these quantities, therefore its average value is also zero and the second moments read

$$\left\langle \left(q_j(0) - \frac{\gamma_j}{\omega_j^2} x(0) \right)^2 \right\rangle = \frac{k_{\text{B}}T}{\omega_j^2}, \quad \left\langle p_j(0)^2 \right\rangle = k_{\text{B}}T.$$

In this case there are no correlations between the initial values. Then direct calculations lead to a second-kind fluctuation–dissipation theorem,

$$\langle F_p(t) F_p(t') \rangle = k_{\text{B}}TK(t - t').$$

The noise in this case is a Gaussian random variable, since it is a linear combination of Gaussian variables. With the dynamical properties discussed above necessary to get a memory function with the properties of a delta function, then the noise is white or Markovian, so that all the properties of a Langevin equation are fulfilled.

Another interesting example of microscopic derivation of a Langevin equation is due to R J Rubin [26]. He considered a one-dimensional harmonic lattice in which one particle of mass M is much heavier than the others with mass m. Each particle j has displacement and velocity x_j and v_j, respectively, where j goes from 0 to $N - 1$. Periodic boundary conditions dictate $x_N = x_0$. The Hamiltonian is

$$H = \frac{M}{2} v_0^2 + \sum_{j=1}^{N-1} \frac{m}{2} v_j^2 + \sum_{j=0}^{N-1} \frac{K}{2} (x_j - x_{j+1})^2,$$

leading to the following equations of motion

$$[m + (M - m)\delta_{j0}]\ddot{x}_j = K(x_{j+1} - 2x_j + x_{j-1}).$$

The linearity of the equations of motion implies that the position and velocity at time t are linear combinations of initial conditions. Equilibrium averages show that the velocities of different particles are not correlated,

$$\langle x_j v_0 \rangle = 0, \quad \langle v_j v_0 \rangle = \frac{k_B T}{M}\delta_{j0}.$$

By assuming that all initial coordinates and velocities are equal to zero, except $v_0(0)$, applying the Laplace transform of the jth coordinate we get

$$\hat{x}_j(z) = \int_0^\infty dt e^{-zt} x_j(t),$$

and the equations of motion for the given initial conditions read

$$[m + (M - m)\delta_{j0}](z^2\hat{x}_j - \delta_{j0}v_0(0)) = K(\hat{x}_{j+1} - 2\hat{x}_j + \hat{x}_{j-1}).$$

Since we have a harmonic potential energy we can proceed by normal mode decomposition, where the new coordinates q_k are

$$q_k = \frac{1}{\sqrt{N}}\sum_{j=0}^{N-1} x_j \exp\left(-\frac{2\pi i}{N}jk\right).$$

Straightforward calculations lead to

$$\hat{q}_k = \frac{1}{\sqrt{N}}\frac{1}{z^2 + \omega_k^2}\left[\frac{M}{m}v_0(0) - \frac{M - m}{m}z^2\hat{x}_0\right],$$

where the frequencies of the modes are

$$\omega_k^2 = \frac{2k}{m}\left(1 - \cos\frac{2\pi k}{N}\right).$$

By summing over k we get

$$\hat{x}_0 = \frac{1}{N}\sum_k \frac{1}{z^2 + \omega_k^2}\left[\frac{M}{m}v_0(0) - \frac{M - m}{m}z^2\hat{x}_0\right]. \tag{6.33}$$

The sum

$$\hat{\phi}(z) = \frac{1}{N}\sum_{k=0}^{N-1}\frac{1}{z^2 + \omega_k^2}$$

simplifies in the large N limit, by changing variables from k to θ, and replacing the k-sum with a θ-integral:

$$\begin{aligned}\hat{\phi}(z) &\cong \frac{1}{2\pi}\int_0^{2\pi} d\theta \frac{1}{z^2 + (2K/m)(1 - \cos\theta)} \\ &= \frac{1}{z}\frac{1}{\sqrt{z^2 + 4K/m}}.\end{aligned}$$

After having solved equation (6.33) for $\hat{x}_0$, we get

$$\hat{x}_0 = \frac{(1+Q)\hat{\phi}}{1+Qz^2\hat{\phi}} v_0(0),$$

where $Q = \frac{M-m}{m}$. The velocity Laplace transform is $z\hat{x}_0$, therefore the velocity correlation function $C(t) = \frac{\langle v_0(t)v_0(0)\rangle}{\langle v_0^2\rangle}$ has the following transform

$$\hat{C}(z) = \frac{(1+Q)}{1+Qz^2\hat{\phi}(z)} z\hat{\phi}(z).$$

We can insert the approximate (large N) $\hat{\phi}$, obtaining an algebraic function of z,

$$\hat{C}(z) = \frac{1+Q}{Qz + \sqrt{z^2 + 4K/m}}.$$

The expansion for large z gives the short-time behaviour of $C(t)$,

$$\hat{C}(z) = \frac{1}{z} - \frac{2K}{Mz^3} + \cdots.$$

so that C is

$$C(t) = 1 - \frac{K}{M}t^2 + \cdots,$$

which means that the heavy mass inertial motion dominates at short times. Algebraic manipulation shows that a memory function form can be obtained,

$$\hat{C}(z) = \frac{1}{z + \frac{m}{M}\hat{k}(z)}, \qquad \hat{k}(z) = \sqrt{z^2 + 4K/m} - z.$$

The inverse transform leads to

$$\frac{d}{dt}C(t) = -\frac{m}{M}\int_0^t dsk(s)C(t-s),$$

where $k(s)$ (which is independent of M) depends upon time in the following way

$$k(s) = \frac{4K}{m}\frac{J_1(2s\sqrt{K/m})}{s},$$

where we denote by $J_1(x)$ the Bessel function of order 1. The above formula indicates a convolution with a memory kernel that has a short lifetime, here $O(\sqrt{\mathrm{m/K}})$. The coefficient in front of the integral can be very small, in this case of the order of m/M. When this goes to zero (heavy mass limit), the time derivative of $C(t)$ is small, and one can assume $C(t-s) \sim C(t)$ which is a Markovian approximation so that the integral can be extended to infinity,

$$\frac{d}{dt}C(t) \approx -\frac{m}{M}\int_0^\infty dsk(s)C(t).$$

The integral is simply

$$\int_0^\infty dsk(s) = \sqrt{4K/m} = \frac{1}{t_0}.$$

Then one has an exponential time decay for the velocity correlation function, with the relaxation timescale of the order of $t_0 M/m$,

$$C(t) \approx \exp\left(-\frac{m}{M}\frac{t}{t_0}\right).$$

For more details about the complete calculation of the integrals, see the work by Rubin [26].

Sometimes it is possible to invert a Laplace transform numerically: figure 6.2 shows $C(t)$ from the numerical inversion for $M = 10$ m (curve A), and the exponential approximation $\exp(-t/10)$ (curve B). Note that, apart from small values of t, the exponential is a fair approximation. When N is finite we should apply two adjustments: first, the amplitude of the velocity correlation function is changed by terms of order of $1/N$; second, when the particle at $j = 0$ moves, it triggers sound waves that—when the system has periodic boundary conditions—will eventually come back to the massive particle, leading to an 'almost periodic' motion with recurrences. If N is large, however, recurrences occur at times much longer than the exponential relaxation time.

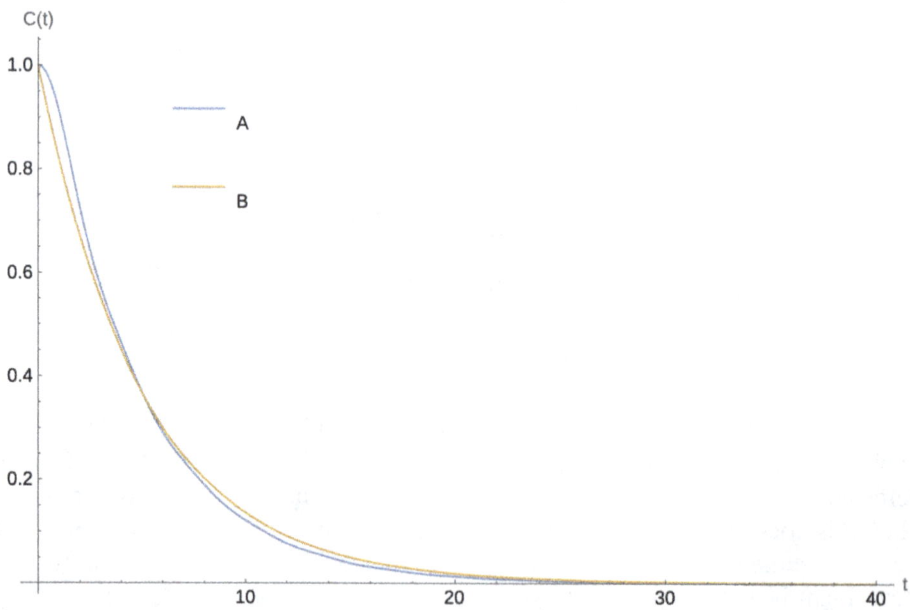

Figure 6.2. The velocity correlation function $C(t)$ as a function of time t. Curve A is the result of the numerical inversion of the Laplace transform when $M = 10$ m and $K/m = 1$. Curve B is the exponential $\exp(-t/(10t_0))$ where $1/t_0 = 2$.

6.3.4 Projection methods and generalized Langevin equations

Here we discuss a general and abstract method to derive the stochastic equation of motion for an arbitrary observable, for instance the position of a single particle [27], under conservative dynamics, which implies microscope time-reversibility. We consider the full phase space Ω for N 'molecules'. The point in the phase space representing the system at time t is $\omega_t \equiv (\mathbf{r}_1(t), \mathbf{r}_2(t), \ldots, \mathbf{r}_N(t), \mathbf{p}_1(t), \mathbf{p}_2(t), \ldots, \mathbf{p}_N(t))$, and it obeys Hamiltonian dynamics with potential $V(\mathbf{r}_1, \mathbf{r}_2, \ldots, \mathbf{r}_N)$. A generic observable evolves through the Liouville operator

$$L = \sum_{n=1}^{N} \left\{ \frac{\mathbf{p}_n}{m_n} \cdot \nabla_{\mathbf{r}_n} - [\nabla_{\mathbf{r}_n} V(\mathbf{r}_1, \mathbf{r}_2, \ldots, \mathbf{r}_N)] \cdot \nabla_{\mathbf{p}_n} \right\},$$

for instance the phase space point obeys $\dot{\omega}_t = L\omega_t$. We will use a notation where sometimes it will be useful to keep track of the initial conditions at time 0 in the operator at time t, i.e. we will call $A(\omega_0, t) = A(\omega_t) = e^{tL}A(\omega_0)$.

The idea is to introduce a projection operator P that reduces the complete phase space (containing fast and slow variables) onto a subspace called the 'relevant' subspace (typically containing only the slow variables): it is a linear, idempotent operator, i.e. for arbitrary scalars c_1, c_2, it fulfils the properties $P(c_1 A_t + c_2 B_t) = c_1 PA_t + c_2 PB_t$ and $P^2 = P$. The operator $Q = 1 - P$, where 1 denotes the identity operator, projects onto the complementary subspace. The operators P and Q can be used to decompose the evolution equation for the observable, for instance we can consider a 'position' observable A_t, and therefore decompose the 'velocity' evolution $\ddot{A}_t = L\dot{A}_t$, in the following way

$$\ddot{A}_t = e^{tL}(P + Q)L\dot{A}_0 = e^{tL}PL\dot{A}_0 + e^{tL}QL\dot{A}_0. \tag{6.34}$$

The operator

$$\Phi(t) = e^{tL}Q$$

propagates the part of an observable that lies in the complementary subspace in time. Since

$$\frac{d}{dt}e^{tL}Q = e^{tL}LQ = e^{tL}QLQ + e^{tL}PLQ, \tag{6.35}$$

for $\Phi(t)$ we find

$$\dot{\Phi}(t) = \Phi(t)LQ + e^{tL}PLQ. \tag{6.36}$$

Equation (6.36) is an inhomogeneous differential equation of first order. Using $\Phi(0) = Q$, the solution reads

$$\Phi(t) = Qe^{tLQ} + \int_0^t du e^{uL}PLQe^{(t-u)LQ}.$$

By using $Qe^{tLQ} = e^{tQL}Q$ and the substitution $s = t - u$, we get

$$\Phi(t) = e^{tL}Q = e^{tQL}Q + \int_0^t ds e^{(t-s)L}PLe^{sQL}Q. \tag{6.37}$$

Replacing $e^{tL}Q$ in equation (6.34) by equation (6.37) leads to the GLE for A_t in terms of a general projection P:

$$\ddot{A}_t = e^{tL}PL\dot{A}_0 + \int_0^t dse^{(t-s)L}PLF^R(s) + F^R(t), \tag{6.38}$$

$$F^R(t) \equiv e^{tQL}QL\dot{A}_0 = Qe^{tLQ}L\dot{A}_0. \tag{6.39}$$

The function $F^R(t)$ stays in the complementary subspace for all times and is an explicit function of the initial state of the entire system. The first term on the rhs of equation (6.38) represents the time evolution of the part of $\ddot{A}_t = L\dot{A}_t$ which lies in the relevant subspace and reflects a deterministic force. The second term is due to the relevant part of $LF^R(t)$ and describes dissipative effects. Clearly, the explicit form of equation (6.38) depends on the specific form of the projection operator P. In the following we discuss the GLEs generated by the Mori projection P_M and by the Zwanzig projection P_Z.

6.3.4.1 Mori projection

The Mori projection is applied on an observable A_t, projecting it onto the (generic) observables B_0 and $\dot{B}_0$, which are called the projection functions. It is given by [28]

$$P_M A_t = \frac{\langle A_t, B_0 \rangle}{\left\langle B_0^2 \right\rangle} B_0 + \frac{\langle A_t, \dot{B}_0 \rangle}{\left\langle \dot{B}_0^2 \right\rangle} \dot{B}_0, \tag{6.40}$$

where the inner product is defined by

$$\langle A_t, B_{t'} \rangle \equiv \int_\Omega d\omega_0 \rho_{\text{eq}}(\omega_0) A(\omega_0, t) B(\omega_0, t'),$$

where $\rho_{\text{eq}}(\omega_0) = e^{-\beta H(\omega_0)}/Z$ is the canonical Boltzmann distribution with the inverse thermal energy $\beta = 1/k_B T$ and the partition function $Z = \int d\omega_0 e^{-\beta H(\omega_0)}$. The projection in equation (6.40) maps any observable A_t onto the subspace of all functions linear in the observables B_0 and $\dot{B}_0$. In addition to being linear and idempotent, P_M is self-adjoint with respect to the inner product, i.e. for two arbitrary observables A_t, $C_{t'}$, one has

$$\langle P_M A_t, C_{t'} \rangle = \langle A_t, P_M C_{t'} \rangle,$$

and

$$\left\langle P_M A_t, Q_M C_{t'} \right\rangle = 0,$$

as follows directly from equation (6.40) and from the idempotence of P; therefore, all functions $P_M A_t$ and $Q_M C_{t'}$ are orthogonal. For $P = P_M$ and choosing the projection functions to be $B_t = A_t$ and $\dot{B}_t = \dot{A}_t$, i.e. projecting onto the observable of interest itself, equation (6.38) takes the form [28, 29]

$$\ddot{A}_t = -\mathrm{K}A_t - \int_0^t ds \Gamma^M(s) \dot{A}_{t-s} + F^R(\omega_0, t), \tag{6.41}$$

$$\mathrm{K} = \frac{\left\langle \dot{A}_0^2 \right\rangle}{\left\langle A_0^2 \right\rangle}, \quad \Gamma^M(t) = \frac{\langle F^R(0), F^R(t) \rangle}{\left\langle \dot{A}_0^2 \right\rangle}, \tag{6.42}$$

where $\Gamma^M(s)$ is the memory friction kernel obtained from the Mori projection. Equation (6.41) is an exact decomposition of the Liouville equation into three terms: the first term is a generalized force due to a potential of quadratic form; the second term accounts for linear friction and includes the memory kernel $\Gamma^M(s)$, which is related through equation (6.42) to the second moment of the random force $F^R(t)$, defined in equation (6.39).

An explicit expression for the memory function can be derived for simple models, but this becomes challenging for realistic systems and practical applications. This difficulty arises primarily because the fluctuating term $F^R(t)$ depends explicitly on the initial state of the entire system, rather than behaving as a stochastic process with zero mean and a fixed second moment. Moreover, even assuming a Gaussian form for $F^R(t)$ is inadequate, as $F^R(t)$ inherently incorporates all nonlinearities present in A_t. This makes it a poor approximation for nonlinear systems and highlights a fundamental limitation of the Mori projection scheme when applied in practice.

6.3.4.2 Zwanzig projection

Contrary to the Mori projection, the Zwanzig projection P_Z of an observable A_t is nonlinear in the projection functions B_0 and $\dot{B}_0$[26]

$$P_Z A_t = \langle A_t \rangle_{B_0, \dot{B}_0} = \frac{\langle \delta[B(\hat{\omega}_0) - B_0] \delta[\dot{B}(\hat{\omega}_0) - \dot{B}_0], A(\hat{\omega}_0, t) \rangle}{\langle \delta[B(\hat{\omega}_0) - B_0] \delta[\dot{B}(\hat{\omega}_0) - \dot{B}_0] \rangle}, \tag{6.43}$$

where $\hat{\omega}_0$ represents the phase space variables which are integrated away; therefore, the δ function inside the inner product realizes a conditional average over the hypersurface of the phase space such that B and $\dot{B}$ are equal to B_0 and $\dot{B}_0$ at initial time. The Zwanzig projection thus is a conditional average and is linear, idempotent, and self-adjoint, similarly to the Mori projection. The resulting GLE from the Zwanzig projection is best illustrated by choosing the observable of interest to be the momentum of a single particle, $\dot{A}_0 = \mathbf{p}_0$, and the projection functions as the position and the linear momentum of the same particle, i.e. $B_0 \to \mathbf{r}_0$, $\dot{B}_0 \to \mathbf{p}_0$. With this, equation (6.38) becomes [30]

$$\dot{\mathbf{p}}_t = -\nabla_{\mathbf{r}_t} U_{\mathrm{PMF}}(\mathbf{r}_t) + \mathbf{F}^R(\omega_0, t) + \int_0^t ds \left[\left(\frac{\nabla_{p_s}}{\beta} - \frac{\mathbf{p}_s}{m} \right) \right]^{\mathrm{T}} \cdot \Gamma^Z(t - s, \mathbf{r}_s, \mathbf{p}_s), \tag{6.44}$$

where $U_{\mathrm{PMF}}(\mathbf{r}) = -k_{\mathrm{B}} T \ln \mathscr{P}(\mathbf{r})$ denotes the potential of mean force (PMF), where the normalized probability that $\mathbf{r}_t$ has the value $\mathbf{r}$ is given by $\mathscr{P}(\mathbf{r}) \equiv \langle \delta(\mathbf{r}_t - \mathbf{r}) \rangle$. This creates in the GLE a force on the particle that tends to establish the equilibrium positional distribution. This is the main advantage over the Mori projection, since this ensures the correct equilibrium behaviour once we switch to a stochastic description and replace the fluctuating force $F^R(t)$ by a Gaussian stochastic variable with zero mean [31]. The memory friction kernel Γ^Z is a 3×3 matrix that, as a result of the conditional average, is a function of particle position $\mathbf{r}_s$ and particle momentum $\mathbf{p}_s$ with the matrix elements defined by

$$\beta\Gamma_{ij}^Z(t - s, \mathbf{r}_s, \mathbf{p}_s) = \left\langle F_i^R(0), F_j^R(t - s)\right\rangle_{\mathbf{r}_s, \mathbf{p}_s}.$$

This is the main drawback of the Zwanzig GLE, since the position and momentum dependence is difficult to be understood in applications. One typically invokes the assumption that the memory function is independent of position and momentum, i.e. $\Gamma^Z(t - s, \mathbf{r}_s, \mathbf{p}_s) \approx \Gamma^{\text{app}}(t - s)$, giving the following, amply used in literature, GLE:

$$\dot{\mathbf{p}}_t = -\nabla_{\mathbf{r}_t} U_{\text{PMF}}(\mathbf{r}_t) - \int_0^t ds\Gamma^{\text{app}}(t - s) \cdot \frac{\mathbf{p}_s}{m} + \mathbf{F}^R(t).$$

In applications such an approximate GLE can accurately reproduce the full system dynamics, for instance in protein folding simulations [32], but the assumption on which it is based is not easy to be validated.

6.4 Symmetries and coarse-graining

Time reversal symmetry holds in the microscopic Hamilton equations. The coarse-graining procedures discussed above reduce the description to a few 'slow' degrees of freedom (e.g. the dynamics of a massive intruder, such as a colloid in a fluid of molecules); it is then natural to wonder if the time-reversal symmetry is preserved: that symmetry, in a stochastic process, is replaced by a statistical time-reversal symmetry which is the detailed balance when the process is Markovian. How is this symmetry verified for a Langevin equation, which is Markovian and continuous in space–time? In chapter 5, we have mentioned how entropy production can be computed in Langevin equations; in the first subsection below we discuss the simplest problem of time-reversibility for this kind of continuous Markov processes. A different question may also arise. What happens if one starts with a microscopic description that *breaks* time-reversal symmetry, i.e. where external driving forces are present? That can happen—for instance—in an equilibrium fluid where a massive tracer is dragged by an external drive. In principle, one can apply some of the reduction methods discussed above (for instance, the van Kampen size expansion that does not require microscopic time-reversibility) and—hopefully—end up with a Langevin equation containing external driving forces. The question then is: how can we be sure that the amount of dissipation present in the microscopic description is faithfully reproduced by the mesoscopic one? This issue is discussed in the second subsection.

6.4.1 The generalized equilibrium conditions

For a stochastic process described by the Langevin or Fokker–Planck equation, being Markovian, the time-reversal symmetry is equivalent to detailed balance, which for continuous processes takes the form of a lack of irreversible currents. The Fokker–Planck equation can be written as a continuity equation:

$$\frac{\partial f(\mathbf{z}, t)}{\partial t} = -\nabla \cdot \mathbf{J}(\mathbf{z}, t),$$

where $\mathbf{J}$ is the current probability which can be decomposed into a reversible and an irreversible part:

$$\mathbf{J}(\mathbf{z}, t) = \mathbf{J}^{irr}(\mathbf{z}, t) + \mathbf{J}^{rev}(\mathbf{z}, t),$$

according to the following definitions

$$J_i^{irr}(\mathbf{z}, t) = \frac{1}{2}[A_i(\mathbf{z}, t) + \varepsilon_i A_i(\varepsilon\mathbf{z}, t)]f(\mathbf{z}, t) - \sum_j \partial_j D_{ij} f(\mathbf{z}, t),$$

$$J_i^{rev}(\mathbf{z}, t) = \frac{1}{2}[A_i(\mathbf{z}, t) - \varepsilon_i A_i(\varepsilon\mathbf{z}, t)]f(\mathbf{z}, t),$$

where $\varepsilon_i = \pm 1$ denotes the parity of the variable z_i and $A_i(\mathbf{z}, t)$ is the drift (mean force) acting on degree i. When D_{ij} does not depend upon $\mathbf{z}$, then $\mathbf{J}^{irr} = 0$ (in the steady state) is a condition equivalent to detailed balance. A general discussion of detailed balance for Fokker–Planck equations can be found in references [8, 33].

We show how this condition takes the form of the Einstein relation using a peculiar starting point where the dynamics is a generalization of Hamiltonian mechanics, i.e. the phase space is made of positions and momenta Q, P and the drift A_i comes from the Hamiltonian equations of motion; however, the kinetic energy of the Hamiltonian has a more general form, rather than the usual term proportional to the momentum squared P^2:

$$H(P, \{p\}, Q, \{q\}) = K(P) + \sum_n \tilde{K}(p_n) + U(Q) + \sum_n V_I(Q, q_n) + \sum_{n,n'} \tilde{V}(q_n, q_n'), \quad (6.45)$$

where (P, Q) denote the canonical variables of a 'heavy' particle and $(\{p_n\}, \{q_n\})$ indicate the 'light' particles. We have denoted by $\tilde{V}$ and V_I the potential for the interaction between light particles and that for the interaction heavy-light particles, respectively, while U is the external potential confining the heavy particle; K, $\tilde{K}$ are the kinetic energies of the heavy and light particles, respectively. The use of a general kinetic part has been introduced in models of statistical mechanics where negative temperatures appear, such as those used to describe point vortices in an inviscid fluid or trapped atomic systems [34–38]

The evolution equations for (P, Q) are

$$\dot{Q} = \partial_P H = \partial_P K(P), \quad (6.46a)$$

$$\dot{P} = -\partial_Q H = -\partial_Q U(Q) - \sum_n \partial_Q V_I(Q, \{q_n\}). \quad (6.46b)$$

The coarse-graining procedures described above cast the term $-\sum_n \partial_Q V_I$ into a 'viscous term'—only function of the variables P, Q—and a noisy term. This is a generalization of the Klein–Kramers equation for a generic form of $K(P)$, i.e. including cases where the kinetic term is different from P^2 and therefore one may also have ranges with inverse temperature $\beta < 0$. Our candidate is a couple of equations of the kind:

$$\dot{Q} = \partial_P K(P), \tag{6.47a}$$

$$\dot{P} = -\partial_Q U(Q) + B(P, Q, t), \tag{6.47b}$$

where $B(P, Q, t)$ is the effective force due to the interaction with the rest of the system, assumed here to take the simplified form:

$$B(P, Q, t) = \Gamma(P) + \sqrt{2D_P}\xi(t), \tag{6.48}$$

where $\xi(t)$ is a Gaussian white noise, with $\langle\xi(t)\rangle = 0$ and $\langle\xi(t)\xi(t')\rangle = \delta(t - t')$ and $D_P > 0$.

Even before applying the detailed balance condition, a first argument to determine the function $\Gamma(P)$ immediately emerges in the overdamped limit, i.e. when the inertial term $\dot{P}$ can be neglected. In that case the only way to have a closed equation for $\dot{Q}$ is to impose

$$\Gamma(P) = c\partial_P K(P), \tag{6.49}$$

with c some constant to be found. With such a choice, in fact, setting to 0 the left hand side of equation (6.47b), one gets

$$\dot{Q} = \partial_P K(P) = \frac{\Gamma(P)}{c} = \frac{1}{c}\partial_Q U(Q) - \frac{\sqrt{2D_P}}{c}\xi(t), \tag{6.50}$$

which has the steady probability density $f_Q(Q) \sim \exp[cU(Q)/D_P]$. Such a density must be consistent with equilibrium, which implies $c = -\beta D_P$. In summary one has

$$f_Q(Q) \sim \exp[-\beta U(Q)], \tag{6.51a}$$

$$\Gamma(P) = -D_P\beta\partial_P K(P). \tag{6.51b}$$

If inertia cannot be neglected, we must take the Fokker–Planck equation associated with equations (6.47) and (6.48), in the steady state, which reads

$$\partial_Q J_Q(Q, P) + \partial_P J_P(Q, P) = 0 \tag{6.52}$$

$$J_Q(Q, P) = f(Q, P)\partial_P K(P) \tag{6.53}$$

$$J_P(Q, P) = -f(Q, P)\partial_Q U(Q, P) + \Gamma(P)f(Q, P) - D_P\partial_P f(Q, P), \tag{6.54}$$

where $f(Q, P)$ is the stationary probability density.

Now we impose the detailed balance condition mentioned before, i.e. that the irreversible current vanishes, leading to:

$$\Gamma(P)f(Q, P) - D_P\partial_P f(Q, P) = 0, \tag{6.55}$$

which can be solved by factorization, i.e. $f(Q, P) = f_Q(Q)f_P(P)$, and therefore we have $f_P(P) \sim \exp[-\beta K(P)]$ and to equations (6.51). Of course in the most common case, i.e. when $K(P) = P^2/(2M)$, one recovers $\Gamma(P) = -D_P\beta P/M$, that is the usual viscous term $-\gamma V$ with viscosity satisfying the Einstein relation $\gamma = \beta D_P$ and therefore it can only be $\beta > 0$. Interestingly, one always has $\Gamma(P) = -D_P\beta\dot{Q}$, i.e. somehow the 'velocity' $\dot{Q}$ sees no consequences of the different shape of the kinetic term $K(P)$. Moreover, in all cases one obviously has $f_P(P) \sim \exp[-\beta K(P)]$. It is clear that in cases where $\beta < 0$ boundary conditions on P must be consistent with the normalization of $f_P(P)$. The positional part of the probability density, $f_Q(Q)$, can be found by using the stationary condition.

6.4.2 Loss of symmetry: dissipation without fluctuations

Parts of this section have been reprinted with permission from [39] copyright EPLA, 2015.

Here we briefly discuss, qualitatively, the problem of coarse-graining a microscopic model which, for some reason, does not conserve energy: the simplest one is the presence of an external perturbation. Consider as a starting point the usual Hamiltonian system as composed of two interacting subsystems Σ and Θ: a point in the complete phase ('zero level') space is $\Gamma_0 = (\Gamma_\Sigma, \Gamma_\Theta)$. An external perturbation is applied to some of the degrees of freedom of Σ: the nature and formal details of the perturbation are not important for the general discussion, but examples are discussed in reference [39]. We consider a probabilistic description where the phase space position Γ_0 is distributed with some probability density $P_0(\Gamma_0, t)$ at time t. This density obeys an equation of the kind

$$\frac{\partial P_0(\Gamma_0, t)}{\partial t} = [L_0(\Gamma_0) + L_{ext}(\Gamma_\Sigma, t)]P_0(\Gamma_0, t), \tag{6.56}$$

where L_0 is the Liouville operator associated with the total Hamiltonian $H_0(\Gamma_0)$ and L_{ext} represents the external perturbation, which can be deterministic, stochastic, time-dependent or not, etc. The first level of coarse-graining, C1, is similar to those considered above, i.e. it consists in focusing on $\Gamma_1 \equiv \Gamma_\Sigma$ alone, by replacing equation (6.56) with the following equations:

$$\frac{\partial P_1(\Gamma_1, t)}{\partial t} = [L_1(\Gamma_1) + L_{ext}(\Gamma_1, t)]P_1(\Gamma_1, t), \tag{6.57a}$$

$$L_1(\Gamma_1) = L_H(\Gamma_1) + L_T(\Gamma_1). \tag{6.57b}$$

In equation (6.57), the L_H operator is the Liouville operator associated with the Hamiltonian $H(\Gamma_1)$ of the system Σ alone, i.e. the internal dynamics of the system of interest, and L_T is the operator describing—in some simplified form—its coupling to Θ. When L_T is a stochastic operator and, for instance, memory effects can be neglected, it takes the form of a Markovian master equation operator, and if only the continuous part dominates (as in the limit of large mass for a tracer), L_T becomes a Fokker–Planck operator. Its transition rates, in order to reflect the invariance of L_0 under time reversal, satisfy detailed balance with respect to the Gibbs measure defined by Hamiltonian H and temperature T.

When the external perturbation L_{ext} acts on space–time scales much larger than those dictated by L_H and L_T, it is convenient and common to scale up the description to a 'macroscopic' level, what we call the C2 coarse-graining. This operation is usually achieved by phenomenological considerations. Only those degrees of freedom which are relevant at large scales, Γ_2, are retained and the evolution of their probability takes the form

$$\frac{\partial P_2(\Gamma_2, t)}{\partial t} = [L_2(\Gamma_2) + L_{ext}(\Gamma_2, t)]P_2(\Gamma_2, t). \tag{6.58}$$

The L_2 operator represents the contraction of microscopic Hamiltonian (L_H) and thermostat (L_T) parts, and it is often non-conservative and deterministic. Indeed, the aim of C2 is to get a fair description of the trajectories at a macroscopic resolution where energy is dissipated and fluctuations are usually negligible. The result is that L_2 fairly accounts for dissipation, but may violate detailed balance, a required symmetry in the absence of L_{ext}. For instance, a typical choice is to *neglect* the noise part of thermal origin, because it is exceedingly small with respect to other forces. An example is given by macroscopic solid friction, e.g. Coulomb friction models, where thermal fluctuations are not taken into account. Another example is the description of macroscopic electrical currents in circuits: the dynamics contains dissipative terms but (unless in the presence of very small currents) the thermal electrical Johnson–Nyquist noise is neglected. In the reality, the huge difference of energy scales (between dissipation and noise), makes the probability that the time reversal of a typical trajectory is observed exceedingly small, in the modelling this can become zero: in many common cases [40–42], L_2 does not describe properly those trajectories and strongly twists the definition of entropy production. Generalizations of the fluctuating entropy production have been proposed (see for instance references [42, 43]) where the probability weighing the reversed trajectory, i.e. that appearing at the denominator of equation (5.54), is replaced by a different probability, generated by an *auxiliary dynamics*. Unfortunately, this ad hoc prescription changes the physical meaning (and the accessibility in experiments) of the action functional and usually does not entirely solve the discrepancy. The only way to get an action functional which properly represents the thermodynamics of the system is to include, in L_2, the fluctuations which are conjugate to the modelled dissipation restoring the condition of detailed balance in the absence of L_{ext}. An example of this problem can be seen at work in models of granular gases, which are discussed in chapter 8.

6.5 Inferring Langevin models from data

This last section introduces the reader to a different problem, that is, inferring a model from experimental or numerical data. Here we still consider (generalized) Langevin models. In the first subsection we discuss a procedure that is able to build iteratively the memory kernel of a GLE, but is not able to distinguish an equilibrium from a nonequilibrium system, as it always gives the noise correlation in terms of the fluctuation–dissipation theorem, i.e. the resulting model is always at equilibrium. The second subsection in contrast presents a procedure which is able to infer the

Langevin parameters (friction and noise amplitude) from the data including also out-of-equilibrium systems, but assumes that the model is Markovian, i.e. that the data exhaust all the relevant degrees of freedom of the system, which is of course far from the general situation.

6.5.1 Inferring a generalized Langevin equation at equilibrium

Procedures to construct the memory function from data have been discussed in references [27, 44]. We summarize here the latter. The idea is to consider a different generalized Langevin equation with respect to both Mori and Zwanzig projections called, by the authors, a hybrid projection scheme. The equation is the following:

$$\ddot{A}_t = -\frac{1}{M(A_t)}\frac{d}{dA_t}[U_{\mathrm{PMF}}(A_t) + k_{\mathrm{B}}T \ln M(A_t)] \tag{6.59}$$

$$-\int_0^t ds\Gamma^{\mathrm{L}}(s)\dot{A}_{t-s} + \int_0^t ds\Gamma^{\mathrm{NL}}(A_{t-s}, s) + F^R(t), \tag{6.60}$$

where: (i) the potential mean force $U_{\mathrm{PMF}}(A_t)$ appears explicitly in the equation of motion, similarly to the Zwanzig projection scheme. (ii) An inhomogeneous effective mass $M(A_t)$ appears defined through

$$\left\langle \dot{A}_t^2 \right\rangle_{A_t} \equiv k_{\mathrm{B}}T/M(A_t),$$

and it gives rise to a drift term. If $M(A_t)$ is constant, i.e. if the variance of $\dot{A}_t$ is independent of A_t then this drift term vanishes. For an observable A_t that is a linear combination of positions, then the effective mass is constant. Even for certain nonlinear observables, such as distances in position space, it can be shown that the generalized mass is constant [27]. (iii) The memory kernel $\Gamma^{\mathrm{L}}(s)$ is the unconditional average over the random-force correlations

$$\Gamma^{\mathrm{L}}(s) = \frac{\langle F^R(0), F^R(s)\rangle}{\left\langle \dot{A}_0^2 \right\rangle},$$

similarly to the Mori projection, and therefore only depends on time. It thus describes the linear friction contribution. Note that here the equilibrium hypothesis has been exploited. (iv) The memory friction function Γ^{NL} has the following expression

$$\Gamma^{\mathrm{NL}}(A_t, s) = \frac{d}{dA_t}D(A_t, s) - \beta D(A_t, s)\frac{d}{dA_t}U_{\mathrm{PMF}}(A_t),$$

where the conditional correlation function between the time derivative of the observable at the initial time, $\dot{A}_0$, and the random force $F^R(s)$, has been introduced: $D(A_t, s) = \langle \dot{A}_0, F^R(s)\rangle_{A_t}$.

Let us stress that both the noise and the memory kernels appearing in equation (6.59) can be reconstructed from an experimental or numerical trajectory of the observable A_t;

the detailed procedure is described in reference [27]. If the observable A_t has at time $t = i\Delta t$ a value in the interval I_α, then we write $A_i \in I_\alpha$; $\sum_{A_i \in I_\alpha}$ denotes the sum over all times i for which A_i is in the interval I_α, which is used to compute conditional averages needed in the following. N_{traj} denotes the total length of the A_t trajectory used. The sums run from $i = 0$ to $N_{\text{traj}} - j - 1$, because for a given j, the iterative scheme has only determined the random force at times up to $N_{\text{traj}} - j - 1$. The iterative scheme works as follows: first, note that at $t = 0$, the random force is given by $F^R(\omega_i, 0) = \ddot{A}_i + [1/M(A_i)]d/dA_i[U_{\text{PMF}}(A_i) + k_B T \ln M(A_i)]$. That is, it contains contributions from the accelerations and the forces due to the effective potential, taken over all possible initial configurations at $i\Delta t$, for $i = 0, 1, 2, \ldots, N_{\text{traj}} - 1$. This, together with $\dot{A}_i$, can be obtained directly from a given trajectory of the observable A_t. Then, $F^R(\omega_i, 0)$, A_i, and $\dot{A}_i$ are inserted into the iterative scheme (see below) to compute $F^R(\omega_i, 1)$ for $i = 0, 1, 2, \ldots, N_{\text{traj}} - 2$. $F^R(\omega_i, 1)$ is then used to compute $F^R(\omega_i, 2)$ for $i = 0, 1, 2, \ldots, N_{\text{traj}} - 3$ and so forth. While computing $F^R(\omega_i, j)$, the memory friction functions $\Gamma^{\text{L}}(j)$ and $\Gamma^{\text{NL}}(A, j)$ are computed simultaneously. The scheme at step j reads

$$
\begin{aligned}
F^R(\omega_i, j+1) = {} & F^R(\omega_{i+1}, j) + \Delta t \Gamma^{\text{L}}(j)\dot{A}_{i+1} \\
& - \Delta t \Gamma^{\text{NL}}(A_{i+1}, j),
\end{aligned}
$$

$$
\Gamma^{\text{L}}(j) = \frac{\sum_{i=0}^{N_{\text{traj}}-j-1} F^R(\omega_i, 0) F^R(\omega_i, j)}{\sum_{i=0}^{N_{\text{traj}}-j-1} \dot{A}_i^2},
$$

$$
\begin{aligned}
\Gamma^{\text{NL}}(A_{i+1}, j) = {} & \left[\frac{D(\alpha+1, j) - D(\alpha-1, j)}{2\Delta A} - \beta D(\alpha, j) \right. \\
& \left. \times \frac{U_{\text{PMF}}(\alpha+1) - U_{\text{PMF}}(\alpha-1)}{2\Delta A} \right]_{A_{i+1} \in I_\alpha}.
\end{aligned}
$$

The sums in the denominator extend over the same interval as in the numerator to increase the numerical stability [6, 37]. If one needs to compute the memory friction functions, then the computation of $F^R(\omega_i, j)$ can be stopped as soon as the memory functions have dropped to zero. For instance, if the memory functions decay to zero after N_{mem} time steps, then one can abort the computation of the random force at $F^R(\omega_i, N_{\text{mem}})$. At that point, $N_{\text{traj}} - N_{\text{mem}} - 1$ distinct random-force trajectories of length N_{mem} each have been generated.

6.5.2 Inferring a Langevin equation, out of equilibrium

Parts of this section have been reprinted from [47] CC BY 4.0.

A much more direct approach can be described if one wishes to build a Langevin equation without memory from data. The method is based upon the straightforward definitions of the parameters of the Langevin equations; however, there are some rather severe difficulties both at a conceptual and technical level [45–47]. The advantage of this method over the previous one, apart from a simpler implementation, is the fact that equilibrium is not assumed, i.e. the drift and diffusion functions do not necessarily satisfy a fluctuation–dissipation theorem. However, it is limited to data which can be described by a Markov process.

Let us assume that we know that the slow 'good' variables are the component of a vector $\mathbf{X}(t) \in \mathbb{R}^N$. As it has been made clear in the present chapter, this is a rather subtle point, and the choice of the proper $\mathbf{X}$ is not easy at all. In general, if this choice is not appropriately done, one ends up with a non-Markovian description. Let us also suppose that we have measured, e.g. in an experiment or a simulation, a long time series of these variables, $\{\mathbf{X}(t)\}$. With these assumptions, one can determine the N Langevin equations ($n \in [1, N]$)

$$\frac{dX_n}{dt} = F_n(\mathbf{X}) + \sqrt{2D_n(\mathbf{X})}\,\eta_n(t),$$

where $\eta_n(t)$ are white noises, i.e. Gaussian processes with $\langle \eta_n(t) \rangle = 0$ and $\langle \eta_n(t)\eta_{n'}(t') \rangle = \delta_{nn'}\delta(t - t')$, following the definitions found in textbooks on stochastic processes [6, 8, 33], that is, in terms of the statistical features of

$$\Delta X_n(\Delta t) = X_n(t + \Delta t) - X_n(t).$$

In fact the drifts and diffusion coefficients are given by the following formulae:

$$F_n(\mathrm{X}) = \lim_{\Delta t \to 0} \frac{1}{\Delta t} \langle \Delta X_n(\Delta t) | \mathrm{X}(t) = \mathrm{X} \rangle,$$

$$D_n(\mathrm{X}) = \lim_{\Delta t \to 0} \frac{1}{2\Delta t} \langle (\Delta X_n(\Delta t) - F_n(\mathrm{X})\Delta t)^2 | \mathrm{X}(t) = \mathrm{X} \rangle.$$

The limit $\Delta t \to 0$ must be considered in a proper physical sense, i.e. smaller than any typical time, but not too small [46]. Here the mean force $\mathbf{F}$ is not necessarily linear and the noise may be multiplicative, i.e. D can depend upon $\mathbf{X}$.

The practical procedure to extract the $\{F_n\}$ and $\{D_n\}$ from data is not trivial; however, the main conceptual trouble is given by the absence of a general method for choosing the 'right' variables, an aspect that is too often overlooked. Usually the 'proper variables' are unknown and one can use the time series of just one observable or a few ones. Such a problem is quite similar to the phase space reconstruction in dynamical systems, or even harder because of the presence of noise that hugely amplifies the dimensionality of phase space. There are no automatic protocols for this choice, and typically to obtain some good results it is necessary to possess the expertise and/or intuition about the problem under investigation. For a general discussion on the difficulties in building models from data, see reference [48]. Sometimes mathematics can help to anticipate that a certain set of variables is not adequately described by a Markovian model. For instance, see reference [46], in the case of a single scalar variable, the shape of the correlation function may be already sufficient to exclude that such a variable is an equilibrium Markov process, so that it becomes necessary to look for a different set of variables. In chapter 8 we will discuss an application of the procedure sketched here to the case of a granular gas.

References

[1] Givon D, Kupferman R and Stuart A 2004 Extracting macroscopic dynamics: model problems and algorithms *Nonlinearity* **17** R55

[2] Arnold V I 2013 *Mathematical Methods of Classical Mechanics* (Berlin: Springer Science & Business Media)

[3] Spohn H 2012 *Large Scale Dynamics of Interacting Particles* (Berlin: Springer Science & Business Media)

[4] Chibbaro S, Rondoni L and Vulpiani A 2014 *Reductionism, Emergence and Levels of Reality* (Berlin: Springer)

[5] Espanol P 2004 Statistical mechanics of coarse-graining *Novel Methods in Soft Matter Simulations* ed M Karttunen, I Vattulainen and A Lukkarinen (Berlin: Springer) pp 69–115

[6] van Kampen N G 1992 *Stochastic Processes in Physics and Chemistry* (Amsterdam: Elsevier)

[7] Bocquet L 1997 High friction limit of the Kramers equation: the multiple time-scale approach *Am. J. Phys.* **65** 140–4

[8] Gardiner C 2009 *Stochastic Methods* (Berlin: Springer)

[9] Biferale L, Crisanti A, Vergassola M and Vulpiani A 1995 Eddy diffusivities in scalar transport *Phys. Fluids* **7** 2725–34

[10] Majda A J and Kramer P R 1999 Simplified models for turbulent diffusion: theory, numerical modelling, and physical phenomena *Phys. Rep.* **314** 237–574

[11] Crisanti A, Falcioni M and Vulpiani A 1991 Lagrangian chaos: transport, mixing and diffusion in fluids *Riv. Nuovo Cimento (1978–1999)* **14** 1–80

[12] Weinan E and Engquist B 2003 Multiscale modeling and computation *Not. AMS* **50** 1062–70

[13] Weinan E 2011 *Principles of Multiscale Modeling* (Cambridge: Cambridge University Press)

[14] Van den Broeck C, Kawai R and Meurs P 2004 Microscopic analysis of a thermal Brownian motor *Phys. Rev. Lett.* **93** 090601

[15] Costantini G, Marconi U M B and Puglisi A 2007 Granular Brownian ratchet model *Phys. Rev.* E **75** 061124

[16] Puglisi A 2014 *Transport and Fluctuations in Granular Fluids: From Boltzmann equation to Hydrodynamics, Diffusion and Motor Effects* (Berlin: Springer)

[17] Cecconi F, Cencini M and Vulpiani A 2007 Transport properties of chaotic and non-chaotic many particle systems *J. Stat. Mech. Theory Exp.* **2007** P12001

[18] Lorentz H A 1905 The motion of electrons in metallic bodies I *Pro. K. Ned. Akad. Wet.* **7** 438–53

[19] Rayleigh L 1891 Dynamical problems in illustration of the theory of gases *London, Edinburgh Dublin Phil. Mag. J. Sci.* **32** 424–45

[20] Rayleigh J W S 1902 *Scientific Papers: 1887–1892* **vol 3** (Cambridge: Cambridge University Press)

[21] Dorfman J R 1999 *An Introduction to Chaos in Nonequilibrium Statistical Mechanics* (Cambridge: Cambridge University Press)

[22] Alder B J and Wainwright T E 1970 Decay of the velocity autocorrelation function *Phys. Rev.* A **1** 18

[23] Dorfman J R and Cohen E G D 1970 Velocity correlation functions in two and three dimensions *Phys. Rev. Lett.* **25** 1257

[24] Perondi L F and Binder P-M 1993 Mean-squared displacement of a hard-core tracer in a periodic lattice *Phys. Rev.* B **48** 4136

[25] van Kampen N G 1961 A power series expansion of the master equation *Can. J. Phys.* **39** 551–67

[26] Rubin R J 1960 Statistical dynamics of simple cubic lattices. Model for the study of Brownian motion *J. Math. Phys.* **1** 309–18

[27] Ayaz C, Scalfi L, Dalton B A and Netz R R 2022 Generalized Langevin equation with a nonlinear potential of mean force and nonlinear memory friction from a hybrid projection scheme *Phys. Rev.* E **105** 054138

[28] Mori H 1965 Transport, collective motion, and Brownian motion *Prog. Theor. Phys.* **33** 423–55

[29] Zwanzig R 2001 *Nonequilibrium Statistical Mechanics* (Oxford: Oxford University Press)

[30] Darve E, Solomon J and Kia A 2009 Computing generalized Langevin equations and generalized Fokker-Planck equations *Proc. Natl Acad. Sci.* **106** 10884–9

[31] Chorin A J, Hald O H and Kupferman R 2000 Optimal prediction and the Mori-Zwanzig representation of irreversible processes *Proc. Natl Acad. Sci.* **97** 2968–73

[32] Ayaz C, Tepper L, Brünig F N, Kappler J, Daldrop J O and Netz R R 2021 Non-Markovian modeling of protein folding *Proc. Natl Acad. Sci.* **118** e2023856118

[33] Risken H 1996 *Fokker-Planck equation* (Berlin: Springer)

[34] Onsager L 1949 Statistical hydrodynamics *Il Nuovo Cimento (1943–1954)* **6** 279–87

[35] Ramsey N F 1956 Thermodynamics and statistical mechanics at negative absolute temperatures *Phys. Rev.* **103** 20

[36] Braun S, Ronzheimer J P, Schreiber M, Hodgman S S, Rom T, Bloch I and Schneider U 2013 Negative absolute temperature for motional degrees of freedom *Science* **339** 52–5

[37] Cerino L, Puglisi A and Vulpiani A 2015 A consistent description of fluctuations requires negative temperatures *J. Stat. Mech. Theory Exp.* **2015** P12002

[38] Baldovin M, Puglisi A, Sarracino A and Vulpiani A 2017 About thermometers and temperature *J. Stat. Mech. Theory Exp.* **2017** 113202

[39] Cerino L and Puglisi A 2015 Entropy production for velocity-dependent macroscopic forces: the problem of dissipation without fluctuations *Europhys. Lett.* **111** 40012

[40] Kim K H and Qian H 2004 Entropy production of Brownian macromolecules with inertia *Phys. Rev. Lett.* **93** 120602

[41] Gnoli A, Puglisi A and Touchette H 2013 Granular Brownian motion with dry friction *Europhys. Lett.* **102** 14002

[42] Munakata T and Rosinberg M L 2014 Entropy production and fluctuation theorems for Langevin processes under continuous non-Markovian feedback control *Phys. Rev. Lett.* **112** 180601

[43] Spinney R E and Ford I J 2012 Nonequilibrium thermodynamics of stochastic systems with odd and even variables *Phys. Rev. Lett.* **108** 170603

[44] Carof A, Vuilleumier R and Rotenberg B 2014 Two algorithms to compute projected correlation functions in molecular dynamics simulations *J. Chem. Phys.* **140** 124103

[45] Friedrich R, Peinke J, Sahimi M and Tabar M R R 2011 Approaching complexity by stochastic methods: from biological systems to turbulence *Phys. Rep.* **506** 87–162

[46] Baldovin M, Puglisi A and Vulpiani A 2018 Langevin equation in systems with also negative temperatures *J. Stat. Mech. Theory Exp.* **2018** 043207

[47] Baldovin M, Puglisi A and Vulpiani A 2019 Langevin equations from experimental data: the case of rotational diffusion in granular media *PLoS One* **14** e0212135

[48] Baldovin M, Cecconi F, Cencini M, Puglisi A and Vulpiani A 2018 The role of data in model building and prediction: a survey through examples *Entropy* **20** 807
[49] Castiglione P, Falcioni M, Lesne A and A Vulpiani 2009 Coarse-graining equations in complex systems *Chaos and Coarse Graining in Statistical Mechanics* (Cambridge University Press) ch 7
[50] Ayaz C, Scalfi L, Dalton B A and Netz R R 2022 Generalized Langevin equation with a nonlinear potential of mean force and nonlinear memory friction from a hybrid projection scheme *Phys. Rev. E* **105** 054138

Chapter 7

Applications: from climate to causation

If we study the history of science we see produced two phenomena which are, so to speak, each the inverse of the other. Sometimes it is simplicity which is hidden under what is apparently complex; sometimes, on the contrary, it is simplicity which is apparent, and which conceals extremely complex realities.

Henri Poincaré

7.1 Langevin equation for the climate dynamics

As already discussed in the previous chapters, in systems with very different characteristic times it is necessary to treat the 'slow dynamics' in terms of effective equations. Historically, such a topic was born in the realm of statistical mechanics with the seminal contribution of Langevin; however, the method is rather general and had been used in a large class of systems. In this section we show how the Langevin equation can be successfully used in the context of climate dynamics.

Focusing over the last 10^6 years, climate records can be obtained by the ice cores: the most striking feature of the paleoclimate signal is a roughly periodic behaviour with a period of around 10^5 years [1]. Remarkably, as noted in the 1940s by M Milankovitch, there is a rather strong correlation between the temperature, as well as the CO_2, and the flux of the energy from the Sun [2].

Following the approach of Langevin, one can describe the observed fluctuations in terms of a stochastic equation. Such an idea was introduced by K Hasselman who was the first to approach the large scale and long time dynamics of the climate evolution in terms of stochastic processes [3]. For this seminal contribution, Hasselmann received the Nobel prize in 2021.

Let us illustrate the idea, considering a simplified description for just one climate variable, the temperature T, which, in this case, should be considered the global

doi:10.1088/978-0-7503-6229-0ch7

average temperature. A minimal model was introduced by Budyko and Sellers who proposed that long time features of the Earth climate can be described by a radiation balance [1, 4]

$$C\frac{dT}{dt} = R_{\text{in}}(1 - \alpha) - I(T) + \sqrt{2\varepsilon}\,\eta, \tag{7.1}$$

where C is the thermal capacity, R_{in} is the incoming solar radiation, α is the Earth albedo[1] and $I(T)$ is the out-coming infrared radiation. Following Hasselmann, we also include the noisy term which originates from the fast dynamics. The infrared emission $I(T)$ is a growing function of T given by the Stefan–Boltzmann law and a crucial role is due to the albedo α: for very low temperature the Earth is covered by ice and α is large, whereas for T large enough, as in the present climate, $\alpha \simeq 0.3$.

With a sensible shape of $\alpha(T)$ we have that the deterministic drift $R_{\text{in}}(1 - \alpha) - I(T)$ is zero for three values of the temperature: $T = T_0$ and $T = T_0 \pm \Delta T$, where $T_0 + \Delta T$ corresponds to the present climate state and $T_0 - \Delta T$ to the interglacial one. Introducing the quantity $\phi = (T - T_0)/(\Delta T)$, for $|\phi|$ not too large, say $O(1)$, a good approximation of (7.1) is given by the following stochastic differential equation

$$\frac{d\phi}{dt} = \frac{1}{\tau_s}\phi(1 - \phi^2) + \sqrt{2\varepsilon}\,\eta. \tag{7.2}$$

In the previous model we have neglected the possibility that ice covers the Earth completely. In the absence of the noise the above system has two stable states, while in the presence of noise one has transitions between $T_0 - \Delta T$ and $T_0 + \Delta T$. We can determine the strength of the noisy term ε and the characteristic time τ_s in a simple way: the fluctuations in the present climate are described by $\delta\phi = \phi - 1$, and for small $\delta\phi$ we can linearize the previous equation obtaining

$$\frac{d\delta\phi}{dt} = -\frac{2}{\tau_s}\delta\phi + \sqrt{2\varepsilon}\,\eta(t); \tag{7.3}$$

from the above equation we obtain $\langle \delta\phi^2 \rangle = (\varepsilon\tau_s)/2$, as well as $\langle \delta\phi(t)\delta\phi(0) \rangle = \langle \delta\phi^2 \rangle e^{-t/\tau_s}$: therefore, we can estimate both τ_s and ε from the present climate data.

7.1.1 Exit time and stochastic resonance

Let us open a brief parenthesis on the exit time problem for stochastic processes: given an initial condition $x(0) = x_*$ in a domain Ω, one looks for the time τ such that $x(t)$ exits for the first time from Ω. Of course τ is a random variable and a natural question is to evaluate the mean value $\langle \tau \rangle$, which depends on Ω and x_*. In the one-dimensional problem Ω is an interval $[x^{(1)}, x^{(2)}]$, and $\langle \tau \rangle$ can be computed solving a differential equation [5]. In the following we prefer to adopt an alternative approach

[1] The albedo is the fraction of solar radiation reflected back to space. The value of α depends on many factors (such as ice, clouds, deserts and forests): in simple models one assumes that α is a function of the temperature T.

due to Kramers [6]; we consider the case of a one-dimensional Langevin equation, i.e. a particle moving (in the overdamped limit) in a potential $U(x)$ and subject to noise. Such an issue is interesting in several problems of physics and chemistry, e.g. for determining the transition rate in chemical reactions involving an energy barrier, or to study how a particle initially in a given metastable state, i.e. in a local potential minimum higher than the absolute one, jumps in the stable state. Consider the Langevin equation

$$\frac{dx}{dt} = -\frac{dU(x)}{dx} + \sqrt{2\varepsilon}\,\eta, \tag{7.4}$$

for instance with the double well potential $U(x)$ (see figure 7.1)

$$U(x) = -\frac{1}{2}\kappa x^2\left(x_0^2 - x^2\right) - Ax, \qquad \kappa, A > 0. \tag{7.5}$$

Assuming $A \ll x_0^3$, we see that the potential has two local minima close to $\pm x_0$, and a local maximum x_M close to 0. The value of the potential at the point $x_- \simeq -x_0$ is larger than its value at the point $x_+ \simeq +x_0$: therefore, if the system is initially placed at x_-, we expect it to cross the barrier at the origin and to reach eventually x_+; in the following it will be clear that the precise shape of the potential is not particularly relevant. We now evaluate the escape rate from this metastable state, that is, the probability per unit time that the system leaves the metastable region on the left of the origin. The Fokker–Planck equation for this system is

$$\frac{\partial p(x, t)}{\partial t} = \frac{\partial}{\partial x}\left[\frac{dU(x)}{dx}p(x, t) + \varepsilon\frac{\partial p(x, t)}{\partial x}\right]; \tag{7.6}$$

introducing the probability current

$$J(x, t) = -\frac{dU(x)}{dx}p(x, t) - \varepsilon\frac{\partial p(x, t)}{\partial x}, \tag{7.7}$$

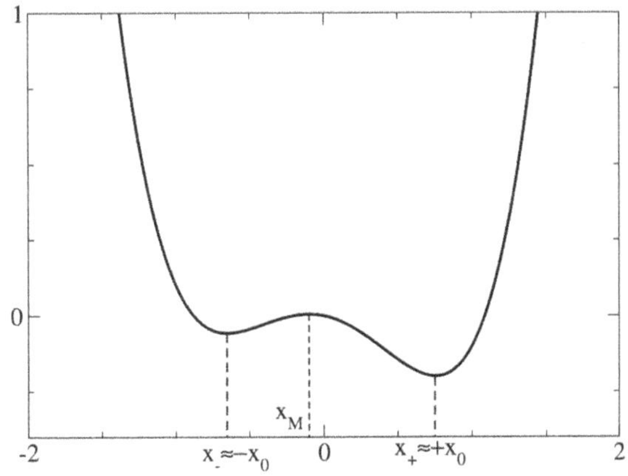

Figure 7.1. Bistable potential equation (7.5), for $k = 1$, $x_0 = 1$ and $A = 0.1$.

the Fokker–Planck equation can be written as

$$\frac{\partial}{\partial t}p(x, t) + \frac{\partial}{\partial x}J(x, t) = 0.$$

It is easy to verify that

$$J(x, t) = -\varepsilon e^{-U(x)/\varepsilon}\frac{d}{dx}[e^{U(x)/\varepsilon}\, p(x, t)]. \tag{7.8}$$

We start from x_- at $t = 0$, namely $p(x, 0) = \delta(x - x_-)$, and if we linearize the Langevin equation around x_-, for the variable $z = x - x_-$ we have

$$\frac{d}{dt}z = -\frac{z}{\tau_c} + \sqrt{2\varepsilon}\,\eta,$$

where $1/\tau_c = U''(x_-)$. It is easy to realize that, in a time $O(\tau_c)$, $p(x, t)$ will relax in the well on the left in a metastable situation, while on the right $p(x, t)$ is basically zero. In other words, denoting by $p_s(x)$ the stationary solutions of the Fokker–Planck equation $p_s(x) \propto e^{-U(x)/\varepsilon}$, we have $p(x, t) \simeq p_{\rm meta}(x)$ with $p_{\rm meta} \simeq C\, p_s(x)$ for $x < x_M$ and $p_{\rm meta} \simeq 0$ for $x > x_M$, the constant C being fixed by the normalization, see in the following. In this metastable state, at varying x, J is basically constant on the left and zero on the right; given arbitrarily x_1 and x_2 on the left, from equation (7.8) one has

$$J \cdot \int_{x_1}^{x_2} dx e^{U(x)/\varepsilon} = -\varepsilon[e^{U(x)/\varepsilon}\, p_{\rm meta}(x)]_{x_1}^{x_2}. \tag{7.9}$$

Let us take $x_1 = x_-$ and x_2 slightly larger than the maximum x_M of $U(x)$[2]; since $p_{\rm meta}(x_M)$ is very small, we have

$$J \simeq \frac{\varepsilon\, e^{U(x_-)/\varepsilon} p_{\rm meta}(x_-)}{\int_{x_-}^{x_M} dx e^{U(x)/\varepsilon}}. \tag{7.10}$$

We can obtain $p_{\rm meta}(x_-)$ noting that for $x < x_M$ it is proportional to the stationary distribution for x close to x_-; in addition, for small ε, we can use the Laplace method, i.e. we assume $U(x) = U(x_-) + \frac{1}{2}\omega_-^2(x - x_-)^2$ for x close to x_-, where $\omega_-^2 = \left.\frac{d^2U}{dx^2}\right|_{x_-}$. By imposing the normalization condition:

$$\int_{-\infty}^{x_M} dx\; p_{\rm meta}(x)\, dx = 1, \tag{7.11}$$

since

$$p_{\rm meta}(x) = \frac{e^{-U(x)/\varepsilon}}{\int_{-\infty}^{x_M} dx e^{-U(x)/\varepsilon}} = p_{\rm meta}(x_-) e^{-[U(x) - U(x_-)]/\varepsilon},$$

[2] Such a choice is quite natural: once the particle has crossed the maximum then it will go quickly toward x_+.

one has

$$\int_{-\infty}^{x_M} dx\, p_{\text{meta}}(x_-)\, e^{-[U(x)-U(x_-)]/\varepsilon} \simeq p_{\text{meta}}(x_-)\left(\frac{2\pi\varepsilon}{\omega_-^2}\right)^{1/2} = 1. \tag{7.12}$$

We have extended the integration limits to infinity, assuming that $p_{\text{meta}}(x)$ is concentrated near x_-. The integral in the denominator of equation (7.10) is dominated by the contribution around $x_M \simeq 0$, where we have $U(x) \simeq -\frac{1}{2}\omega_0^2 x^2$, with $\omega_0^2 = \left.\frac{d^2U}{dx^2}\right|_{x_0}$, and thus, as shown by a straightforward integration, it is equal to $(2\pi/\omega_0^2)^{1/2}$. By collecting all these results, we obtain

$$J \simeq \frac{\omega_0\, \omega_-}{2\pi}\, e^{-\Delta U/\varepsilon}, \tag{7.13}$$

where $\Delta U = U(0) - U(x_-)$ is the height of the energy barrier. This characteristic behaviour proportional to the exponential of minus the barrier height (in units of ε) is known as the Arrhenius law[3].

There is no need to assume that the point x_+ is more stable than x_-. Suppose, for instance, that $U(x_+) = U(x_-)$: as long as the barrier ΔU is large enough, the system will switch from x_- to x_+ or back, at random times separated, on average, by $1/J$. Such behaviour is shown in figure 7.2. Of course the exit time τ is a random independent variable, and its probability distribution is roughly Poissonian:

$$p(\tau) \simeq \frac{1}{\langle\tau\rangle} e^{-\frac{\tau}{\langle\tau\rangle}} \text{ with } \langle\tau\rangle = \frac{1}{J}.$$

Let us now come back to our climate problem, i.e. equation (7.2). Using a realistic estimate of τ_s and ε, the average transition time between the glacial and the interglacial climate state is of order 50 000 years. On the other hand, since the transition times are

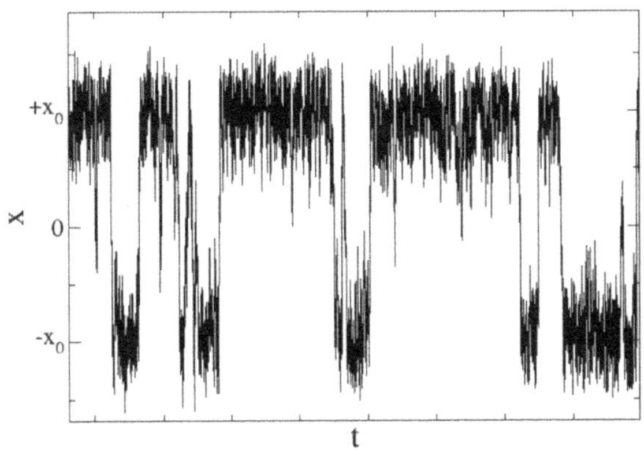

Figure 7.2. Behaviour of a Brownian particle in a bistable potential.

[3] Arrhenius obtained his law from an analysis of experimental data for chemical reactions.

random variables, there is no way for equation (7.2) to show the oscillations roughly periodic with a period around 10^5 years, observed in climate records.

The next step in the modelling of paleoclimate is to take into account an extra ingredient. The serb mathematician M Milankovitch in the first half of the 20th century made an accurate analysis of the Earth orbital variation, and discovered that the global incoming radiation from the Sun was roughly modulated by a 10^5 years cycle in close correspondence to the observed glacial/interglacial transition. This modulation is extremely weak and if we apply the Milankovitch forcing to equation (7.1) we deduce only a small change of about 0.2 degrees: too small to explain the interglacial/glacial transition which is $O(10)$ degrees. Let us note that the Milankovitch forcing is *in phase* with the observed climate records and this suggests the possibility of a resonance-like behaviour.

Including the time-dependent behaviour of the solar constant due to the roughly periodic variation of the Earth eccentricity, the final equation reads [4]:

$$\frac{d\phi}{dt} = \frac{1}{\tau_s}\phi(1 - \phi^2) + \sqrt{2\varepsilon}\,\eta(t) + A(t), \tag{7.14}$$

where now $A(t) = A_0 \cos(\omega t)$, with $\omega = 2\pi/T$, being $T = O(10^5)$ years, and A_0, in the reduced units of equation (7.14), is order 0.1. Once we introduce the effect of the external Milankovitch forcing, the behaviour of the transition mechanism between the stable states changes completely.

Let us now reconsider the exit time problem in a non-stationary potential, e.g. we assume that the quantity A that appears in equation (7.5), instead of being constant, is sinusoidally modulated with angular frequency ω, i.e. $A(t) = A_0 \cos \omega t$, as for the Milankovitch forcing. Then, when $t = 0$, the point x_- is metastable and the point x_+ is stable, and the escape rates from x_- and x_+ are, respectively, given by

$$\begin{aligned} J_- &\simeq \frac{\kappa x_0^2}{\sqrt{2}\pi} \exp\left[-\frac{\kappa}{\varepsilon}\left(\frac{x_0^4}{4} - A_0 x_0\right)\right], \\ J_+ &\simeq \frac{\kappa x_0^2}{\sqrt{2}\pi} \exp\left[-\frac{\kappa}{\varepsilon}\left(\frac{x_0^4}{4} + A_0 x_0\right)\right]. \end{aligned} \tag{7.15}$$

When $t = \pi/\omega$, the two points swap their stability. Consider now the case such that

$$J_+ < \frac{\omega}{\pi} < J_-; \tag{7.16}$$

then, if the system is in the temporarily metastable state, say x_-, it has time to overcome the barrier and go to the stable one, x_+. After a half-period, x_+ becomes itself metastable, and the system has time to cross the barrier in the reverse direction. In other words, if we apply to the system a small periodic modulation of amplitude $A_0 \ll x_0^3$ and angular frequency ω, in the case the strength of the noise, ε, is too small, the system does never have the time to escape from the metastable state during the half-period. If ε is too large, the system has the time to make a number of comings and goings from one minimum to the other in each half-period. In both cases, the resulting behaviour exhibits oscillations with frequency close to ω. But, if ε satisfies the inequalities

$$\frac{\kappa(x_0^4/4 - A_0x_0)}{\ln(\sqrt{2}\,\pi\omega/\kappa x_0^2)} < \varepsilon < \frac{\kappa(x_0^4/4 + A_0x_0)}{\ln(\sqrt{2}\,\pi\omega/\kappa x_0^2)}, \tag{7.17}$$

the system will have time to do exactly one crossing, on average, per half-period, and the trajectory will exhibit approximate periodicity with the angular frequency ω: this behaviour is shown in figure 7.3.

Thus, surprisingly, a moderate amount of noise is able to reveal a weak modulation of the potential. This phenomenon, called stochastic resonance (SR), appears in many different contexts, from signal analysis to neuroscience, biology and physiology; there have been a number of experimental confirmations of the phenomenon, for a review see reference [7]. The stochastic resonance was also referred to in the nomination of the Nobel prize to G Parisi in 2021.

The term 'resonance' is due to the fact that in a plot of output signal-to-noise ratio (SNR) as a function of input noise intensity one has a single maximum for some nonzero ε, see figure 7.4 [7]. Such behaviour is similar to that observed in frequency

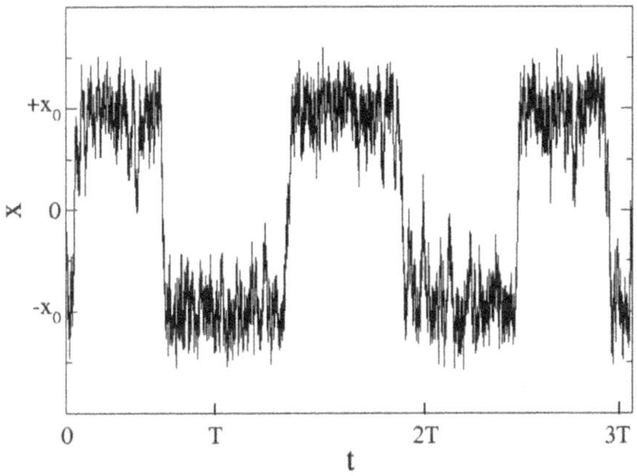

Figure 7.3. Behaviour of a Brownian particle in a bistable potential modulated with period T, exhibiting stochastic resonance.

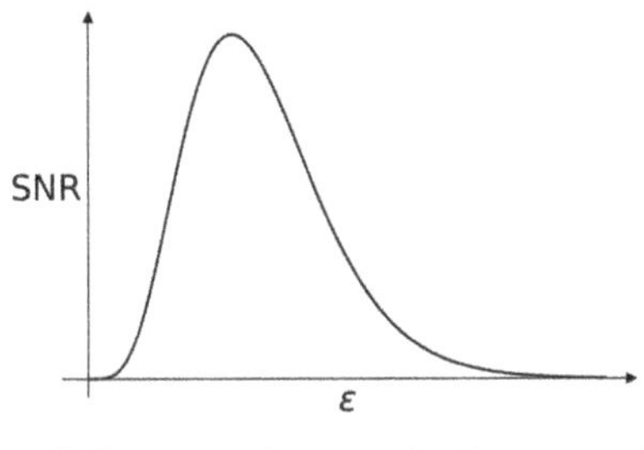

Figure 7.4. Typical curve of output SNR versus noise strength ε for systems showing stochastic resonance. For small and large ε, the output SNR is very small, while some intermediate nonzero ε provides the maximum output SNR.

dependent systems that have a maximum SNR, or output response, for some resonant frequency. However, in the case of SR, the resonance is noise induced.

7.2 Data analysis via information theory

In experimental investigations, typically we have access to the system only through the measurement of a few observables, therefore, the treatment of time records has an obvious practical relevance. For such an aim one can use methods and ideas of a powerful theory introduced at the end of the 1940s by C Shannon: information theory [8].

7.2.1 Shannon entropy

For the sake of self-consistency we briefly recall the basic concepts in information theory, in particular the Shannon entropy [9]. Consider a source that can output m different symbols; denote by s_t the symbol emitted at time t and by $P(W_N)$ the probability that a given block of length N, $W_N = (s_1, s_2, \dots, s_N)$, is emitted:

$$P(W_N) = P(s_1, s_2, \dots, s_N); \tag{7.18}$$

we assume that the source is statistically stationary, i.e. $\forall t \in \mathbb{N}$

$$P(s_1, \dots, s_N) = P(s_{t+1}, \dots, s_{t+N}).$$

The mathematical foundation of the information theory is based on the following fundamental result: given a set of events whose probabilities of occurrence are $(p_1, p_2, \dots, p_n)$, under rather natural assumptions and consistency requirements, one can prove that [9]

$$H = -\lambda \sum_{i=1}^{n} p_i \log p_i \quad \text{with } \lambda > 0,$$

is the unique quantity which measures the average uncertainty about one outcome or, else, the average information that is supplied by one occurrence. More specifically, it is possible to show that $H(p_1, \dots, p_n)$ is the unique function with the 'good natural properties' for the characterization of the amount of information, namely:

(a) H is zero if there is a sure event, i.e. $p_i = \delta_{i,k}$;

(b) H is maximal for equiprobable events, $p_i = 1/n$;

(c) in the case of two probabilistic mutually independent sources A and B, one has $H(A; B) = H(A) + H(B)$, where $H(A; B) = -\lambda \sum_{i=1}^{n} \sum_{j=1}^{m} p_{ij} \log p_{ij}$ is the joint entropy of A and B, and p_{ij} the joint probability;

(d) in the general case one has $H(A; B) = H(A) + H(B|A)$ where $H(B|A) = -\lambda \sum_{i=1}^{n} \sum_{j=1}^{m} p_{ij} \log p_{ij}/p_i$ is the entropy of the source B under the condition that the source A is known;

(e) $H(B|A) \leqslant H(B)$; the interpretation of this inequality is obvious: the knowledge of the outcome of A cannot increase our uncertainty on that of B. Of course the value of λ is not relevant and in the following we take $\lambda = 1$.

Let us now introduce the N-block entropies

$$H_N = -\sum_{\{W_N\}} P(W_N) \ln P(W_N), \tag{7.19}$$

that measures the information content of the N-block ensemble, and the differential entropies

$$h_N = H_{N+1} - H_N, \tag{7.20}$$

whose meaning is the average information supplied by observing the $(N + 1)$-th symbol, provided the N previous ones are known. It is possible to show that the h_N are non-increasing quantities: $h_{N+1} \leqslant h_N$; i.e. the knowledge of a longer past history cannot increase the uncertainty on the next outcome. Therefore, for a stationary source the following limits there exist, are equal, and define the Shannon entropy h_{Sh}:

$$h_{\mathrm{Sh}} = \lim_{N\to\infty} \frac{H_N}{N} = \lim_{N\to\infty} h_N. \tag{7.21}$$

In the case of a periodic source, with period T, $h_{\mathrm{Sh}} = 0$; for a non-correlated source (i.e. emitting independent identically distributed symbols) one has $h_N = h_1 = H_1 = h_{\mathrm{Sh}}$. For a kth order Markov process $h_N = h_{\mathrm{Sh}}$ for all $N \geqslant k$; e.g. in a Markov chain one has

$$h_{\mathrm{Sh}} = -\sum_{i,j} P_i^{\mathrm{inv}} P_{i\to j} \ln P_{i\to j},$$

where P_i^{inv} is the invariant distribution and $P_{i\to j}$ are the transition probabilities.

The Shannon entropy is a measure of the 'surprise' the source emitting the sequences can reserve to us, and it quantifies the asymptotic uncertainty about the further emission of a symbol in a very long sequence. We stress that h_{Sh} is a statistical property of the source, however, because of the ergodicity it can be obtained by analyzing just one single long sequence, and it can also be viewed as a property of each one of the typical sequences.

Let us mention a very important theorem due to Shannon and McMillan that applies to stationary ergodic sources:

If N is large enough, the ensemble of N-long sequences can be separated in two classes, $\Omega_1(N)$ and $\Omega_0(N)$ such that all the blocks $W_N \in \Omega_1(N)$ have (roughly) the same probability $P(W_N) \sim \exp(-Nh_{\mathrm{Sh}})$ and

$$\sum_{W_N \in \Omega_1(N)} P(W_N) \to 1 \text{ for } N \to \infty, \tag{7.22}$$

while

$$\sum_{C_N \in \Omega_0(N)} P(C_N) \to 0 \text{ for } N \to \infty. \tag{7.23}$$

We can explain the main relevance of the theorem as follows: an m-states process admits in principle m^N possible sequences of length N, but the number of the sequences effectively observable, $N_{\text{eff}}(N)$, those ones in $\Omega_1(N)$, also called *typical* sequences, is

$$N_{\text{eff}}(N) \sim \exp(Nh_{\text{Sh}}); \tag{7.24}$$

of course if $h_{\text{Sh}} < \ln m$ we have $N_{\text{eff}} \ll m^N$.

Let us observe that equation (7.24) is somehow the equivalent, in information theory, of the Boltzmann law in statistical thermodynamics: $S \propto \ln W$, W being the number of possible microscopic configurations and S the thermodynamic entropy; this justifies the name 'entropy' for h_{Sh}, which in both cases represents a measure of the exponential proliferation of possibilities. In this analogy one should understand that the physical entropy S is an extensive quantity (of order N) and therefore it corresponds to Nh_{Sh}, i.e. h_{Sh} is an intensive entropy (an entropy per symbol).

7.2.2 The Kolmogorov–Sinai entropy and the ε-entropy

The information theory treats sources emitting discrete symbols, therefore for the case of continuous variables, it is necessary to adapt the approach: this can be done introducing the Kolmogorov–Sinai (KS) entropy [10, 11]. Consider a trajectory, $\mathbf{x}(t)$, generated by a deterministic system, sampled at the times $t_j = j\,\tau$, with $j = 1, 2, 3,...,$ and a finite partition $\mathscr{A}$ of the phase space. If the system evolves in a bounded region, all the trajectories visit a finite number of different cells, each one identified by a symbol. With the finite number of symbols $\{s\}_{\mathscr{A}}$ enumerating the cells of the partition, the time-discretized trajectory $\mathbf{x}(t_j)$ determines a sequence $\{s_1, s_2, s_3, ...\}$, whose meaning is clear: at the time t_j the trajectory is in the cell labelled by s_j, see figure 7.5.

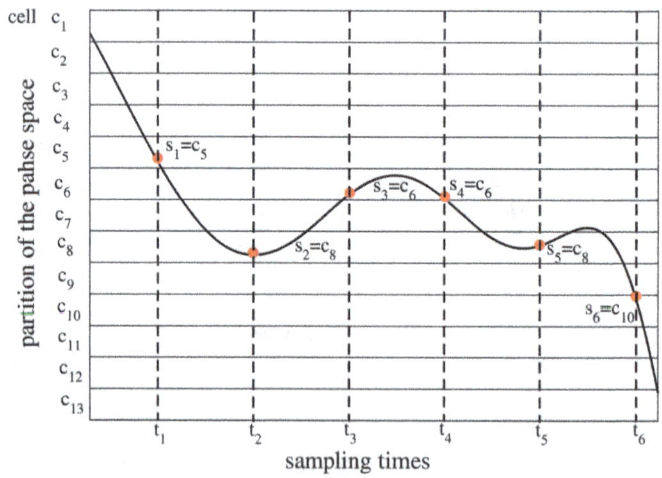

Figure 7.5. A one-dimensional trajectory $x(t)$, sampled at times $t_j = j\tau$, with $j = 1, \ldots, 6$, and the finite partition $\mathscr{A}$ of the phase space in cells, identified by the symbols $\{s\}_{\mathscr{A}} = \{c_1, \ldots, c_{13}\}$. The time-discretized trajectory $x(t_j)$ (red dots) determines the sequence $\{c_5, c_8, c_6, c_6, c_8, c_{10}\}$. The subsequences of length $N\tau$, e.g. with $N = 4$, define the blocks $W_1^4(\mathscr{A}) = (c_5, c_8, c_6, c_6)$, $W_2^4(\mathscr{A}) = (c_8, c_6, c_6, c_8)$, $W_3^4(\mathscr{A}) = (c_6, c_6, c_8, c_{10})$.

To each subsequence of length $N\tau$ one can associate a block of length N: $W_j^N(\mathscr{A}) = (s_j, s_{j+1}, \dots, s_{j+N-1})$. If the system is ergodic from the observed frequencies of the blocks one obtains the probabilities from which one calculates the block entropies $H_N(\mathscr{A})$:

$$H_N(\mathscr{A}) = -\sum_{\{W^N(\mathscr{A})\}} P[W^N(\mathscr{A})] \ln P[W^N(\mathscr{A})]. \tag{7.25}$$

The entropy per unit time of the trajectory with respect to the partition $\mathscr{A}$, $h(\mathscr{A})$, is defined as follows:

$$h_N(\mathscr{A}) = \frac{1}{\tau}[H_{N+1}(\mathscr{A}) - H_N(\mathscr{A})], \tag{7.26}$$

$$h(\mathscr{A}) = \lim_{N\to\infty} h_N(\mathscr{A}) = \frac{1}{\tau} \lim_{N\to\infty} \frac{1}{N} H_N(\mathscr{A}). \tag{7.27}$$

The KS-entropy is defined as the supremum of $h(\mathscr{A})$ over all possible partitions [10]

$$h_{KS} = \sup_{\mathscr{A}} h(\mathscr{A}). \tag{7.28}$$

It is not simple at all to determine h_{KS} according to the above definition. A more tractable way to define h_{KS} is based upon considering the partition $\mathscr{A}_\varepsilon$ made up by a grid of cubic cells of edge ε, from which one has

$$h_{KS} = \lim_{\varepsilon\to 0} h(\mathscr{A}_\varepsilon). \tag{7.29}$$

We expect that $h(\mathscr{A}_\varepsilon)$ becomes independent of ε when $\mathscr{A}_\varepsilon$ is fine enough. Since the computation of h_{KS} involves the limit of arbitrary fine resolution and infinite times, it turns out that, for most systems, it is not possible to compute h_{KS}. Nevertheless, by relaxing the strict requirement of reproducing a trajectory with arbitrary accuracy, one can introduce the ε-entropy which measures the amount of information for reproducing a trajectory with accuracy ε in phase space.

We consider a continuous variable $\mathbf{x}(t) \in \mathbb{R}^d$, which represents the state of a d-dimensional system; we discretize time by introducing an interval τ and we consider the new variable

$$\mathbf{X}^{(m)}(t) = (\mathbf{x}(t), \mathbf{x}(t+\tau), \dots, \mathbf{x}(t+(m-1)\tau)). \tag{7.30}$$

Introduce now a partition of the phase space in $\mathbb{R}^d$, using cells of edge ε in each of the d directions; each $\mathbf{X}^{(m)}(t)$ can be coded into a block of length m, out of a finite alphabet:

$$\mathbf{X}^{(m)}(t) \longrightarrow W^m(\varepsilon, t) = (i(\varepsilon, t), i(\varepsilon, t+\tau), \dots, i(\varepsilon, t+m\tau-\tau)), \tag{7.31}$$

where $i(\varepsilon, t+j\tau)$ labels the cell in $\mathbb{R}^d$ containing $\mathbf{x}(t+j\tau)$. From the time evolution of $X^{(m)}(t)$ one obtains, under the hypothesis of ergodicity, the probabilities $P(W^m(\varepsilon))$ of the admissible blocks $\{W^m(\varepsilon)\}$. The (ε, τ)-entropy per unit time, $h(\varepsilon, \tau)$, is defined as follows:

$$h_m(\varepsilon, \tau) = \frac{1}{\tau}[H_{m+1}(\varepsilon, \tau) - H_m(\varepsilon, \tau)], \tag{7.32}$$

$$h(\varepsilon, \tau) = \lim_{m\to\infty} h_m(\varepsilon, \tau) = \frac{1}{\tau} \lim_{m\to\infty} \frac{1}{m} H_m(\varepsilon, \tau), \tag{7.33}$$

where H_m is the entropy of blocks of length m:

$$H_m(\varepsilon, \tau) = -\sum\nolimits_{\{W^m(\varepsilon)\}} P(W^m(\varepsilon)) \ln P(W^m(\varepsilon)). \tag{7.34}$$

We underline that the ε-entropy is not a mere way to have a proxy of the 'true' entropy, i.e. $h_{\rm KS}$. In systems with many different characteristic times, the shape of $h(\varepsilon)$ has a precise interest and it depends on the dynamics ruling the evolution [11, 12]. For instance in fully developed turbulence, in a suitable range of ε, related to the so called inertial range, one has $h(\varepsilon) \sim \varepsilon^{-3}$, and the exponent -3 follows from the multifractal structure of the turbulence [11].

7.2.3 Information and entropy production

Let us briefly discuss how the approach used for the Shannon entropy, as well the Kolmogorov–Sinai entropy, and the ε-entropy, can be useful for the computation of the entropy production introduced in chapter 5.

For a given resolution ε and sampling time τ we can introduce an ε, τ- entropy production $\Sigma(\varepsilon, \tau)$ which is determined by the the probabilities of the blocks

$$W_N(\varepsilon, \tau) = (i_1(\varepsilon, \tau), i_2(\varepsilon, \tau), \ldots, i_{N-1}(\varepsilon, \tau), i_N(\varepsilon, \tau)),$$

and also of the reverse blocks

$$W_N^I(\varepsilon, \tau) = (i_N(\varepsilon, \tau), i_{N-1}(\varepsilon, \tau), \ldots, i_2(\varepsilon, \tau), i_2(\varepsilon, \tau)),$$

in the limit of large N.

The protocol for the computation of $\Sigma(\varepsilon, \tau)$ is conceptually simple, however, one has to face severe limitations when applying it. The origin of the difficulties are clear and already well known from the 1980s to researchers working in chaotic dynamical systems: e.g. only in few simple cases can the computation of the block entropy H_N be carried on, because, in general, it needs an enormous amount of data, namely exponentially large with N [11].

A simple way to obtain a proxy $\Sigma_p(\varepsilon, \tau)$ of $\Sigma(\varepsilon, \tau)$ is taking the saturation value for $N = 2$; this corresponds to assuming that the system is Markovian, so one has

$$\Sigma_p(\varepsilon, \tau) = \frac{1}{\tau} \sum_{I,J} P_i^{inv}(\varepsilon, \tau) P_{i\to j}(\varepsilon, \tau) \ln \frac{P_{i\to j}(\varepsilon, \tau)}{P_{j\to i}(\varepsilon, \tau)}.$$

We conclude this section noting that, in data analysis, the space where the state vectors of the system live is not known. Moreover, usually only a scalar variable $u(t)$ can be measured. In such a situation, as a special case of equation (7.30), one considers the vectors $\mathbf{Y}^m(t) = (u(t), u(t+\tau), \ldots, u(t+m\tau-\tau))$, that live in $\mathbb{R}^m$. In a deterministic system the vector $\mathbf{Y}^m(t)$ allows us a reconstruction of the original phase space (that is, reconstruction of topologically equivalent geometrical features of the dynamics), known as delay embedding in the literature [13, 14].

It is natural to wonder about the possibility to extend the above result to stochastic processes. For instance, an interesting question is whether partial information, e.g. instead of the whole vector $\mathbf{X}(t)$ just one component, say $X_1(t)$, allows us the computation of the entropy production, or, at least, to have some qualitative information. Unfortunately, in the case of Gaussian stochastic processes the scenario is rather different from that existing for deterministic systems, and it is possible to show a no-go theorem: given a stationary Gaussian stochastic process $\mathbf{X}(t) \in \mathbb{R}^D$, the proxy of the entropy production computed with the variable $X_1(t)$ is always zero[4]. This means that even if the vector $\mathbf{X}(t)$ describes a non-equilibrium system and the true Σ is positive, the incomplete knowledge of the state does not allows us to understand the equilibrium, or non-equilibrium, character of the system under investigation (see reference [15], where the example of the Brownian gyrator introduced in chapter 3 is discussed).

7.3 Causation

Let us introduce the problem of the detection of causation: given two variables $\{x_t\}$ and $\{y_t\}$, one wants to understand whether, and how much, the variable $\{x_t\}$ influences $\{y_t\}$. As an important example of causation we can mention, in the context of the climate change, the debated relationship between temperature and CO_2.

The modern study of causation begins with David Hume; we cannot enter into a detailed philosophical discussion [16], however, we want to mention the following interesting remark by Bertrand Russell: *[Hume] supposes the law to state that there are propositions 'A causes B' where A and B are classes of events; the fact that such laws do not occur in any well-developed science appears unknown to philosophers.* Let us note that, at a fundamental level, from the third law of Newton we have that in Nature there is no separation between cause and effect. This is a consequence of the fact that in physics, both classical and quantum, there are only interactions, and these are always mutual; for a nice general discussion about causation see reference [16].

In the following we ignore such a delicate aspect and focus on the practical problem: how to understand if A $\to$ B? And how to quantify the 'degree of causation'? The first natural idea, according to the latin saying *cum hoc ergo propter hoc* (with this, therefore because of this), is to *look at the correlation* between x and y. On the other hand, it is well known that correlation does not imply causation [17]; among the many hilarious examples we mention the strong correlation between: (a) country's chocolate consumption and Nobel Prize victories; (b) shark attacks and icecream sales. Of course in the last case the origin of the strong correlation is rather obvious: when the temperature increases more people want icecream and to go swimming, and the number of shark attacks increases with the number of swimmers.

One of the first, and most popular, attempts to go beyond the correlation is due to Granger, and it is based on the idea that the variable $x^{(j)}$ influences $x^{(k)}$ if the knowledge of the past history of $x^{(j)}$ enhances the ability to predict future values of

[4] The same result holds for any signal $y(t) = \sum_j a_j X_j(t)$, independently of the choice of $a_1, \dots, a_m$.

$x^{(k)}$ [18]. Another approach based on the information theory, uses the transfer entropy [19], namely a degree of information exchange from $x^{(j)}$ to $x^{(k)}$ which quantifies the loss of information about $x^{(k)}$ that one measures if $x^{(j)}$ is ignored.

Remarkably, the Granger causality and the transfer entropy have been shown to be equivalent in linear Gaussian autoregressive systems. In more recent studies, some authors introduced protocols which use methods and ideas borrowed from the theory of information and dynamical systems [19].

It is appropriate to underline the difference between two methodologies in the approach to the problem, namely interventional and observational causation. We say that two variables can be assumed to be in a cause–effect relationship if an external action on one of them results in a change of the observed value of the second: we call interventional this physics-inspired definition of the cause–effect relation. The observational approach is based on tests to determine whether, and to what extent, the knowledge of a certain variable is useful to the actual determination of future values of another [20].

Let us compare the two approaches discussing the measurement of the electrical current passing trough a resistor, whose extremities are connected to an external electric potential $v(t)$. If the amperometer is affected by some noise $\eta(t)$ independent of $v(t)$, the measured current is given by

$$J_{\text{meas}}(t) = J_{\text{true}}(t) + \eta(t) = Gv(t) + \eta(t),$$

where J_{true} is the (unknown) value of the current and G is the electrical conductance. A good estimator of the interventional causality between $v(t)$ and the measured J_{meas} will only depend on the conductance G since this parameter establishes to which extent an external action on $v(t)$ will influence the observed value of the current, J_{meas}; note that G does not depend on the intensity of the noise. Conversely, from an observational perspective, the amplitude of the noise also plays a role, since our ability to predict future values of J_{meas} given $v(t)$, crucially depends on it. If the noise is small, the knowledge of $v(t)$ will suffice to give a good estimate of J_{meas}; conversely, if it is large, the information about $v(t)$ is almost useless.

7.3.1 Causation and linear response theory

Let us now show how linear response theory can allow us to understand causal links (in the interventional sense) from time series of data [20]. We start assuming that the process under investigation is of Markov type, and the state of the system is given by the vector $\mathbf{x}_t = (x_t^{(1)}, x_t^{(2)}, \dots, x_t^{(n)})$, where t is a (discrete) time index. We say that $x^{(j)}$ influences $x^{(k)}$ if a perturbation on the variable $x^{(j)}$ at time 0, $x_0^{(j)} \to x_0^{(j)} + \delta x_0^{(j)}$, induces, on average, a change on $x_t^{(k)}$, with $t > 0$. In formulae, we will say that $x^{(j)}$ has an influence on $x^{(k)}$ if

$$R_t^{kj} = \frac{\overline{\delta x_t^{(k)}}}{\delta x_0^{(j)}} \neq 0 \qquad \text{for some } t > 0, \tag{7.35}$$

where the over-line represents an average over many realizations of the experiment.

The idea to use the linear response theory is reminiscent of the framework developed by J Pearl, in which causation is detected by observing the effects of an action on the system: *Interventions and counterfactuals are defined through a mathematical operator called do(x), which simulates physical interventions by deleting certain functions from the model, replacing them with a constant X = x while keeping the rest of the model unchanged* [21].

In chapter 4 we discussed in detail the computation of the response functions, in particular we saw that, provided that the stationary probability distribution $P_s(\mathbf{x})$ is known, it is possible to compute the response functions from suitable correlation functions. On the other hand, the determination of $P_s(\mathbf{x})$ from data is a non-trivial task, in particular for high-dimensional systems.

First we consider the case of linear stochastic processes of the form

$$\mathbf{x}_{t+1} = A\mathbf{x}_t + B\eta_t, \tag{7.36}$$

where A and B are constant $n \times n$ matrices, and η is an n dimensional uncorrelated (non necessarily Gaussian) noise. In such a class of systems the following relation between the response matrix and the correlation matrix with entries $C_t^{kj} = \langle x_t^{(k)} x_0^{(j)} \rangle$ holds:

$$R_t = C_t C_0^{-1}, \tag{7.37}$$

where C_0^{-1} is the inverse of C_0. The above relation has been discussed in chapter 4 in the case of linear Langevin equations, but it also holds for a generic uncorrelated noise.

From the previous relation we have that if we know the proper variables which describe the system, i.e. the vector $\mathbf{x}_t$, it is possible to understand causation from correlations. Of course the result holds also in cases with continuous time. In order to compare the effect of different 'causes', it may be useful to rescale correlations and responses with the standard deviations of the corresponding variables:

$$\tilde{C}_t^{kj} = \frac{1}{\sigma_k \sigma_j} C_t^{kj}, \quad \tilde{R}_t^{kj} = \frac{\sigma_j}{\sigma_k} R_t^{kj}. \tag{7.38}$$

Let us discuss in an explicit case the results on causation comparing the response theory approach and the other methods; consider the following system

$$x_{t+1} = ax_t + \varepsilon y_t + b\eta_t^{(x)} \tag{7.39a}$$

$$y_{t+1} = ax_t + ay_t + b\eta_t^{(y)} \tag{7.39b}$$

$$z_{t+1} = ax_t + az_t + b\eta_t^{(z)} \tag{7.39c}$$

where $\eta^{(x)}$, $\eta^{(y)}$, $\eta^{(z)}$ are independent Gaussian processes with zero mean and unitary variance, while a, ε and b are constant parameters.

The case $\varepsilon = 0$ is a minimal example in which the behaviour of two quantities, y and z, is influenced by a common-causal variable x; as a consequence, y and z are correlated even if there is no causal relationship between them.

The same mechanism may be identified in many cases in which surprising correlations arise, such as that between the number of Nobel laureates of a country and its chocolate consumption per year: in this specific case, both quantities may be expected to be roughly influenced by the gross domestic product of the nation.

According to our approach, in order to decide whether there is a causal relation between y and z, one perturbs y at time 0 and measures the average variation δz_t for $t > 0$, see figure 7.6. For $\varepsilon = 0$, not surprisingly, $\tilde{R}_t^{zy} = 0$ for any t. The situation changes if we allow for a small feedback $\varepsilon \neq 0$ from y to x, which will result in a causal link between y and z. The corresponding response function correctly reveals that the behaviour of z starts to be influenced by a perturbation of y after two time steps, and that the intensity of such causal influence roughly scales with ε.

We note that none of the above conclusions could have been drawn from the simple analysis of the correlation functions. In addition, both Granger and transfer entropy approaches would lead to a null causal dependence of z from y for every choice of ε; indeed, once the trajectory of x is known, the knowledge of y does not add any useful information to predict future values of z.

Of course the above discussed approach for linear systems is not completely general and it is natural to wonder about the effects of nonlinear contributions to the dynamics. Numerical tests show that the Gaussian approximation (7.37) still gives meaningful information about the causal relations between the variables of the system [20].

Let us summarize differences and advantages of the approach to causation via linear response theory with respect to other approaches:

- it is easy to use;
- at variance with Granger causality and transfer entropy, which are based on the idea of prediction and information, the method built on linear response

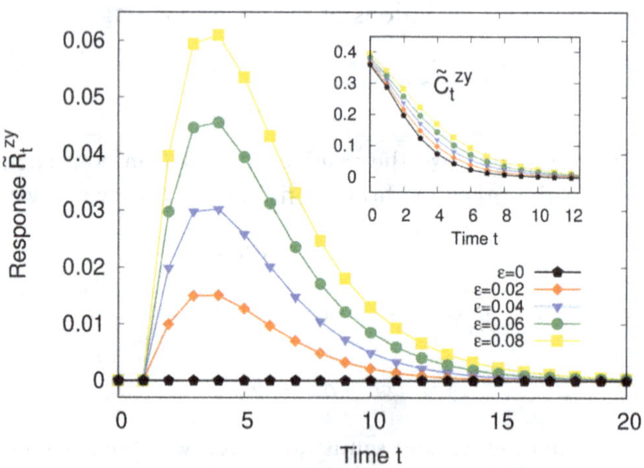

Figure 7.6. Response of z when y is perturbed. The inset plots the corresponding correlations. Several values of ε are considered; in all cases, $a = 0.5$, $b = 1$. Each plot has been obtained with an average over 10^5 trajectories; responses have been computed inducing an initial perturbation $\delta y_0 = 0.01$. Reprinted figure with permission from reference [20], copyright (2020) by the American Physical Society. CC BY 4.0.

has an objective nature, i.e. it is founded, as in Pearl's approach, on the understanding about how the system behaves after a certain operation;
- it is able to find the 'causation link' even in non-trivial cases;
- it allows for a global, as well time-dependent, characterization of causation.

7.3.2 Difficulties in the treatment of data

Let us briefly discuss the rather common situation in which we do not know the whole state vector **x** of the system. We can understand the difficulties one has to face, discussing the case where we only have access to the time series of a limited number of variables, e.g. when the unique information we have comes from y and z of the model (7.39).

A first obvious attempt is to consider $\Gamma_t^{(2)} = (y_t, z_t)$ as the proper vector to describe the system, but this leads to wrong results: for instance, the proxy response computed by applying equation (7.37) to $\Gamma_t^{(2)}$ turns out to be nonzero even in the case $\varepsilon = 0$; also for $\varepsilon > 0$ the results are rather inaccurate, see figure 7.7. The reason of this failure is that $\Gamma_t^{(2)}$ is not able to completely describe the state of the system.

It is tempting, following the approach of Takens, to build a vector of dimension $2d$ by looking at the previous times, e.g.

$$\Gamma_t^{(2d)} = (y_t, y_{t-1}, \dots, y_{t-d+1}, z_t, z_{t-1}, \dots, z_{t-d+1}),$$

and then to use the protocol based on the fluctuation–dissipation theorem, as if $\Gamma_t^{(2d)}$ would describe the whole system we are interested in [13]. Figure 7.7 shows how upon increasing d there is not an improvement of the result.

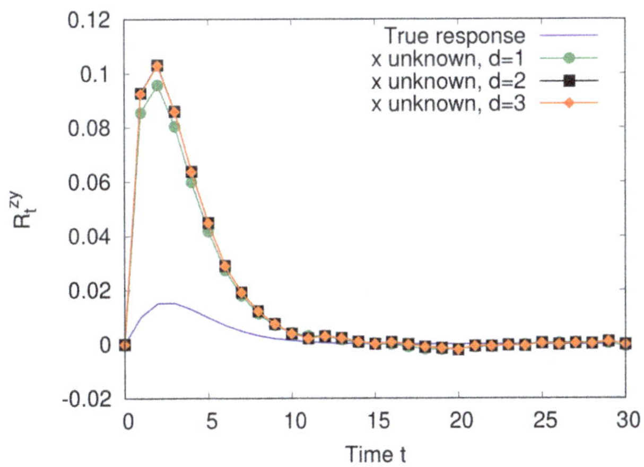

Figure 7.7. The response function of model (7.39) with $a = 0.5$, $b = 1.0$, $\varepsilon = 0.02$. The blue solid line shows the actual response function measured from simulations; the other curves, marked with different symbols, represent the results obtained from the only knowledge of correlations between y and z, as if x was not part of the system, with different embedding dimensions. All curves have been obtained with an average over 10^5 trajectories. Reprinted figure with permission from reference [20], Copyright (2020) by the American Physical Society. CC BY 4.0.

The failure of the approach based on the attempt to generalize the Takens methods to stochastic processes is due to the following negative result: *it is possible to show that, at variance with deterministic dynamics, the 'embedding' procedure for stochastic processes is not able to reconstruct a finite-dimensional vector whose dynamics is equivalent to that of the original system* [20].

In other words, it is possible to infer causation relations from time series of a stochastic dynamics only if we know all the variables which are relevant to the dynamics: in the above specific case not only y and z, but also the knowledge of x is necessary for a good description of the causal relations.

7.3.3 Toward an understanding of causal relations of paleoclimate dynamics

Although, as previously discussed, in general an incomplete knowledge of the vector which describes the system does not allow for a good understanding of causation, at least in some cases there is the possibility to arrive to some results [22]. Let us consider again the paleoclimate dynamics, and assume that it is described by the $2d$ vector $\mathbf{x} = (x_1, x_2) =$ (Temperature, CO_2) whose dynamics has the shape

$$\frac{d\mathbf{x}}{dt} = \mathscr{A}\mathbf{x} + \mathbf{f}(t) + \text{noise},$$

where $\mathbf{f}(t)$ indicates the astronomical and tectonic forcing.

We consider the case such that the matrix $\mathscr{A}$ and $\mathbf{f}$ are not known, and we have only the time series of $x_1(t)$ and $x_2(t)$.

In figure 7.8 we can note the clear correlation between the temperature and CO_2, as well the role of the flux of energy by the Sun, in agreement with the intuition of Milankovitch. Of course, since x_1 and x_2 are both driven by the same forcing their strong correlation is not necessarily significative, therefore in order to master the 'internal dynamics', we have to 'remove' the effects of the forcing on x_1 and x_2.

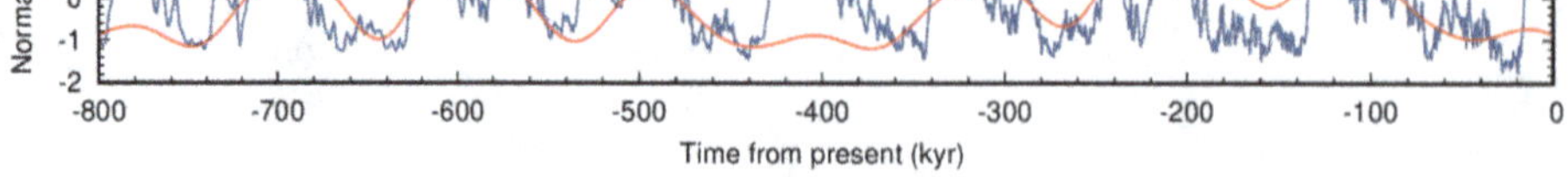

Figure 7.8. The deviation from average of temperature and CO_2 as a function of time (blue curves), normalized by the standard deviation. The integrated annual insolation at the 65th north parallel, presenting the typical timescales of the external driving, is also reported (red curves). Reprinted by permission from [22] Springer Nature. CC BY 4.0.

Our aim is to understand the causation relations $x_1 \to x_2$ and $x_2 \to x_1$ only from the time series of temperature and CO_2; we know that due to the negative result on the embedding technique, in the general case such a problem is not well posed. However, there is at least a case where it is possible to use the approach to causation in terms of linear response, namely if the characteristic time τ_I of the internal dynamics (given by $\mathscr{A}$) is much smaller than the typical time τ_f of the forcing.

Let us introduce the protocol to remove the effect of the forcing: we perform a filtering procedure

$$\mathbf{x}_S(t) = (G_\tau{}^*\mathbf{x})(t) = \int G_\tau(t - t')\mathbf{x}(t')\, dt',$$

where the filter $G_\tau(t - t')$ has a maximum at $t = t'$, is very small at large values of $|t - t'|/\tau$ and $\int_{-\infty}^{\infty} G_\tau(t - t')\, dt' = 1$, e.g. $G_\tau(t - t') = \frac{1}{\sqrt{2\pi\tau^2}} e^{-\frac{(t-t')^2}{2\tau^2}}$. If $\tau \ll \tau_f$ we can introduce the fast variables

$$\mathbf{x}_F(t) = \mathbf{x}(t) - \mathbf{x}_S(t).$$

In the case of timescale separation, $\tau_I \ll \tau \ll \tau_f$, it is possible to detect the internal dynamics, and one can show that

$$\frac{d\mathbf{x}_F}{dt} = \mathscr{A}\mathbf{x}_F + \text{noise}.$$

Therefore, one can use the previously discussed approach in terms of fluctuation/response relation to understand the causation relations $x_1 \to x_2$ and $x_2 \to x_1$ only from the time series [20].

In figure 7.9 we show the results: there is fair evidence that the causation $CO_2 \to T$ is more relevant than the $T \to CO_2$; we can say that the filtering procedure, at least in this case, is able to 'disentangle' the variables from the common forcing [20].

For comparison we also show the results obtained with the transfer entropy, which is an indicator of the information which a given time-dependent signal $x_1(t)$ provides about a second variable $x_2(t)$: basically one measures how much information is lost about the distribution of $x_2(t)$ when the knowledge of $x_1(t)$ is ignored. For a Markovian system, the transfer entropy $T_{1\to 2}$ with lag t is defined as

$$T_{1\to 2}(t) = H(x_2|x_2)(t) - H(x_2|x_1, x_2)(t),$$

where

$$H(y|x)(t) = -\int dxdyP(x, 0; y, t)\ln P(x, 0; y, t) + \int dxP(x)\ln P(x)$$

is the conditional Shannon entropy. Here $P(x)$ represents the marginal probability density function for the variable x, while $P(x, 0; y, t)$ is the joint distribution of x at time 0 and y at time t. If the dynamics is Gaussian the above relations can be simplified: it can be shown that $T_{i\to j}$ is a function of two-point correlation functions [19].

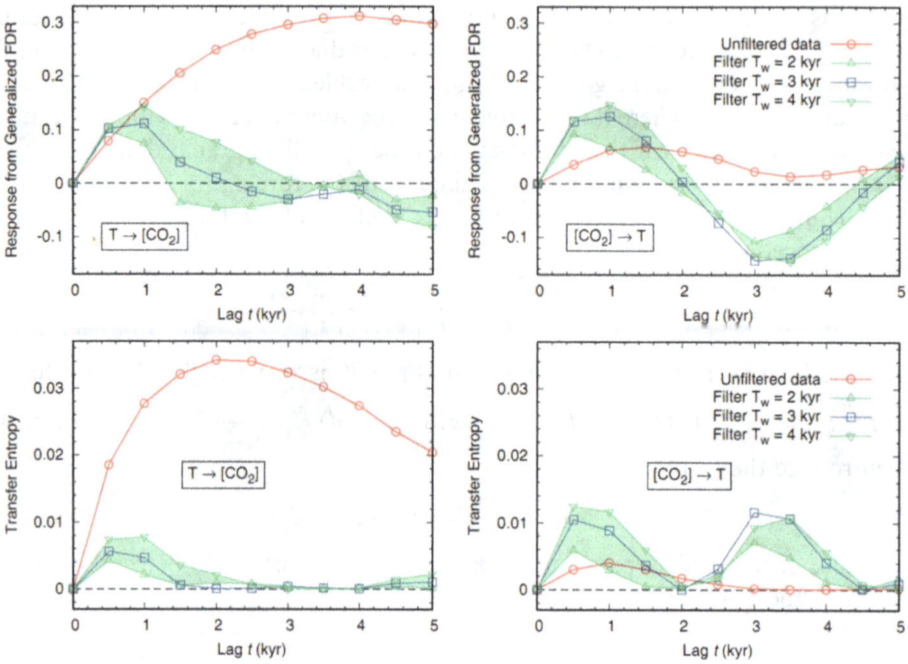

Figure 7.9. In the top panels the response function, computed according to the generalized fluctuation–dissipation theorem, is plotted. The analytical formula is given by the the linear response matrix, where x and y are the normalized CO_2 and T signals shown in figure 7.8. Panel (c) refers to the effect of T on CO_2, while panel (d) shows the opposite relation. Red circles represent the results of a direct application of the method on raw data, apparently suggesting that $T \to CO_2$ is stronger than $CO_2 \to T$. The response on data filtered over $\tau = 3$ kyr window (bluesquares) instead indicates that the impact of CO_2 on T becomes larger. The result is robust with respect of τ variations by one kyr (green up/down triangles). Reprinted by permission from [22] Springer Nature. CC BY 4.0.

Let us conclude this section with a recap of the main results.

- In spite of the (correct) common-wisdom statement *correlation does not imply causation*, if we know the state of systems, i.e. the vector $\mathbf{x} = (x^{(1)}, \ldots, x^{(N)})$, a proper employ of time correlations and fluctuation–response theory allows us to understand the causal relations between the variables of a multi-dimensional linear Markov process.
- The fluctuation–response formalism can be used both to find the direct causal links between the variables of a system and to introduce a degree of causation, cumulative in time, whose physical interpretation is straightforward.
- In the absence of the complete knowledge of $\mathbf{x}$, e.g. if one only has access to the time series of some components $\{x_t^{(i)}\}$, in general it is not possible to use the embedding methodology.
- In some cases where the variables are driven by a common external forcing, in the presence of timescale separation, it is possible to treat time series and to obtain some information about the causal relations.

We conclude this chapter stressing again that in the treatment of time series we have to face a difficult problem, which is quite common in all the attempts dealing with data, i.e. the selection of the proper variables. In a few words, we can say that only if we are able to select the 'good' variables is there the possibility for a fair understanding.

References

[1] Imkeller P and Von Storch J-S (ed) 2001 *Stochastic Climate Models* (Berlin: Springer Science & Business Media)

[2] Milankovitch M K 1941 Kanon der erdbestrahlung und seine anwendung auf das eiszeitenproblem *R. Serb. Acad. Spec. Publ.* **133** 1–633

[3] Hasselmann K 1976 Stochastic climate models part I. Theory *Tellus* **28** 473–85

[4] Benzi R, Parisi G, Sutera A and Vulpiani A 1982 Stochastic resonance in climatic change *Tellus* **34** 10–6

[5] Gardiner C 2009 *Stochastic Methods* (Berlin: Springer)

[6] Kramers H A 1940 Brownian motion in a field of force and the diffusion model of chemical reactions *Physica* **7** 284–304

[7] Gammaitoni L, Hänggi P, Jung P and Marchesoni F 1998 Stochastic resonance *Rev. Mod. Phys.* **70** 223

[8] Shannon C E 1948 A mathematical theory of communication *Bell Syst. Tech. J.* **27** 379–423

[9] Khinchin A Y 2013 *Mathematical Foundations of Information Theory* (North Chelmsford, MA: Courier Corporation)

[10] Eckmann J-P and Ruelle D 1985 Ergodic theory of chaos and strange attractors *Rev. Mod. Phys.* **57** 617

[11] Boffetta G, Cencini M, Falcioni M and Vulpiani A 2002 Predictability: a way to characterize complexity *Phys. Rep.* **356** 367–474

[12] Gaspard P and Wang X-J 1993 Noise, chaos, and (ε, τ)-entropy per unit time *Phys. Rep.* **235** 291–343

[13] Takens F 2006 Detecting strange attractors in turbulence *Dynamical Systems and Turbulence, Warwick 1980: Proc. Symp. held at the University of Warwick 1979/80* ed D Rand and L S Young (Berlin: Springer) 366–81 pp

[14] Kantz H and Schreiber T 2004 *Nonlinear Time Series Analysis* (Cambridge: Cambridge University Press)

[15] Lucente D, Baldassarri A, Puglisi A, Vulpiani A and Viale M 2022 Inference of time irreversibility from incomplete information: linear systems and its pitfalls *Phys. Rev. Res.* **4** 043103

[16] Aurell E and Del Ferraro G 2016 Causal analysis, correlation-response, and dynamic cavity *J. Phys.: Conf. Ser.J. Phys.* **699** 012002

[17] Calude C S and Longo G 2017 The deluge of spurious correlations in big data *Found. Sci.* **22** 595–612

[18] Granger C W J 1969 Investigating causal relations by econometric models and cross-spectral methods *Econometrica: J. Econometric Soc.* **37** 424–38

[19] Bossomaier T, Barnett L, Harré M and Lizier J T 2016 *An Introduction to Transfer Entropy: Information Flow in Complex Systems* (Berlin: Springer)

[20] Baldovin M, Cecconi F and Vulpiani A 2020 Understanding causation via correlations and linear response theory *Phys. Rev. Res.* **2** 043436
[21] Pearl J 2012 The do-calculus revisited arXiv preprint arXiv: 1210.4852
[22] Baldovin M, Cecconi F, Provenzale A and Vulpiani A 2022 Extracting causation from millennial-scale climate fluctuations in the last 800 kyr *Sci. Rep.* **12** 15320

Chapter 8

Granular and active matter

Perhaps, by staring at the sand as sand, words as words, we can come close to understanding how and to what extent the world that has been ground down and eroded can still find in sand a foundation and model[a].

Collection of sand, Italo Calvino (translated by M McLaughlin)

8.1 Introduction

Parts of this section have been reprinted from [6] copyright (2015), with permission from Springer Nature.

Granular matter is distinct from conventional molecular matter primarily due to the size of its elementary constituents. Grains, being macroscopic, typically have a linear dimension rarely smaller than 0.1 mm. This scale allows them to be governed by the principles of classical mechanics, complemented by dissipative interactions. When grains collide, the kinetic energy associated with their centres of mass is converted into internal energy, i.e. heat, which is then rapidly dissipated into the environment. Essentially, a portion of the kinetic energy is lost from the system's description. These dissipative interactions have several implications, the most fundamental of them being the breakdown of time-reversal symmetry. Additionally, the mass of a grain is of the order of 10^{20} times that of a typical molecule, meaning the kinetic or potential energy of a grain is vastly larger than molecular thermal energy. Consequently, the ambient temperature has a negligible effect on the dynamics of grains, allowing them to be effectively considered as being at $T = 0$.

[a] 'Forse fissando la sabbia come sabbia, le parole come parole, potremo avvicinarci a capire come e in che misura il mondo triturato ed eroso possa ancora trovarvi fondamento e modello.'

In the kinetic theory of granular gases, the grains themselves act as the 'microscopic degrees of freedom', and a 'granular temperature' is introduced based on the kinetic energy of the grains. To initiate and sustain the motion of granular particles, some form of 'thermostat' is required. The most common method of energy injection into a granular system is through the application of forces to the container, often by moving (e.g. vibrating) it. The container's motion is then transferred to the grains through collisions with the container walls. Furthermore, for many practical purposes, grains can be treated as rigid bodies, meaning the volume occupied by one grain is excluded from the volume available to the others. When the total occupied volume becomes a significant fraction of the total available volume, this exclusion leads to important consequences, including geometrical frustration, strong spatial correlations, the dominance of collisional transport over streaming transport, and an increase in recollisions within kinetic equations, leading to the breakdown of molecular chaos, among other effects.

Given these essential characteristics of granular matter, it is customary to refer to two extreme 'limit states': granular solids and granular gases. Real granular materials typically exist in an intermediate state between these two extremes, depending on external conditions, available volume, driving intensity, degree of inelasticity, presence of interstitial fluids, and other factors. The experimental and theoretical tools used to study granular solids, such as elastoplastic continuum models, can be quite different from those applied to granular gases, where kinetic theory is more commonly employed.

Over the past 40 years, three main categories of granular problems have been identified in the literature: (1) stable or metastable granular issues encompass the study of internal force distribution and correlation analysis within a pile or silo of grains, the characterization of sound propagation in densely packed arrays, the slow compaction dynamics observed under tapping (where grains can remain in a metastable state, far from the minimum attainable packing fraction, in the absence of vibration), and the examination of the time and size distributions of avalanches in a pile that has reached its critical slope [1]; (2) slow granular flows, where particles remain in prolonged contact with their neighbours, interacting through friction: this is known as the 'quasi-static' regime of granular flow, and it is typically studied using modified plasticity models based on Coulomb's friction criterion [2]; and finally, (3) the rapid-flow regime, corresponding to high-speed flows [3, 4]. This is also known as the granular fluid regime.

In the last category, discussed more in detail in this chapter, particles move more freely and independently, rather than in large, many-particle blocks. In this regime, a particle's velocity can be decomposed into the mean velocity of the bulk material and a seemingly random component that describes its motion relative to the mean. The analogy between the random motion of granular particles and the thermal motion of molecules in kinetic theory is strong. Building on this analogy, the mean-square value of these random velocities is referred to as 'granular temperature'—a term first introduced by Ogawa [5]. As the stationary velocity of the flow increases (due to greater external driving forces), shear work generated by internal friction raises the granular temperature and pressure, leading to a decrease in the volume fraction occupied by the grains [4]. This suggests that rapid flows tend to be dilute, making theoretical approaches from kinetic theory, as well as hydrodynamic descriptions, applicable and sometimes successful. A variety of fluid-like experiments have been conducted on granular systems, including those involving Couette cells, inclined channels, and rotating drums, revealing nonlinear constitutive relations. High-amplitude vibrations can induce interesting convection phenomena, such as size and density segregation. Patterns, like two-dimensional standing waves, can form on the free surface of a vibrated granular layer. The study of simulated models has introduced new questions regarding the constitutive behaviour in rapid flows. Recent experiments and numerical studies have focused on this area, measuring velocity probability distribution functions and discovering that, in many cases, these distributions are non-Gaussian. The examination of internal stress fluctuations and velocity structure factors has provided additional insights, leading to refinements in granular kinetic theories. A debate has emerged regarding the limits of applying hydrodynamic formalism in this context [6].

A privileged experimental setup for granular fluids is vibro-fluidization. When a granular medium is subjected to vertical vibration, particularly under the influence of gravity, a range of intriguing phenomena can be observed. As mentioned earlier, slow vibrations of a container filled with grains lead to a very gradual compaction of the material. However, when the amplitude of the vibration is sufficiently strong, specifically when

$$\Gamma = \frac{a_{\max}}{g} > 1,$$

where g is the gravity acceleration and $a_{\max}$ is the maximum acceleration of the vibrating plate (e.g. $a_{\max} = A\omega^2$ for a plate vibrating harmonically with amplitude A and frequency ω), the granular medium exhibits several remarkable behaviours. We list below some of the most interesting ones.

Convection and segregation phenomena are widely documented in granular media contained within a shaken box [7]. Faraday [8] was one of the first to observe such effects. The geometry of the container can significantly influence the nature of convection. For instance, in a cylindrical container, grains near the walls might move downward while those in the centre move upward, whereas in an inverted cone, the direction of convection may reverse. Typically, larger grains, regardless of their density, tend to rise to the surface, leading to material segregation [9–13].

Another extensively studied phenomenon is the formation of patterns on the surface of vibrated layers of grains. Depending on various factors—such as the amplitude and frequency of vibration, the shapes and sizes of the grains, the size of the container, and the depth of the granular bed—different types of standing waves can emerge, resulting in unexpected and fascinating textures [14–17].

The formation of clusters has been extensively studied in experiments where steel balls roll on a smooth surface, which can be either horizontal or inclined and may include a vibrating side [18, 19]. These experiments focused on a monolayer of grains, ensuring a true 2D setup. In both horizontal and inclined scenarios, at sufficiently high global densities, the density distribution shifts from Poissonian to exponential, indicating strong clustering. High-density clusters have also been observed in vibrated cylindrical pistons [20–22]. As the number of particles in the cylinder increases, a transition occurs from gas-like behaviour to collective solid-like behaviour. A similar transition has been noted in fluidized beds [23], where vertically shaken granular monolayers transition from gas-like motion to a coexistence between a crystallized state and a surrounding gas as the vibration amplitude decreases.

The Leidenfrost effect in granular media appears in a stationary configuration observed under gravity and vertical shaking: a granular 'drop' at high packing fraction floats above a more dilute granular gas, which remains in direct contact with the vibrating boundary [24]. Such an effect is analogous to the liquid Leidenfrost effect, known since the 18th century, where a liquid drop skitters across a surface much hotter than its boiling point due to a thin vapour layer that insulates the drop from direct contact with the surface.

A systematic study of vertically vibrated quasi-two-dimensional containers (where the length and height are much larger than the depth) has led to the development of a robust phase diagram [25]. The experiment revealed a variety of phenomena, including bouncing beds, undulations, the granular Leidenfrost effect, convection rolls, and granular gas. These phenomena, and the transitions between them, are governed by several key control parameters: the shaking's maximum acceleration (Γ), the number of bead layers (F), the inelasticity parameter ($1 - r^2$, where r is the restitution coefficient, see later for a definition), and the aspect ratio (the ratio between the container's length and the height of the granular media at rest). More recent experiments have provided interesting insights into the effect of amplitude and frequency of vibration upon the linear and nonlinear response to external forces [26, 27].

Recently, the investigation of random lasers has been extended to shaken granular lasers [28, 29]. In general, random lasers involve pumping light through a scattering and amplifying random medium, resulting in an emitted light spectrum with random peaks that depend on various parameters of the scattering system. In a shaken granular laser, a cell containing glass or steel spheres ($\sim$1 mm in diameter) dispersed in a 'gain medium' such as a rhodamine solution is vibrated. Shaking allows for direct control over the statistical properties of the emitted spectra, with different shaking parameters leading to different stationary regimes, characterized by either more dilute or denser granular assemblies.

Experimental efforts have also focused on validating kinetic theory by studying hydrodynamic and kinetic fields (e.g. packing fraction profiles, granular temperature profiles, self-diffusion, and velocity statistics) in vertically vibrated boxes or vertical slices (2D setups) [30–33]. These studies have highlighted the challenges of imposing boundary conditions on existing kinetic models due to the presence of non-hydrodynamic boundary layers [34]. This has led to hypotheses about the scaling of granular temperature as a function of vibration amplitude [35, 36]. For more recent experimental results, see [37].

The question of the form of velocity distributions in granular media has prompted numerous experiments following numerical studies on granular rapid dynamics, which suggested deviations from Gaussian distributions. In [38], the velocity distributions were studied on an inclined plane at varying angles, revealing non-Gaussian statistics with enhanced high-energy tails. It was observed that as the angle of inclination increased, the distributions tended toward a Maxwellian form. In the experiments reported in [23, 39], a horizontally arranged granular monolayer subjected to vertical vibrations (with horizontal velocity measurements) revealed that in the presence of clustering, the velocity distributions deviate from the Gaussian norm, instead exhibiting nearly exponential tails. Another experiment [40], using a similar monolayer with vertical vibration, confirmed the theoretical predictions made in reference [41] regarding the high-energy tails of cooling and driven granular gases. The results showed exponential tails for the cooling gas and $\exp(-\text{const}\ v^{3/2})$ tails for the driven gas. More recently, in [42], velocity fluctuations in a vertically vibrated monolayer were measured, again yielding a velocity distribution with $\exp(-\text{const}\ v^{3/2})$ tails.

Several experiments have focused on the nonequilibrium properties of granular materials. Notably, two experiments have verified the breakdown of energy equipartition [43] and measured fluctuations in internal energy flow [44]. In the latter, the authors claimed to observe the Gallavotti–Cohen fluctuation theorem [45] (see chapter 5), although subsequent theoretical work questioned this claim [46]. More recently, experiments have measured the energy exchanges between a granular gas and a harmonic oscillator, both in the stationary state [47] and during periodic cycles of varying forces [48].

Experiments similar to those in reference [38] have revealed strong correlations between particle velocities [49]. More recent studies have measured velocity structure factors in a monolayer of spheres moving on a vertically vibrating horizontal rough plate, achieving good agreement with fluctuating granular hydrodynamics [50, 51]. The results showed that the measured velocity correlations are characterized by a correlation length that increases with the packing fraction. We recall that the packing fraction is the ratio between the volume occupied by grains and the total available volume. For instance, if N identical grains of volume V_g are confined in a total volume V, then the packing fraction is simply $\phi = NV_g/V$.

Linear response has been investigated in experiments where a Brownian rotator is suspended in a granular gas and subjected to a small torque. The first experiment [52] explored a very dense system over long timescales, finding that the fluctuation–dissipation theorem (FDT), similar to that in equilibrium, holds with an effective temperature. A more recent experiment in dilute and moderately dense configurations [53] highlighted the interplay between fast and slow timescales, which generates nonequilibrium correlations that grow with density, leading to the breakdown of the Einstein relation. The general problem of modelling fast and slow scales for the rotator dynamics in this kind of experiments has been deepened in [54, 55] and recently the issue of characterizing the irreversibility of such a dynamics has been studied in [56].

A series of experiments under gravity and vertical vibration, conducted in containers separated into communicating chambers (compartments), demonstrated the tendency of granular fluids to violate many entropic trends observed in equilibrium fluids [57]. Examples include phase separation (dense versus dilute) between compartments and the spontaneous segregation of mixtures into different compartments. These scenarios are often likened to the realization of a Maxwell demon experiment.

Conceptually similar to the Maxwell demon phenomenon mentioned above, the granular ratchet offers a striking illustration of the rectification of unbiased fluctuations under nonequilibrium conditions. While models have been proposed since reference [58], the first experimental realization was achieved in reference [59]. More recently, an experiment in fair quantitative agreement with kinetic theory was conducted, yielding surprising insights into the crucial role of Coulomb friction [60]. These effects will be discussed in some detail in section 8.3.3.

8.2 Granular kinetic theory

Parts of this section have been reprinted from [6] copyright (2015), with permission from Springer Nature.

Many models have been proposed to describe binary inelastic collisions, including both soft spheres and hard spheres. The most commonly used model in granular gas literature is also the simplest: the gas of inelastic smooth hard spheres with a fixed restitution coefficient[1]. The dynamics of this system is defined by the following relations:

$$m_1\mathbf{v}_1' + m_2\mathbf{v}_2' = m_1\mathbf{v}_1 + m_2\mathbf{v}_2 \tag{8.1a}$$

$$(\mathbf{v}_1' - \mathbf{v}_2')\cdot\hat{\mathbf{n}} = -r(\mathbf{v}_1 - \mathbf{v}_2)\cdot\hat{\mathbf{n}}, \tag{8.1b}$$

where the primes denote post-collisional velocities, $\hat{\mathbf{n}}$ is the unit vector along the line joining the centres of the grains, and $0 \leqslant r \leqslant 1$ is the restitution coefficient. In this model, collisions are instantaneous and occur at contact. When $r = 1$, the gas behaves elastically, following the collision rules for hard spheres. When $r = 0$, the gas is perfectly inelastic, meaning particles exit the collision with no relative velocity in the $\hat{\mathbf{n}}$ direction.

The transformation that gives post-collisional velocities from pre-collisional ones is:

$$\mathbf{v}_1' = \mathbf{v}_1 - (1+r)\frac{m_2}{m_1+m_2}((\mathbf{v}_1 - \mathbf{v}_2)\cdot\hat{\mathbf{n}})\hat{\mathbf{n}} \tag{8.2a}$$

$$\mathbf{v}_2' = \mathbf{v}_2 + (1+r)\frac{m_1}{m_1+m_2}((\mathbf{v}_1 - \mathbf{v}_2)\cdot\hat{\mathbf{n}})\hat{\mathbf{n}}. \tag{8.2b}$$

[1] Hard spheres denotes a kind of particles with a well-defined diameter: the spheres interact only in the exact moment they touch, updating instantaneously their velocities. Smooth hard spheres have the additional property of smoothness, which means that they do not exchange any rotational degree of freedom. Later we mention the so-called rough hard spheres, which can exchange angular velocity as a consequence of roughness (microscopic asperities on the surface).

For reverse transformations, where post-collisional velocities give pre-collisional ones with exchanged primes, the relation is:

$$\mathbf{v}_1 = \mathbf{v}_1' - \left(1 + \frac{1}{r}\right)\frac{m_2}{m_1 + m_2}((\mathbf{v}_1' - \mathbf{v}_2') \cdot \hat{\mathbf{n}})\hat{\mathbf{n}} \tag{8.3a}$$

$$\mathbf{v}_2 = \mathbf{v}_2' + \left(1 + \frac{1}{r}\right)\frac{m_1}{m_1 + m_2}((\mathbf{v}_1' - \mathbf{v}_2') \cdot \hat{\mathbf{n}})\hat{\mathbf{n}}. \tag{8.3b}$$

As seen, the inverse transformation is equivalent to changing the restitution coefficient $r \to 1/r$. Of course, in the case of a perfectly inelastic gas ($r = 0$), no inverse transformation exists.

Various extensions of this model have been proposed and explored in the literature. To account for tangential frictional forces during collisions, one can include the rotational degree of freedom for particles, introducing a variable $\boldsymbol{\omega}_i$ for each grain. The simplest model incorporating rotational motion is the rough hard spheres gas, where post-collisional translational and angular velocities are given by:

$$\mathbf{v}_{1,2}' = \mathbf{v}_{1,2} \mp \frac{1 + r}{2}\mathbf{v}_n \mp \frac{q(1 + \beta)}{2q + 2}(\mathbf{v}_t + \mathbf{v}_r) \tag{8.4a}$$

$$\boldsymbol{\omega}_{1,2}' = \boldsymbol{\omega}_{1,2} + \frac{1 + \beta}{\sigma(1 + q)}[\hat{\mathbf{n}} \times (\mathbf{v}_t + \mathbf{v}_r)]. \tag{8.4b}$$

Here, q is the dimensionless moment of inertia defined by $I = qm(\sigma/2)^2$, σ the diameter, with $q = 1/2$ for disks and $q = 2/5$ for spheres. The term $\mathbf{v}_n = ((\mathbf{v}_1 - \mathbf{v}_2) \cdot \hat{\mathbf{n}})\hat{\mathbf{n}}$ represents the normal relative velocity component, $\mathbf{v}_t = \mathbf{v}_1 - \mathbf{v}_2 - \mathbf{v}_n$ is the tangential velocity component due to translational motion, and $\mathbf{v}_r = -\sigma(\boldsymbol{\omega}_1 - \boldsymbol{\omega}_2)$ is the tangential velocity component due to particle rotation. The tangential restitution coefficient β, which ranges from -1 to $+1$, also appears in these equations. When $\beta = -1$, tangential effects are absent (i.e. rough spheres behave like smooth spheres), while $\beta = +1$ indicates a perfectly rough surface. Energy conservation occurs for $\beta = \pm 1$ when $r = 1$.

Other collision models have been introduced, motivated by a detailed analysis of the collision process. In these models, the restitution coefficient r (or the coefficients r and β in more detailed descriptions) depends on the relative velocity of the colliding particles. Notably, as the relative velocity approaches zero, collisions tend to become more elastic. This refined description, known as the 'viscoelastic' model, is significant in various aspects of the statistical mechanics of granular gases. A notable kinetic instability in cooling (and sometimes driven) granular gases is the so-called inelastic collapse, where the local collision rate diverges due to a few particles becoming trapped very close to each other. Simulations using the viscoelastic model have shown that this instability is mitigated, suggesting it is an artefact of the fixed restitution coefficient idealization. Below is an expression for the leading term in the velocity dependence of the normal restitution coefficient r in the viscoelastic model (the viscoelastic theory also provides velocity-dependent expressions for the tangential restitution coefficient):

$$r = 1 - C_1|(\mathbf{v}_1 - \mathbf{v}_2) \cdot \hat{\mathbf{n}}\,|^{1/5} + \cdots \tag{8.5}$$

where C_1 depends on the physical properties of the spheres, such as mass, density, radius, Young's modulus, and viscosity [61].

8.2.1 The granular Boltzmann equation

The derivation of the Boltzmann equation for inelastic particles cannot follow the discussion sketched in chapter 2, since that was based upon conservative interactions, see equation (2.25). Instead one has to use a derivation of the Boltzmann equation built specifically for hard-core interactions where a binary collision operator appears. That operator can be adapted to inelastic collisions that obey equations (8.2) and (8.3). This procedure is not described in detail here, one can look into [6, 62, 63]. By deriving from this the Bogoliubov–Born–Green–Kirkwood–Yvon hierarchy and applying the molecular chaos assumption to the first equation, one obtains the Boltzmann equation for granular gases:

$$\left(\frac{\partial}{\partial t} + L_1^0\right)P(\mathbf{r}_1, \mathbf{v}_1, t) = N\sigma^2 Q(P, P) \tag{8.6}$$

$$Q(P, P) = \int d\mathbf{v}_2 \int_{\mathbf{V}_{12}\cdot\hat{\mathbf{n}}>0} d\hat{\mathbf{n}}|\mathbf{V}_{12}\cdot\hat{\mathbf{n}}|\left[\frac{1}{r^2}P(\mathbf{r}_1, \mathbf{v}_1', t)P(\mathbf{r}_1, \mathbf{v}_2', t) - P(\mathbf{r}_1, \mathbf{v}_1, t)P(\mathbf{r}_1, \mathbf{v}_2, t)\right], \tag{8.7}$$

where L_i^0 is defined in equation (2.26a), while the primed velocities are defined in equations (8.3). A key difference from the elastic case is the presence of the factor $\frac{1}{r^2}$ in front of the gain term in the collision integral. This factor is the main source of imbalance between the gain and loss terms and underlies the violation of time-reversal symmetry and the H-theorem (see the discussion at the end of section 8.2.2).

This equation was first studied in the spatially homogeneous case (i.e. with no spatial gradients, $L_1^0 = 0$), by Goldshtein and Shapiro [64], and by Ernst and van Noije [65]. In this case, the equation is given by:

$$\frac{\partial}{\partial t}P(\mathbf{v}_1, t) = n\sigma^2 Q(P, P). \tag{8.8}$$

To facilitate the analysis under the assumption of spatial homogeneity, it is useful to define a rescaled distribution function:

$$NP(\vec{r}, \vec{v}, t) = \frac{n}{v_T^3}\tilde{f}\left(\frac{\vec{v}}{v_T}\right), \tag{8.9}$$

where $T(t) = \frac{m\langle\vec{v}^2\rangle}{3} = \frac{1}{2}mv_T^2(t)$ (assuming $k_B = 1$), $\vec{c} = \frac{\vec{v}}{v_T}$. Here, v_T is the thermal velocity. Under this scaling, the collision integral transforms as $N^2Q \to n^2 v_T^{-2}\tilde{Q}$, where

$$\tilde{Q} = \int d\vec{c}_2 \int_+ d\hat{n}\,|\vec{c}_{12}\cdot\hat{n}|\left[\frac{1}{r^2}\tilde{f}(\vec{c}_1', t)\tilde{f}(\vec{c}_2', t) - \tilde{f}(\vec{c}_1)\tilde{f}(\vec{c}_2)\right], \tag{8.10}$$

where the second integral is over the region $\vec{c}_{12}\cdot\hat{n} > 0$.

The primary contribution to the time derivative of the temperature arises from inelastic collisions. In homogeneous scenarios, where collisions reduce kinetic energy proportionally to the kinetic energy itself, it is expected that $\dot{T} \propto T$. The exact expression for this rate is:

$$\begin{aligned}\frac{d}{dt}\left(\frac{3}{2}nT\right)\bigg|_{\text{coll}} &= \int d\vec{v}\, \frac{mv^2}{2}\sigma^2 N^2 Q(P, P) \\ &= \sigma^2 n^2 v_T \frac{mv_T^2}{2}\int d\vec{c}_1 c_1^2 \tilde{Q} = -\sigma^2 n^2 v_T T \mu_2,\end{aligned} \tag{8.11}$$

where

$$\mu_p = -\int d\vec{c}_1 c_1^p \tilde{Q}. \tag{8.12}$$

Thus, the time evolution of the temperature due to collisions is given by:

$$\frac{dT}{dt}\bigg|_{\text{coll}} = -\zeta(t)T, \tag{8.13}$$

where the cooling rate $\zeta(t)$ is:

$$\zeta(t) = \frac{2\sqrt{2}}{3} n\sigma^2 \mu_2 \sqrt{\frac{T}{m}}. \tag{8.14}$$

The computation of μ_2, and thus ζ, requires the knowledge of the distribution $\tilde{f}(\vec{c}, t)$.

To address this point, it is useful to introduce a polynomial expansion that is applicable in both standard and granular kinetic theories. This expansion is particularly helpful for describing small deviations from the Maxwellian distribution, which are typical in homogeneous granular gases as well as in dilute gases with spatial inhomogeneities. The expansion is given by:

$$\tilde{f}(\vec{c}) = f_{MB}(\vec{c})\left[1 + \sum_{p=1}^{\infty} a_p S_p(c^2)\right], \tag{8.15}$$

where the Maxwellian distribution is:

$$f_{MB}(c) = \pi^{-3/2}\exp(-c^2). \tag{8.16}$$

The polynomials S_p, known as Sonine polynomials (specifically associated Laguerre polynomials $S_p^{(m)}$ with $m = d/2 - 1$), form a complete set of orthogonal functions:

$$\int d\vec{c}\, f_{MB}(c) S_p(c^2) S_{p'}(c^2) = \frac{2(p + 1/2)!}{\sqrt{\pi}p!}\delta_{pp'} = \mathcal{N}_p \delta_{pp'}. \tag{8.17}$$

In homogeneous granular systems, good agreement is often found by truncating the expansion at $p = 2$. For $d = 3$, the first Sonine polynomials are:

$$S_0(x) = 1, \tag{8.18}$$

$$S_1(x) = -x + \frac{3}{2}, \tag{8.19}$$

$$S_2(x) = \frac{x^2}{2} - \frac{5x}{2} + \frac{15}{8}. \tag{8.20}$$

It can be verified that:

$$\langle c^2 \rangle = \frac{3}{2}(1 - a_1), \tag{8.21}$$

and

$$\langle c^4 \rangle = \frac{15}{4}(1 + a_2). \tag{8.22}$$

Note also that:

$$N\int d\vec{v}\, \frac{mv^2}{2} P(\vec{r}, \vec{v}, t) = \frac{mv_T^2}{2} n \int d\vec{c}\, c^2 \tilde{f}(\vec{c}) = \langle c^2 \rangle \frac{mv_T^2}{2} n, \tag{8.23}$$

and

$$N\int d\vec{v}\, \frac{mv^2}{2} P(\vec{r}, \vec{v}, t) = n\frac{m\langle v^2 \rangle}{2} = \frac{3}{2} nT = \frac{3}{2} n \frac{mv_T^2}{2}. \tag{8.24}$$

Thus, $\langle c^2 \rangle = \frac{3}{2}$ implies $a_1 = 0$, making a_2 the first non-trivial coefficient. The equations for a_2 can be determined once a specific model or boundary conditions are specified. The explicit expression for μ_2 is:

$$\mu_2 = -\int d\vec{c}_1 c_1^2 \int d\vec{c}_2 \int_+ d\hat{n}\, |\vec{c}_{12} \cdot \hat{n}| \left[\frac{1}{r^2} \tilde{f}(\vec{c}_1', t) \tilde{f}(\vec{c}_2', t) - \tilde{f}(\vec{c}_1) \tilde{f}(\vec{c}_2) \right]. \tag{8.25}$$

Using the Sonine expansion truncated at $p = 2$, we finally obtain:

$$\mu_2 = \sqrt{2\pi}(1 - r^2)\left(1 + \frac{3}{16} a_2 + O(a_2^2)\right). \tag{8.26}$$

This formula, put in equation (8.14), gives the homogeneous cooling rate, as discussed in the following.

The homogeneous cooling state (HCS) represents the simplest regime in granular systems, characterized by spatial homogeneity and the absence of any external energy input. The system is initialized with a non-trivial velocity distribution, and as it evolves, the kinetic energy dissipates over time due to inelastic collisions.

In this regime, the rescaled distribution function introduces an additional term in the time derivative:

$$\frac{\partial(NP)}{\partial t} = \frac{n}{v_T^3} \frac{\partial \tilde{f}}{\partial t} + \left(-\frac{3n}{v_T^4} \tilde{f} + \frac{n}{v_T^3} \frac{\partial \tilde{f}}{\partial c_1} \frac{\partial c_1}{\partial v_T} \right) \frac{dv_T}{dt}. \tag{8.27}$$

This leads to the following time evolution equation:

$$\frac{1}{v_T}\frac{\partial \tilde{f}}{\partial t} - \frac{1}{v_T^2}\frac{\partial(\vec{c}_1\tilde{f})}{\partial \vec{c}_1}\frac{dv_T}{dt} = \sigma^2 n\tilde{Q}. \tag{8.28}$$

Considering the relation $\dot{T}(t) = -\zeta(t)T(t)$ and the expression for $\zeta(t)$, it follows that:

$$\left.\frac{1}{v_T^2}\frac{dv_T}{dt}\right|_{\text{coll}} = \frac{1}{2v_T T}\frac{dT}{dt} = -\frac{1}{3}\sigma^2 n\mu_2, \tag{8.29}$$

which is time-independent.

Typically, it is assumed that a scaling function exists such that $\tilde{f} \to \tilde{f}_{HC}$ with $\frac{\partial \tilde{f}_{HC}}{\partial t} = 0$. If such a function exists, it must satisfy the following kinetic equation:

$$\frac{\mu_2}{3}\frac{\partial(\vec{c}_1\tilde{f}_{HC})}{\partial \vec{c}_1} = \tilde{Q}. \tag{8.30}$$

This equation defines the HCS in kinetic theory. The temperature evolution in this state follows:

$$T(t) = \frac{T(0)}{\left(1 + \dfrac{\zeta(0)t}{2}\right)^2}, \tag{8.31}$$

which is known as Haff's law [66].

Using the Sonine polynomial approximation, truncated at the second term, the cooling rate $\zeta(t)$ is given by:

$$\zeta(t) = \frac{4\sqrt{\pi}}{3}n\sigma^2\sqrt{\frac{T(t)}{m}}(1 - r^2)\left(1 + \frac{3}{16}a_2 + O(a_2^2)\right) = \frac{1 - r^2}{3}\omega_c(t), \tag{8.32a}$$

where $\omega_c(t)$ is the collision frequency:

$$\omega_c = 4\sqrt{\pi}n\sigma^2\sqrt{\frac{T(t)}{m}}\left(1 + \frac{3}{16}a_2 + O(a_2^2)\right). \tag{8.33}$$

Following Haff's law, equation (8.31), it becomes evident that:

$$\omega_c \sim \frac{1}{1 + \zeta(0)t/2}, \tag{8.34}$$

implying that the cumulative number of collisions scales as $\sim \ln(1 + \zeta(0)t/2)$. This observation suggests introducing a new timescale:

$$\tau(t) = \tau_0 \ln(1 + \zeta(0)t/2), \tag{8.35}$$

where τ_0 is a constant. The time derivative with respect to this new timescale is:

$$\frac{\partial}{\partial t} = \frac{\tau_0\zeta(0)/2}{1 + \zeta(0)t/2}\frac{\partial}{\partial \tau}. \tag{8.36}$$

This leads to:

$$\frac{1}{v_T(t)}\frac{\partial}{\partial t} = \frac{\tau_0\zeta(0)/2}{v_T(0)}\frac{\partial}{\partial \tau}. \tag{8.37}$$

Finally, in terms of the new timescale, the evolution equation becomes:

$$\frac{\partial \tilde{f}}{\partial \tau} + \frac{n\sigma^2\mu_2}{3}\frac{\partial(\vec{c}_1\tilde{f})}{\partial \vec{c}_1} = \sigma^2 n\tilde{Q}, \tag{8.38}$$

which is equivalent to the Boltzmann equation for particles subject to an effective force:

$$\vec{F} = \frac{n\sigma^2\mu_2\vec{c}}{3}, \tag{8.39}$$

implying a positive viscosity.

This equivalence holds as long as the system remains homogeneous. However, the HCS becomes unstable when subjected to long-wavelength perturbations [6].

Ernst and van Noije [65] provided estimates for the tails of the velocity distribution using an asymptotic method initially developed by Krook and Wu [67]. This method assumes that for a fast particle, the dominant contributions to the collision integral arise from collisions with thermal (bulk) particles, and that the gain term in the integral is negligible compared to the loss term. The loss term in the Boltzmann equation can be approximated by:

$$-\int d\vec{c}_2 \int_+ d\hat{n}\, |\vec{c}_{12}\cdot\hat{n}|\tilde{f}(\vec{c}_1)\tilde{f}(\vec{c}_2) \approx -\pi c_1\tilde{f}(\vec{c}_1). \tag{8.40}$$

If $\tilde{f}$ is isotropic, then $\vec{c}\cdot\nabla_{\vec{c}}\tilde{f} = c\frac{d\tilde{f}}{dc}$. Therefore, for large c, we find:

$$\mu_2\tilde{f} + \frac{1}{3}\mu_2 c\frac{d\tilde{f}}{dc} = -\pi c\tilde{f}, \tag{8.41}$$

which leads to the asymptotic behaviour:

$$\tilde{f} \sim \exp\left(-\frac{3\pi}{\mu_2}c\right). \tag{8.42}$$

Given that $\mu_2 \sim (1 - r^2)$, this estimate is valid for $c > 1/(1 - r^2)$.

The one-dimensional inelastic Boltzmann equation for Maxwell molecules, discussed in section 2.4.2, is given by:

$$\partial_\tau P(v, \tau) + P(v, \tau) = \beta\int du\, P(u, \tau)P(\beta v + (1 - \beta)u, \tau), \tag{8.43}$$

where $\beta = 2/(1 + r)$ and τ represents the number of collisions per particle.

Interestingly, equation (8.43) can be seen as the master equation for the inelastic version of a process originally introduced by Ulam [68]. In this process, a random pair of particles is selected, and their scalar velocities are updated according to the rule specified in equation (8.2). This model was first studied by Ben-Naim and Krapivsky [69], who derived the evolution of the moments of the velocity distributions. They observed that at large times, the moments scale as $\langle v^n \rangle \sim \exp(-\tau q_n)$, where the decay rates q_n are not proportional to $nq_2/2$ but depend nonlinearly on n.

This multiscaling behaviour suggests that a rescaled asymptotic distribution f of the form $P(v, \tau) \to f(v/v_0(\tau))/v_0(\tau)$ may not exist for large τ, where $v_0^2(\tau) = \int v^2 P(v, \tau) dv = E(\tau)$. However, the multiscaling property of moments indicates that the moments of the rescaled distribution $\int x^n f(x) dx = \langle v^n \rangle / v_0^n$ diverge asymptotically for $n \geqslant 3$. This divergence does not exclude the possibility of an asymptotic distribution with power-law tails.

One can study equation (8.43) using the Fourier transform:

$$\partial_\tau \hat{P}(k, \tau) + \hat{P}(k, \tau) = \hat{P}\left(\frac{k}{1-\beta}, \tau\right)\hat{P}\left(\frac{k}{\beta}, \tau\right). \tag{8.44}$$

This equation admits several self-similar solutions of the form $\hat{P}(k, \tau) = \hat{f}(kv_0(\tau))$, corresponding to the asymptotic rescaled distribution $P(v, \tau) = f(v/v_0(\tau))/v_0(\tau)$. However, many of these solutions do not correspond to physically meaningful velocity distributions [69].

The divergence of higher moments implies a non-analytic structure for $\hat{f}(k)$ near $k = 0$, since $\langle v^n \rangle / v_0^n = (-i)^n \frac{d^n \hat{f}(k)}{dk^n}|_{k=0}$. This non-analyticity serves as a guide in selecting the physically relevant solution, which is:

$$f\left(\frac{v}{v_0(\tau)}\right) = \frac{2}{\pi\left[1 + \left(\frac{v}{v_0(\tau)}\right)^2\right]^2}. \tag{8.45}$$

This distribution corresponds to the self-similar solution $\hat{f}(k) = (1 + |k|)\exp(-|k|)$.

It is noteworthy that equation (8.45) is a solution of equation (8.44) for any $r < 1$, indicating that the asymptotic velocity distribution is independent of the value of r as long as $r < 1$. The discovery of this exact scaling solution [70] sparked considerable interest, leading to numerous subsequent studies by various research groups [71–73]. These studies extended the analysis to higher dimensions and provided rigorous results concerning convergence, uniqueness, and related properties [74].

8.2.2 Granular steady states

As mentioned earlier, a granular system must be driven by some external perturbation, e.g. by shaking, in order to be kept in a steady state. How do we model in the most general way such an external energy injection? The randomly driven granular gas, first introduced in [75, 76], consists of an assembly of N identical hard objects (spheres, disks, or rods) of mass m and diameter σ. For simplicity, we set the Boltzmann constant $k_B = 1$. The grains move within a box of volume $V = L^d$, where L is the length of the sides of the box, and periodic boundary conditions are used. The mean free path, which is calculated exactly for a homogeneous gas of 3D hard spheres with a Maxwellian distribution of velocities, can be estimated as:

$$\lambda = \frac{1}{nS}, \tag{8.46}$$

where $n = N/V$ is the mean number density and S is the total scattering cross-section. It is important to note that S has the dimensions of a surface in $d = 3$ (with $S \sim \sigma^2$), a line in $d = 2$ ($S \sim \sigma$), and no dimensions in $d = 1$ (where the diameter becomes irrelevant).

The gas dynamics result from the interplay of two main physical phenomena: continuous interaction with the surroundings and inelastic collisions among the grains. The interaction with the surroundings is modelled by a Langevin equation, which satisfies the Einstein relation (see, for example, reference [77]), governing the evolution of particle velocities during the free time between collisions. The inelastic collisions are described by the standard inelastic collision rules. For a particle i that is not currently colliding with another particle, the equations of motion *between collisions* are:

$$m\frac{d\mathbf{v}_i(t)}{dt} = -\gamma_b \mathbf{v}_i(t) + \sqrt{2\gamma_b T_b}\,\eta_i(t), \tag{8.47a}$$

$$\frac{d\mathbf{x}_i(t)}{dt} = \mathbf{v}_i(t), \tag{8.47b}$$

where $\tau_b = m/\gamma_b$ and T_b are the characteristic time of the bath and temperature of the bath, respectively. The stochastic function $\eta_i(t)$ is characterized by $\langle \eta_i(t) \rangle = 0$ and correlations $\langle \eta_i^\alpha(t)\eta_j^\beta(t') \rangle = \delta(t - t')\delta_{ij}\delta_{\alpha\beta}$, where α and β are component indices, i.e. $\eta_i(t)$ is a standard white noise process.

In the dynamics of the N particles, as described by equations (8.47) and the inelastic hard-core collision rules, the key parameters are: the coefficient of normal restitution r, which determines the degree of inelasticity; the ratio $\rho = \tau_b/\tau_c$ between the characteristic time of the bath and the mean free time between collisions. Based on these two parameters, three fundamental limits of the model's dynamics can be defined: (1) the elastic limit: $r \to 1^-$; (2) the collisionless limit: $\rho \to 0$ ($\tau_c \gg \tau_b$); (3) the cooling limit: $\rho \to \infty$ ($\tau_c \ll \tau_b$).

The elastic limit is smooth for dimensions $d > 1$, and thus can be treated as equivalent to setting $r = 1$. In this case, collisions conserve energy in both the centre-of-mass frame and the absolute frame, leading to a homogenization of particle positions and a relaxation of the velocity distribution toward a Maxwellian with temperature $T = \langle v^2 \rangle / d = \langle v_x^2 \rangle$. This temperature equals the initial kinetic energy but is modified by the relaxation toward T_b due to the Langevin dynamics (equations (8.47)). In one dimension, however, this mixing effect is absent, as elastic collisions preserve the initial velocity distribution, effectively making the particles non-interacting walkers [75].

In the collisionless limit, where $\tau_c \gg \tau_b$, collisions are rare compared to the characteristic time of the bath. In this scenario, the model behaves as an ensemble of non-interacting Brownian walkers, each governed by equations (8.47). Regardless of r, and in any dimension, the velocity distribution relaxes in a time τ_b to a Maxwellian with temperature $T = \langle v^2 \rangle / d = T_b$ and homogeneous density.

Finally, in the cooling limit, collisions dominate the dynamics, while particles move nearly ballistically between collisions. If $r < 1$, the gas can be considered stationary only over observation times much longer than τ_b, where the effect of external driving (the Langevin equation) becomes significant. For observation times larger than the mean free time τ_c but shorter than τ_b, the gas resembles a cooling granular gas.

In summary, the model of a randomly driven granular gas exhibits two fundamental stationary regimes: (1) a collisionless stationary regime, when $\rho \ll 1$, approaching the collisionless limit. In this regime, after a transient period of order τ_b, the system behaves like an ensemble of non-interacting Brownian particles, with homogeneous density, Maxwellian velocity distribution, and no correlations; (2) a colliding stationary regime, when $\rho \gg 1$, approaching the cooling limit, but observed over timescales longer than τ_b. This regime is characterized by anomalous statistical properties.

In this model, the Boltzmann equation includes additional terms equivalent to the Fokker–Planck operators that govern the evolution of the velocity distribution in a Langevin equation. The the equation is thus written as:

$$\frac{\partial P}{\partial t} = n\sigma^2 Q(P, P) + \frac{\gamma_b}{m}\nabla_{\mathbf{v}} \cdot (\mathbf{v}P) + \frac{\gamma_b T_b}{m^2}\nabla_{\mathbf{v}}^2 P, \tag{8.48}$$

where $Q(P, P)$ is defined in equation (8.6). Using the rescaled distribution definition (8.9) and noting that for the thermal velocity $\dot{v}_T = 0$ (in a statistically stationary state), we obtain:

$$\frac{\partial \tilde{f}}{\partial t} = v_T n\sigma^2 \tilde{Q} + \frac{\gamma_b}{m}\nabla_{\mathbf{c}} \cdot (\mathbf{c}\tilde{f}\,) + \frac{\gamma_b}{2m}\frac{T_b}{T}\nabla_{\mathbf{c}}^2 \tilde{f}\,. \tag{8.49}$$

From this, we can define the granular temperature:

$$T = \frac{m}{d}\langle v^2 \rangle, \tag{8.50}$$

which leads to:

$$\langle v\dot{v} \rangle = \frac{\dot{T}}{2m} = -\frac{\gamma_b}{m}\langle v^2 \rangle + \frac{\gamma_b T_b}{m^2} - \zeta\frac{T}{2m}. \tag{8.51}$$

Setting $\dot{T} = 0$ in the stationary state, we obtain:

$$T - T_b = \zeta \tau_b T, \tag{8.52}$$

which can be solved numerically to find T (recalling that $\zeta \propto (1 - r^2)T^{1/2}$).

And what about the relaxation—from some given initial configuration—toward the steady state? The H-functional, as discussed in section 2.1.1, is monotonically non-increasing for systems governed by the homogeneous elastic Boltzmann equation. However, when collisions become inelastic, the monotonicity of H can no longer be guaranteed. Indeed, numerical simulations have shown that this monotonicity is lost [78]. A recent observation [79] proposes a possible replacement for the Boltzmann H-functional within the framework of the Boltzmann–Fokker–Planck (BFP) model. This model is precisely represented by equation (8.48). Variants of this model have also been explored, where the velocities are discretized, and the Fokker–Planck operator is substituted with a stochastic jump operator. The transition rates of this operator satisfy detailed balance with respect to an equilibrium steady distribution.

The proposed Lyapunov functional is as follows:

$$H_K(t) = \int d\mathbf{v}\, P(\mathbf{v},\, t)\log\frac{P(\mathbf{v},\, t)}{\Pi(\mathbf{v})}, \tag{8.53}$$

where $\Pi(\mathbf{v})$ is the stationary velocity distribution that the system asymptotically approaches. Numerical evidence and some analytical arguments suggest that, for the BFP model, the following relation holds:

$$\frac{dH_K(t)}{dt} \leqslant 0. \tag{8.54}$$

In particular, for the elastic version of the BFP model, the result in equation (8.54) can be rigorously demonstrated. It is noteworthy that while the elastic BFP model exhibits a trivial steady state, its dynamics remain non-trivial. The precise origin of the seemingly exact result in equation (8.54) is still not fully understood, more numerical investigations can be found in reference [80] and a proof for a similar lattice model, in the long time limit, is in reference [81].

8.3 The dynamics of a granular tracer

As we have seen in the previous section, a granular system can be kept in a steady state, representing a model for a gas of 'macroscopic particles', characterized by dissipative interactions. It is then interesting to investigate the features of these systems through the study of the dynamics of a massive tracer, in analogy with the case of Brownian motion described in the first part of the book, to understand to what extent nonequilibrium features intervene to modify the main picture. We will consider first the case of a dilute granular gas and then the effect of an increase of the system density. Our analysis will allow us to unveil some particular behaviours that can be observed due to the intrinsic nonequilibrium nature of these systems, which include non-trivial coupling among variables and motor effects.

8.3.1 Granular Brownian motion

We consider a dilute system where all grains (of radius r and mass m) are coupled to a thermal bath, with interaction time τ_b larger than the inter-particle collision time

τ_c, realizing a steady state. The stationary granular gas is then considered as a 'granular bath' for a massive tracer (of radius R and mass M, with mass ratio $\varepsilon = m/M$) which is also coupled to the external energy source. Therefore the 'Brownian' particle interacts with two baths and its dynamics can be analyzed starting from a linear Boltzmann–Lorentz–Fokker–Planck equation, which can be treated in the large mass limit and approximated by a Langevin equation (our presentation follows reference [82]).

As we will show, the effect of a 'double bath' can be observed in the expressions for the tracer temperature, mobility, and diffusion coefficients, all involving the interplay of both energy sources. Moreover, the large mass limit, together with the molecular chaos assumption (justified by the diluteness of the granular gas) guarantees that an FDT holds in the standard equilibrium form, where the ratio between diffusion and mobility is simply given by the tracer granular temperature.

As described in section 8.2.2, denoting by $\mathbf{V}$ and $\mathbf{v}$ the intruder velocity and the gas velocity, respectively, the coupled Boltzmann equations for the probability distributions $P(\mathbf{V}, t)$ and $p(\mathbf{v}, t)$ read

$$\begin{aligned}\frac{\partial P(\mathbf{V}, t)}{\partial t} &= \int d\mathbf{V}'[W_{tr}(\mathbf{V}|\mathbf{V}')P(\mathbf{V}', t) - W_{tr}(\mathbf{V}'|\mathbf{V})P(\mathbf{V}, t)] + \mathcal{B}_{tr}P(\mathbf{V}, t)\\ \frac{\partial p(\mathbf{v}, t)}{\partial t} &= \int d\mathbf{v}'[W_g(\mathbf{v}|\mathbf{v}')p(\mathbf{v}', t) - W_g(\mathbf{v}'|\mathbf{v})p(\mathbf{v}, t)] + \mathcal{B}_g p(\mathbf{v}, t)\\ &\quad + Q(p, p),\end{aligned} \tag{8.55}$$

where $\mathcal{B}_{tr}$ and $\mathcal{B}_g$ describe the interactions with the thermal bath. The collisions between the tracer and gas particles are given by the transition rates

$$\begin{aligned}W_{tr}(\mathbf{V}|\mathbf{V}') = \chi \int d\mathbf{v}' \int d\hat{n} p(\mathbf{v}', t)\Theta[-(\mathbf{V}' - \mathbf{v}')\cdot\hat{n}](\mathbf{V}' - \mathbf{v}')\cdot\hat{n}\\ \times\, \delta^{(d)}\left\{\mathbf{V} - \mathbf{V}' + \frac{\varepsilon^2}{1+\varepsilon^2}(1+\alpha)[(\mathbf{V}' - \mathbf{v}')\cdot\hat{n}]\hat{n}\right\}\end{aligned} \tag{8.56}$$

and

$$\begin{aligned}W_g(\mathbf{v}|\mathbf{v}') = \frac{\chi}{N} \int d\mathbf{V}' \int d\hat{n} P(\mathbf{V}', t)\Theta[-(\mathbf{V}' - \mathbf{v}')\cdot\hat{n}](\mathbf{V}' - \mathbf{v}')\cdot\hat{n}\\ \times\, \delta^{(d)}\left\{\mathbf{v} - \mathbf{v}' + \frac{1}{1+\varepsilon^2}(1+\alpha)[(\mathbf{v}' - \mathbf{V}')\cdot\hat{n}]\hat{n}\right\},\end{aligned} \tag{8.57}$$

where $\Theta(x)$ is the Heaviside step function, $\delta^{(d)}(x)$ is the Dirac delta function in d dimensions, and $\chi = \frac{g_2(r+R)}{l_0}$, $g_2(r + R)$ being the pair correlation function for a gas particle and an intruder at contact and l_0 the intruder mean free path. We have assumed that the probability $P_2(|\mathbf{x} - \mathbf{X}| = r + R, \mathbf{V}, \mathbf{v}, t)$ that a collision between the tracer and a gas particle occurs, when they have velocities $\mathbf{V}$ and $\mathbf{v}$ and positions $\mathbf{X}$ and $\mathbf{x}$, respectively, is given by the Enskog approximation [61]

$$P_2(|\mathbf{x} - \mathbf{X}| = r + R, \mathbf{V}, \mathbf{v}, t) = g_2(r + R)P(\mathbf{V}, t)p(\mathbf{v}, t), \tag{8.58}$$

which takes a small correction to molecular chaos, via the density correlations near the tracer; the action of the thermal bath is described by

$$\mathcal{B}_{tr}P(\mathbf{V},t) = \frac{\gamma_b}{M}\frac{\partial}{\partial \mathbf{V}}[\mathbf{V}P(\mathbf{V},t)] + \frac{\gamma_b T_b}{M}\Delta_V[P(\mathbf{V},t)] \tag{8.59}$$

$$\mathcal{B}_g p(\mathbf{v},t) = \frac{\gamma_b}{m}\frac{\partial}{\partial \mathbf{v}}[\mathbf{v}p(\mathbf{v},t)] + \frac{\gamma_b T_b}{m}\Delta_v[p(\mathbf{v},t)], \tag{8.60}$$

where Δ_v is the velocity Laplacian operator, and γ_b and T_b represent the viscosity and the temperature of the bath, respectively; finally, $Q(p, p)$ represents the Boltzmann collision operator for the particle–particle interactions, defined in equation (8.7).

The two Boltzmann equations (8.55) are coupled through the terms involving W_{tr} and W_g. However, if the number N of granular particles is large enough, the term W_g can be neglected because of the factor $1/N$ in equation (8.57). This means that the surrounding gas is weakly perturbed by the tracer and one can assume that the probability distribution function $p(\mathbf{v})$ is stationary and Gaussian, with variance T_g/m:

$$p(\mathbf{v}) = \frac{1}{\sqrt{(2\pi T_g/m)^d}}\exp\left[-\frac{m\mathbf{v}^2}{2T_g}\right], \tag{8.61}$$

where T_g is the granular temperature. Substituting equation (8.61) into equation (8.56), and projecting the velocities along the collision direction and the orthogonal one, the integral can be solved [83], yielding in two dimensions

$$\begin{aligned} W_{tr}(\mathbf{V}'|\mathbf{V}) = \chi &\frac{1}{\sqrt{2\pi T_g/m}\,k(\varepsilon)^2} \\ &\times \exp\left\{-m\left[V_n' - V_n + k(\varepsilon)V_n\right]^2/(2T_g k(\varepsilon)^2)\right\}. \end{aligned} \tag{8.62}$$

where $V_n = \mathbf{V}\cdot\hat{n}$ and $k(\varepsilon) = (1+\alpha)\varepsilon^2/(1+\varepsilon^2)$.

Note that, with the decoupling assumption, one has that the dynamics of the tracer alone is Markovian, and therefore the transition rates satisfy detailed balance with respect to a Gaussian invariant probability $P(V)$ [83]. Moreover, we can write the master equation for the tracer

$$\frac{\partial P(\mathbf{V},t)}{\partial t} = L_{gas}[P(\mathbf{V},t)] + L_{bath}[P(\mathbf{V},t)], \tag{8.63}$$

where $L_{gas}[P(\mathbf{V},t)]$ is a linear operator which can be expressed by means of the Kramers–Moyal expansion in the small parameter $\varepsilon = m/M$ (as already discussed in section 6.3.2 and reference [84])

$$L_{gas}[P(\mathbf{V},t)] = \sum_{n=1}^{\infty}\frac{(-1)^n\partial^n}{\partial V_{j_1}\ldots\partial V_{j_n}}D^{(n)}_{j_1\ldots j_n}(\mathbf{V})P(\mathbf{V},t), \tag{8.64}$$

(sum over repeated indices is meant) with

$$D^{(n)}_{j_1 \dots j_n}(\mathbf{V}) = \frac{1}{n!}\int d\mathbf{V}'(V'_{j_1} - V_{j_1}) \dots (V'_{j_n} - V_{j_n}) W_{tr}(\mathbf{V}'|\mathbf{V}). \tag{8.65}$$

The second term in the master equation describes the action of the thermal bath:

$$L_{bath}[P(\mathbf{V}, t)] = \mathcal{B}_{tr} P(\mathbf{V}, t). \tag{8.66}$$

In the limit of large mass M, the interaction between the granular gas and the tracer can be described by an effective Langevin equation. Indeed, keeping only the first two terms of the expansion which are the lowest in the expansion in ε, one gets

$$L_{gas}[P(\mathbf{V}, t)] = -\frac{\partial}{\partial V_i}[D_i^{(1)}(\mathbf{V})P(\mathbf{V}, t)] + \frac{\partial^2}{\partial V_i \partial V_j}[D_{ij}^{(2)}(\mathbf{V})P(\mathbf{V}, t)]. \tag{8.67}$$

Computing these first terms (for details, see reference [82]) we finally obtain the Langevin equation for the tracer

$$M\dot{\mathbf{V}} = -\Gamma \mathbf{V} + \mathcal{E}, \tag{8.68}$$

where $\Gamma = \gamma_b + \gamma_g$ and $\mathcal{E} = \xi_b + \xi_g$, with

$$\gamma_g = M\eta_g = M\omega_0(1+\alpha)\varepsilon^2 = \omega_0(1+\alpha)m \tag{8.69}$$

$$\langle \mathcal{E}_i(t)\mathcal{E}_j(t')\rangle = 2\left[\gamma_b T_b + \gamma_g\left(\frac{1+\alpha}{2}T_g\right)\right]\delta_{ij}\delta(t-t'). \tag{8.70}$$

In the above equation $\omega_0 = \chi\sqrt{2\pi}\left(\frac{T_g}{m}\right)^{1/2}$ represents a collision frequancy.

This result implies that the stationary velocity distribution of the tracer is Gaussian with temperature

$$T_{tr} = \frac{\gamma_b T_b + \gamma_g\left(\frac{1+\alpha}{2}T_g\right)}{\gamma_b + \gamma_g}. \tag{8.71}$$

For the diffusion coefficient one gets

$$D_{tr} = \int_0^\infty dt\langle V_x(t)V_x(0)\rangle = \frac{T_{tr}}{\Gamma} = \frac{\gamma_b T_b + \gamma_g\left(\frac{1+\alpha}{2}T_g\right)}{(\gamma_b + \gamma_g)^2}. \tag{8.72}$$

In summary, this example shows that in the dilute case, the external driving mechanism, characterized by a temperature T_b, and the 'internal' granular bath at temperature $T_g < T_b$, sum up together and yield a linear Langevin dynamics for the tracer, characterized by a tracer temperature T_{tr} given by a weighted sum of T_b and $T'_g = \frac{1+\alpha}{2}T_g$, with $T'_g < T_g$, due to non-conservative interactions. The presented analysis, confirmed by molecular dynamics numerical simulation [82], rests on several assumptions: large mass of the tracer (to justify the truncation of the Kramers–Moyal expansion), low density of the gas (to guarantee the molecular chaos hypothesis), inelasticity not too low (to disregard non-Gaussian corrections), and well separated interaction times of the two baths. In particular, the effects of an increase of gas density will be discussed in the following section.

8.3.2 Tracer in a moderately dense granular fluid: the role of the coupling among variables

We now discuss the effects due to recollisions, i.e. violations of molecular chaos, which cannot be neglected when the density of the granular gas is large. By recollisions we mean the situation where the tracer interacts two (or more) times with the same gas particle before the particle has lost the 'memory' of the previous collisions, namely before the particle has interacted for a sufficiently long time with the other gas particle to make the new interaction with the tracer an independent event. The recollisions introduce some velocity correlations in the system, which clearly appear for instance in the functional form of the autocorrelation function $C(t) = \langle V(t)V(0)\rangle$. Indeed, while in the dilute case the autocorrelation displays a simple exponential decay, in the dense system one observes a more complex structure: depending on the model parameters, $C(t)$ can be characterized by multiple exponential decay, or by oscillations, featuring a negative part which represents a backscattering phenomenon (see figure 8.1). This means that if a fluctuation occurs in one direction, due to the accumulation of fluid particles in front of the tracer, after a certain time a fluctuation in the opposite direction is expected.

In a phenomenological approach, the simplest modification to the Langevin equation to describe such memory effects is represented by the introduction of an exponential memory kernel, with a new timescale τ_1

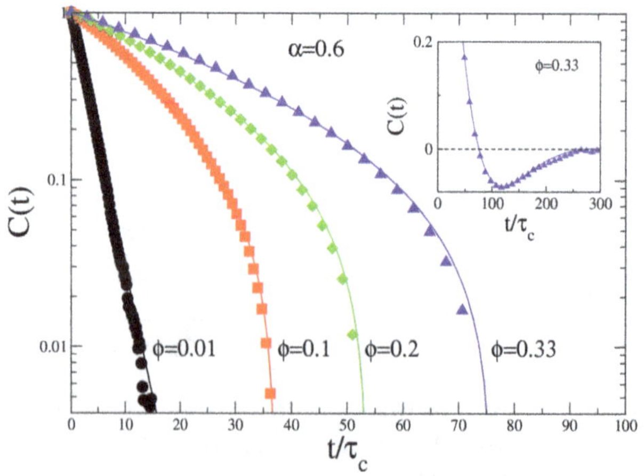

Figure 8.1. Normalized autocorrelation $C(t)/C(0)$ (symbols) for different values of the packing fraction $\phi = 0.01, 0.1, 0.2, 0.33$ at $\alpha = 0.6$. Continuous lines are the best fits obtained with equation (8.79). Inset: the case $\phi = 0.33$ in linear scale. © IOP Publishing. Reproduced with permission. All rights reserved, from reference [85].

$$M\dot{V}(t) = -\int_{-\infty}^{t} dt' \ \Gamma(t - t')V(t') + \xi'(t), \tag{8.73}$$

where

$$\Gamma(t) = 2\gamma_0\delta(t) + \gamma_1/\tau_1 e^{-t/\tau_1}, \tag{8.74}$$

and $\xi'(t) = \xi_0(t) + \xi_1(t)$, with

$$\langle\xi_0(t)\xi_0(t')\rangle = 2T_0\gamma_0\delta(t - t'), \tag{8.75}$$

$$\langle\xi_1(t)\xi_1(t')\rangle = T_1\gamma_1/\tau_1 e^{-(t-t')/\tau_1}, \tag{8.76}$$

and $\langle\xi_1(t)\xi_0(t')\rangle = 0$. In the equilibrium limit $\alpha \to 1$, the parameter T_1 tends to T_0 in order to fulfil the equilibrium fluctuation–dissipation theorem of the second kind $\langle\xi'(t)\xi'(t')\rangle = T_0\Gamma(t - t')$. The dilute limit is recovered if $\gamma_1 \to 0$, and the parameters γ_0 and T_0 coincide with Γ and T_{tr} of the Enskog theory described before. A rigorous derivation of this model should follow the lines described in section 6.3.4, or through perturbative approaches as in the framework of the mode-coupling theory (see for instance reference [86]).

The effect of memory and the role of coupling among variables can be more clearly illustrated mapping equation (8.73) onto a Markovian equivalent model by introducing an auxiliary variable:

$$\begin{aligned} M\dot{V} &= -\gamma_0(V - U) + \sqrt{2T_0\gamma_0}\xi_V \\ \dot{U} &= -\frac{U}{\tau_1} - \frac{\gamma_1}{\gamma_0\tau_1}V + \sqrt{2\frac{T_1\gamma_1}{\gamma_0^2\tau_1^2}}\xi_U, \end{aligned} \tag{8.77}$$

where ξ_V and ξ_U are uncorrelated white noises of unitary variance. The new variable

$$U(t) \propto \gamma_1/(\tau_1\gamma_0)\int_{-\infty}^{t} e^{-\frac{t-t'}{\tau_1}}[V(t') + \xi_1(t')]dt' \tag{8.78}$$

is defined up to a multiplicative factor, and represents a velocity field coupled with the tracer. Physically, $U(t)$ can be interpreted as the local average velocity field of the gas particles surrounding the tracer. Within this picture we expect that $T_0 \sim T_{tr}$ is roughly the temperature of the tracer and τ_1 is the main relaxation time of the average velocity field U around the tracer. Moreover, the parameter γ_1 is the intensity of coupling felt by the gas particles interacting with the intruder, and T_1 is the 'temperature' of the local field U, identified with the bath temperature $T_1 \sim T_b$ (indeed, due to momentum conservation, inelasticity does not affect significantly the average velocity of an ensemble of particles almost only colliding among themselves). In this form, the model belongs to the class of Brownian gyrators, introduced in chapter 5, with the difference that the stochastic variables are here velocities, and therefore odd under time-inversion transformations. Further details can be found in reference [85].

From the linear model (8.73) one can easily obtain the explicit expression for the normalized autocorrelation and response functions, $C(t)/C(0) = f_C(t)$ and $R(t) = f_R(t)$, respectively, with

$$f_{C(R)} = e^{-gt}[\cos(\omega t) + a_{C(R)}\sin(\omega t)]. \tag{8.79}$$

The quantities g, ω, a_C and a_R are known algebraic functions of γ_0, T_0, γ_1, τ_1 and T_1. In particular, the ratio $a_C/a_R = [T_0 - \Omega(T_1 - T_0)]/[T_0 + \Omega(T_1 - T_0)]$, with $\Omega = \gamma_1/[(\gamma_0 + \gamma_1)(\gamma_0/M\tau_1 - 1)]$.

8.3.2.1 Fluctuation dissipation theorem

The memory effects observed at moderately high density are not peculiar to the nonequilibrium properties of the system. Indeed, also molecular fluids can show similar behaviours of the correlation functions [86]. However, in nonequilibrium conditions, couplings among variables arise, that induce 'violations' of the equilibrium form of the FDT. In figure 8.2, we plot the correlation and response functions measured in numerical simulations in a dense case with inelastic collisions: deviations from the equilibrium relation $R(t) \propto C(t)$ are clearly observed. Indeed, as widely discussed in chapter 4, a relation between response and correlation function must now take into account the contribution of the cross-correlation $\langle V(t)U(0)\rangle$, i.e.:

$$R(t) = \sigma_{VV}^{-1}C(t) + \sigma_{UV}^{-1}\langle V(t)U(0)\rangle, \tag{8.80}$$

with the inverse covariance matrix elements $\sigma_{VV}^{-1} = [1 - \gamma_1/M(T_0 - T_1)\Omega_a]$ and $\sigma_{UV}^{-1} = (T_0 - T_1)\Omega_b$, where Ω_a and Ω_b are known functions of the parameters. Note that when interactions are elastic (and $T_0 = T_1$), as well as in the dilute limit ($\gamma_1 \to 0$),

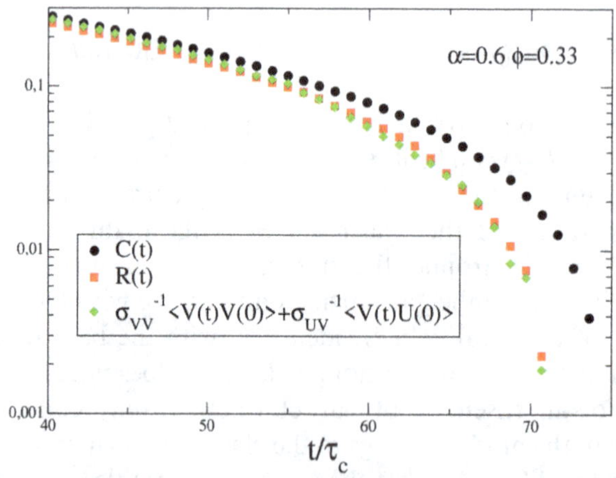

Figure 8.2. Normalized velocity correlation $C(t)/C(0)$ (black circles), response function $R(t)$ (red squares), and generalized FDT equation (8.80) (green diamonds) computed in molecular dynamics numerical simulations of a massive tracer in a granular gas with packing fraction $\phi = 0.33$, coefficient of restitution $\alpha = 0.6$. Reprinted from reference [87], with the permission of AIP Publishing.

one gets $a_C = a_R$ in equation (8.79) and the equilibrium result $C(t) \propto R(t)$ is immediately recovered. In figure 8.2 is shown the very good agreement between the response measured applying the perturbation and the expression in the right-hand side of equation (8.80), where only unperturbed correlators appear.

8.3.2.2 Fluctuation relation

The model (8.73) also allows one to study analytically the fluctuating entropy production, introduced in chapter 5, which quantifies the deviation from detailed balance in the dynamics. Given the trajectory in the time interval $[0,\ t]$, $\{V(s)\}_0^t$, and its time-reversed $\{\overline{V(s)}\}_0^t \equiv \{-V(t-s)\}_0^t$, proceeding along the same lines presented in chapter 5 for the Brownian gyrator, one has

$$\Sigma = \log \frac{P(\{V(s)\}_0^t)}{P(\{\bar{V}(s)\}_0^t)} \approx \gamma_0 \left(\frac{1}{T_0} - \frac{1}{T_1} \right) \int_0^t ds \ \ V(s) U(s). \tag{8.81}$$

This functional vanishes exactly in the equilibrium case, $\alpha = 1$, where equipartition holds, $T_1 = T_0$, and is zero on average in the dilute limit, where $\langle VU \rangle = 0$ (because tracer and gas particles are not coupled). The result (8.81) can be interpreted as follows: the most important source of entropy production is the energy transferred by the 'force' $\gamma_0 U$ on the tracer, weighed by the difference between the inverse temperatures of the two 'thermostats'.

As discussed in chapter 5, the quantities $\text{Prob}(\Sigma = x)$ and $\text{Prob}(\Sigma = -x)$ satisfy the fluctuation relation

$$\log \frac{\text{Prob}(\Sigma = x)}{\text{Prob}(\Sigma = -x)} = x. \tag{8.82}$$

In figure 8.3 we show the numerical results of molecular dynamics simulations, which confirm that at large times equation (8.82) is well verified within the statistical errors [85].

In conclusion, we have discussed a simple (linear) modelization of the dynamics of a tracer in granular fluids, describing nonequilibrium correlators and response

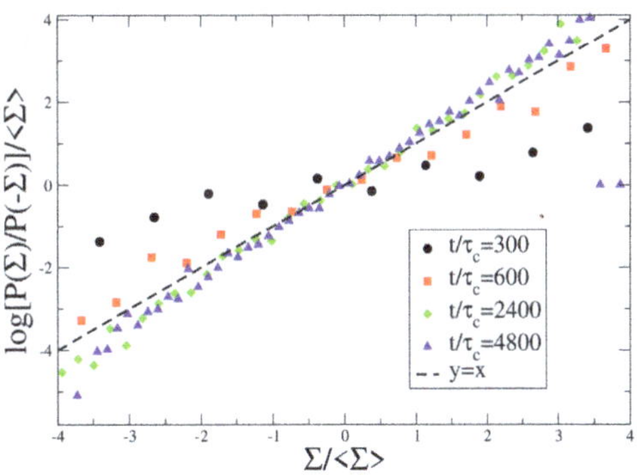

Figure 8.3. Fluctuation relation (8.82) in the system with $\alpha = 0.6$ and $\phi = 0.33$. Adapted from reference [85].

function. This approach is useful to shed light on the role played by recollisions and dynamical memory and their relation with the breakdown of equilibrium properties. In particular, it is interesting to stress that velocity correlations $\langle V(t)U(t')\rangle$ between the tracer and the surrounding velocity field are responsible for both the precise shape of the correlations which appear in the FDT and for the emergence of a finite entropy production, provided that the two fields are *at different temperatures*. Small non-Gaussian corrections, always present in granular fluids, are neglected here in favour of the more important contribution given by memory terms.

8.3.3 Granular Brownian motors

Here we mention another interesting effect that can be realized in granular systems due to their intrinsic nonequilibrium nature: the rectification of unbiased fluctuations. This means that a spatial asymmetric tracer in contact with a granular bath can show a finite average velocity, allowing for the extraction of work from a single (nonequilibrium) bath. The classical setup, introduced by Smoluchowski [88] and studied in detail by Feynman [89], is represented by a rotating ratchet with a pawl mechanism, connected with a wheel in contact with a particle gas. In the case when the particles are molecules at equilibrium at a given temperature, it has been shown that no average direct motion of the ratchet can be obtained [89], unless the system is coupled to two thermostats at different temperatures, in agreement with the second law. However, if the gas is made of macroscopic grains, dissipative interactions introduce nonequilibrium conditions and a finite average motion is possible. In the context of granular systems, several realizations in different geometries have been proposed and studied both theoretically [90–94] and experimentally [95, 96].

We illustrate the following particular case: a massive tracer of mass M, in two dimensions, shaped as an isosceles triangle with base l and angle opposite to the base $2\theta_0$ is immersed in a dilute granular gas (with particles of mass m and unit radius) and is constrained to move along the horizontal axis perpendicular to its base (see figure 8.4). The gas has granular temperature T_g and number density ρ. Computations analogous to those reported in section 8.3.1, based on the linearized Boltmann equation in the form of a master equation and on the Kramers–Moyal expansion with specific transition rates taking into account the geometry of the tracer (for details see reference [90]), lead to the following effective Langevin equation.

$$\dot{V}(t) = -\gamma V(t) + \frac{F}{M} + \Gamma(t). \tag{8.83}$$

The quantities γ, F, $\Gamma(t)$ are functions of the model parameters

$$\gamma = 4\eta\rho l\sqrt{\varepsilon}\sqrt{\frac{T_g}{2\pi M}}(1 + \sin\theta_0), \tag{8.84}$$

where $\varepsilon = m/M$,

$$\frac{F}{M} = -\rho l\frac{T_g}{M}\varepsilon(1 - \sin^2\theta_0)\frac{T_{tr}}{T_g}\left(1 - \frac{T_{tr}}{T_g}\right), \tag{8.85}$$

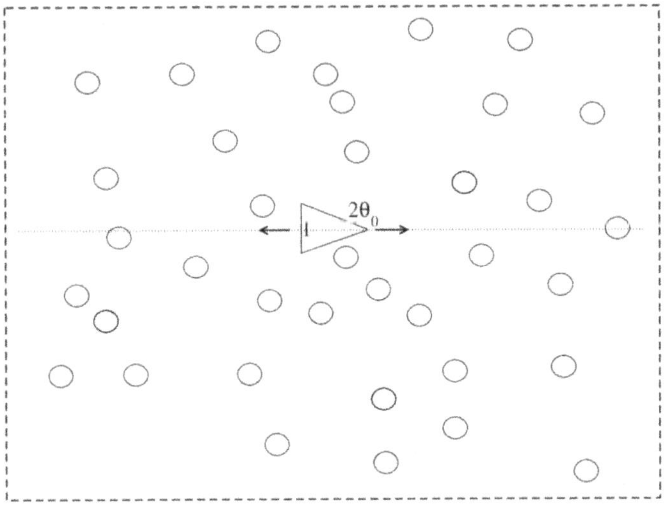

Figure 8.4. The triangular granular Brownian ratchet. Reprinted figure with permission from reference [90], Copyright (2007) by the American Physical Society.

$$\langle \Gamma(t)\Gamma(t')\rangle = \frac{2\gamma T_r}{M}\delta(t - t'), \qquad \langle \Gamma(t)\rangle = 0, \tag{8.86}$$

$$1 - \eta = 1 - \frac{T_{tr}}{T_g} = \frac{1 - \alpha}{2}. \tag{8.87}$$

Therefore, for $\alpha < 1$ the ratchet drifts with an average negative velocity $\langle V(t)\rangle = F/(M\gamma)$. Indeed, the net velocity vanishes linearly with $\varepsilon \to 0$ and is very tiny for massive ratchets. Note that due to equation (8.87) the net driving force is proportional to the temperature difference $T_g - T_r$, so that the tracer and the gas temperatures play a role analogous to the two reservoir temperatures of the standard Brownian ratchet model. The ratchet force vanishes also when the shape becomes symmetric with respect to the inversion of direction of motion, which in this case corresponds to the limit case $\theta_0 \to \pi/2$. A similar general structure for the average drift will be further discussed in chapter 9.

8.4 Hydrodynamics and correlations

Parts of this section have been reprinted from [6] copyright (2015), with permission from Springer Nature.

Fluids can often find themselves in spatially non-homogeneous states, either as a result of nonequilibrium initial conditions—where the system is deliberately set far from its final state, and then observed as it relaxes—or because of the imposed boundary conditions that continuously force the system, maintaining it in a non-equilibrium steady state. In granular fluids there is an intrinsic energy sink that perpetually drives the system out of equilibrium, and an external forcing is necessary in order to maintain it in a stationary state.

The theoretical framework outlined in this chapter is applicable regardless of the source of non-homogeneity, provided the system satisfies the criterion of 'small gradients'. This approach is also useful for describing deviations from homogeneity in the cooling regime, where no external driving force is present.

The Chapman–Enskog procedure is a powerful method for constructing a non-homogeneous solution for the Boltzmann equation under the assumption of weak gradients [97–99]. This method involves several key steps:

- Define densities and fluxes for 'slow' variables: these are quantities that evolve over longer timescales compared to the system's microscopic dynamics.
- Write continuity equations for the 'slow' variables: these equations are universally valid and ensure the conservation of these quantities over time.
- First assumption: the distribution function $P(v, r, t)$ depends on the spatial and temporal coordinates r and t only through the 'slow' variables. This implies that the Boltzmann equation can be approximated by a local Boltzmann equation, supplemented by equations governing the slow parameters.
- Second assumption: the mean free path λ is much smaller than the characteristic length scale of the gradients L, which is comparable to the system's overall size. The ratio $\varepsilon = \lambda/L \ll 1$ is known as the Knudsen number.
- Expand fluxes for small ε: the fluxes are expanded in terms of gradients, retaining only linear terms. The transport coefficients that emerge from this expansion are to be determined.
- Consistency through expansion: the distribution function P and all spatial and temporal derivatives are expanded in powers of ε, such that P is expressed as $P \approx f^{(0)} + \varepsilon f^{(1)} + \varepsilon^2 f^{(2)} + \ldots$. This expansion assumes that the solution P consists of contributions that evolve on different spatial and temporal scales (corresponding to different powers of ε).
- Solving the equations: these expansions are substituted into the Boltzmann equation and the continuity equations for the slow variables, resulting in a hierarchy of equations that can be solved sequentially. Each equation governs the evolution of P at a specific space–timescale.
- Order 0 solution: the zeroth-order solution corresponds to the homogeneous solution (Euler equation for elastic fluids), yielding $f^{(0)}$.
- Order 1 solution: the first-order solution, $f^{(1)}$, is determined through its coefficients in the linear expansion of the gradients. These coefficients relate directly to the transport coefficients.
- Order 2 equations: the second-order equations describe hydrodynamics at the Navier–Stokes level and can be used to find $f^{(2)}$.

The full procedure for molecular gases can be found in classical books of kinetic theory, in the particular granular case one may refer to reference [61]. We sketch the main passages with the aim of a pedagogical introduction. One first assumes that the single-particle distribution function $P(\mathbf{r}, \mathbf{v}, t)$ is defined, and normalized such that integrating over all coordinates and velocities gives the total number of particles N in the fluid. P is assumed to be the solution of the Boltzmann equation.

The particle number density is defined as:

$$n(\mathbf{r},\,t) = \int d^3v\; P(\mathbf{r},\,\mathbf{v},\,t), \tag{8.88}$$

the average molecular velocity is defined as:

$$\mathbf{u}(\mathbf{r},\,t) = \frac{1}{n(\mathbf{r},\,t)} \int d^3v\; \mathbf{v}P(\mathbf{r},\,\mathbf{v},\,t), \tag{8.89}$$

this allows us to introduce the random velocity vector:

$$\mathbf{V}(\mathbf{r},\,t) = \mathbf{v} - \mathbf{u}(\mathbf{r},\,t). \tag{8.90}$$

The vector $\mathbf{V}$ depends on position and time (while $\mathbf{v}$ is independent of both), and has zero average:

$$\int d^3v\; V_i P(\mathbf{r},\,\mathbf{V},\,t) = 0. \tag{8.91}$$

The average fluxes of any molecular quantity $W(\mathbf{v})$ can be expressed as velocity moments of the phase space distribution function:

$$j_W^i(\mathbf{r},\,t) = \int d^3v\; v_i\, W(\mathbf{v}) P(\mathbf{r},\,\mathbf{v},\,t). \tag{8.92}$$

When $W = m$, the flux corresponds to the mass flux:

$$j_m^i = mn(\mathbf{r},\,t)u_i(\mathbf{r},\,t), \tag{8.93}$$

when $W = mv_j$, the flux represents the momentum flux:

$$j_{mv_j}^i = mn(\mathbf{r},\,t)\langle v_i v_j\rangle = mnu_i u_j + mn\langle V_i V_j\rangle. \tag{8.94}$$

This momentum flux is a 3×3 symmetric matrix, including two contributions: the flux due to the bulk (organized) motion and the flux resulting from the random (thermal) motion of the gas particles. The latter is usually referred to as the pressure tensor $\mathscr{P}_{ij} = mn\langle V_i V_j\rangle$.

From the above discussion, two important quantities can be defined: the scalar pressure p and the vector temperature T_i:

$$p = \frac{1}{3}(\mathscr{P}_{xx} + \mathscr{P}_{yy} + \mathscr{P}_{zz}), \tag{8.95}$$

$$\frac{1}{2}k_{\mathrm{B}} T_i = \frac{1}{2}m\langle V_i^2\rangle = \frac{1}{2}\frac{\mathscr{P}_{ii}}{n}. \tag{8.96}$$

In the isotropic case, $T_i = T$, so that $p = nk_{\mathrm{B}}T$, the stress tensor $\mathscr{T}$ can also be defined as:

$$\mathscr{T}_{ij} = \boldsymbol{\delta}_{ij}p - \mathscr{P}_{ij}. \tag{8.97}$$

This tensor expresses the deviation of the pressure tensor from the equilibrium Maxwellian case (where $\mathscr{P}_{ij} = p\delta_{ij}$). Finally, the flux of the quantity $W = mv_j v_k$ is given by:

$$j^i_{mv_j v_k} = mnu_i u_j u_k + u_i \mathscr{P}_{jk} + u_j \mathscr{P}_{ik} + u_k \mathscr{P}_{ij} + \mathscr{Q}_{ijk}. \tag{8.98}$$

Here, $\mathscr{Q}_{ijk} = mn\langle V_i V_j V_k \rangle$ is the generalized heat flow tensor, which describes the transport of random energy $V_j V_k$ due to the thermal motion V_i of the molecules (considering all permutations of i, j, k). In equation (8.98), three contributions can be identified: the first term describes the bulk transport of momentum; the second, third, and fourth terms represent a combination of bulk and random momentum fluxes; and the last term accounts for the transport of random energy due to the random motion itself.

Often, a more intuitive 'classical' heat flow vector is introduced in place of the generalized heat flow tensor:

$$q_i = \frac{\mathscr{Q}_{ikk}}{2} = n\left\langle V_i \frac{mc^2}{2} \right\rangle, \tag{8.99}$$

where $c^2 = \sum_k V_k^2$. By multiplying the Boltzmann equation by 1, v, and v^2, and then integrating over v, one derives the hydrodynamic equations governing the slow variables:

$$\frac{\partial n}{\partial t} + \nabla \cdot (n\vec{u}) = 0, \tag{8.100a}$$

$$\frac{\partial \vec{u}}{\partial t} + \vec{u} \cdot \nabla \vec{u} + \frac{1}{nm} \nabla \cdot \mathscr{P} = 0, \tag{8.100b}$$

$$\frac{\partial T}{\partial t} + \vec{u} \cdot \nabla T + \frac{2}{3n}[\mathscr{P}\colon (\nabla \vec{u}) + \nabla \cdot \vec{q}] + \zeta T = 0. \tag{8.100c}$$

These continuity equations are universally valid. The only additional term in the equation for non-elastic gases is the ζT term, which is zero ($\zeta \equiv 0$) in the case of elastic collisions. Recall that $\zeta(t)$ was previously defined in equation (8.14). The Chapman–Enskog method follows the steps listed above. Among them, the crucial ones are:

1. Spatial scaling: introduce a new spatial scale $r \to \varepsilon r$, where $\varepsilon = \lambda / L$ (with λ being the mean free path and L the macroscopic scale). Consequently, all gradients transform as $\nabla \to \varepsilon \nabla$.

2. Linear approximation: for small ε, the fluxes can be approximated as linear functions of the gradients:

$$\mathscr{P}_{ij} = p\delta_{ij} - \eta\varepsilon\left(\nabla_i u_j + \nabla_j u_i - \frac{2}{3}\delta_{ij} \nabla \cdot \vec{u}\right), \tag{8.101a}$$

$$\vec{q} = -\kappa\varepsilon \nabla T - \mu\varepsilon \nabla n, \tag{8.101b}$$

The primary unknowns here are the transport coefficients η, κ, and μ.

3. Normal solution assumption: assume a 'normal' solution for the distribution function $P(v, r, t) \approx f[V|n(r, t), u(r, t), T(r, t)]$, with $V = v - u$. The time derivative is then expressed as:

$$\frac{\partial f}{\partial t} = \frac{\partial f}{\partial n}\frac{\partial n}{\partial t} + \frac{\partial f}{\partial \vec{u}} \cdot \frac{\partial \vec{u}}{\partial t} + \frac{\partial f}{\partial T}\frac{\partial T}{\partial t}. \tag{8.102}$$

4. Timescale expansion: to ensure consistency with the above expansions, introduce timescales corresponding to different powers of ε, representing variations over distinct spatial scales:

$$\frac{\partial}{\partial t} = \frac{\partial^{(0)}}{\partial t} + \varepsilon\frac{\partial^{(1)}}{\partial t} + \varepsilon^2\frac{\partial^{(2)}}{\partial t} + \cdots. \tag{8.103}$$

5. Space expansion: similarly, expand the spatially non-uniform function f as:

$$f = f^{(0)} + \varepsilon f^{(1)} + \varepsilon^2 f^{(2)} + \cdots. \tag{8.104}$$

Applying this procedure to the granular Boltzmann equation leads—through lengthy calculations—to the granular hydrodynamics equations at the Navier–Stokes order. They correspond to equations (8.100), with constitutive relations given by equations (8.101) and transport coefficients given by

$$\eta = \frac{15}{2(1 + r)(13 - r)\sigma^2}\left(1 + \frac{3}{8}\frac{4 - 3r}{13 - r}a_2\right)\sqrt{mT/\pi}, \tag{8.105}$$

$$\kappa = \frac{75}{2(1 + r)(9 + 7r)\sigma^2}\left(1 + \frac{1}{32}\frac{797 + 211r}{9 + 7r}\right)\sqrt{T/(\pi m)}, \tag{8.106}$$

$$\mu = \frac{750(1 - r)}{(1 + r)(9 + 7r)(19 - 3r)n\sigma^2}(1 + q(r)a_2)\sqrt{T^3/(\pi m)}, \tag{8.107}$$

$$\zeta = \frac{2}{3}n\sigma^2\sqrt{\frac{2T}{m}}\mu_2 + \zeta^{(2)}, \tag{8.108}$$

with $q(r)$ a quite lengthy function of the restitution coefficient r (see reference [100]). In equation (8.108) we have included the contribution $\zeta^{(2)}$ from second-order gradients to the cooling rate: this contribution is necessary for consistency with the rest of the equations (see below). The contribution, however, for low inelasticities is negligible. A detailed discussion can be found in reference [100].

It is important to stress that hydrodynamics for granular fluids is established on a less solid ground than the hydrodynamics for molecular fluids. Critiques have been discussed in several papers, where it is basically understood that for a granular fluid the existence of slow fields obeying hydrodynamic equation is not always guaranteed [6, 101–103].

8.4.1 Langevin equations for hydrodynamic modes

Granular fluids present a fascinating and largely unresolved problem in the study of fluctuations of macroscopic observables. Unlike molecular fluids, granular systems are sometimes composed of a relatively small number of grains, typically in the range of 10^2–10^4, which is far fewer than the number of particles in a molecular fluid. As a result, fluctuations in granular fluids are much more apparent at the macroscopic level. What makes the study of such systems even more challenging and intriguing is the inherently nonequilibrium nature of granular noise. The study of these fluctuations holds potential to bridge gaps between the behaviour of granular systems and that of small molecular systems, such as cell sub-units, biophysical systems, and even micro/nano-mechanical devices, all of which frequently operate out of equilibrium.

A promising theoretical approach to understanding these fluctuations is through fluctuating hydrodynamics. This framework involves examining fluctuations around a system's evolution governed by hydrodynamic equations. Essentially, and in perfect analogy with what we have discussed in chapter 6 of this book, it separates the system's evolution into a set of slowly varying variables and a rapidly relaxing remainder, treated as noise. This method works best when the system is in equilibrium, where time-reversal symmetry imposes strict constraints on the characteristics of the noise. In chapter 4 we already discussed the earliest and most famous examples of this approach: the so-called Einstein relation between diffusivity and mobility in the theory of Brownian motion. Throughout the early 20th century, several other examples followed, culminating in the general theory of linear response, rooted in the equilibrium FDT. Fluctuating hydrodynamics near equilibrium uses these results, notably the Green–Kubo relations, which connect transport coefficients to time-correlated currents or, equivalently, to the amplitude of hydrodynamic noise.

This elegant theoretical framework, however, encounters significant challenges when applied to systems far from equilibrium, such as granular fluids. Despite these obstacles, reasonable assumptions about hydrodynamic noise have allowed researchers to derive fluctuating equations for granular systems. In some instances, these assumptions have been rigorously derived from microscopic models, with varying degrees of success. Usually, as sketched for few cases in the following, specific properties of the noise must be assumed. These assumptions are tested by deriving predictions for fluctuation amplitudes, such as correlations, and comparing them to numerical or experimental results.

A pedagogical introduction to this topic can be obtained by focusing on a particular hydrodynamic mode: the shear mode in two dimensions, defined as

$$U_{\perp}(k,t) = \sum_{j=1}^{N} v_{y,j}(t) e^{-ikx_j(t)},$$

where k is the wave number of the selected mode, and $x_j(t)$ and $v_{y,j}(t)$ represent the position and y-velocity of particle j at time t, respectively. The study of fluctuations of this mode is done by linearizing the hydrodynamic equations around a particular state, for instance the homogeneous one. For both granular and standard (molecular) linearized hydrodynamics, the mode $U_\perp(k, t)$ is decoupled from other modes [104, 105].

To probe the amplitude of fluctuations, the main quantity under investigation is the rescaled autocorrelation function:

$$C_\perp(k, t) = \frac{\langle U_\perp(k, 0) U_\perp^*(k, t) \rangle}{2T_y},$$

measured in a steady state, where $T_\beta = \langle v_\beta^2 \rangle$ represents the granular temperature in the β direction, with β being either x or y. In isotropic systems, such as homogeneous cooling or bulk-driven systems, $T_y / T_x = 1$.

At equilibrium, using the Landau–Lifshitz fluctuating hydrodynamics model and Einstein's fluctuation formula, the shear mode $U_\perp(k, t)$ follows:

$$\partial_t U_\perp(k, t) = -\nu k^2 U_\perp(k, t) + \xi(k, t),$$

where $\nu = \eta / n$ is the kinematic viscosity, and $\xi(t)$ is white Gaussian noise with zero mean. The noise variance is given by:

$$\langle \xi(k, t) \xi(k', t') \rangle = \delta_{k', -k} \delta(t - t') 2TN \nu k^2.$$

For an extensive field $U_\perp(k, t)$, the noise variance is proportional to the number of particles, N, meaning the variance scales with $1/N$ for the intensive field. Solving for the autocorrelation function in this model yields:

$$C_\perp(k, t) = \frac{N}{2} e^{-\nu k^2 t}. \tag{8.109}$$

In the case of inelastic homogeneous cooling, the Einstein formula for noise amplitude does not hold. However, it is possible to apply it in a modified form by using the granular temperature, although additional care is needed. As the system cools, both the kinematic viscosity $\nu(t) \propto \sqrt{T(t)}$ and temperature $T(t)$ decrease over time. By introducing a rescaled time τ, proportional to the cumulative number of collisions, together with a rescaled velocity $w(k, \tau) = U_\perp(k, \tau) / \sqrt{T(\tau)}$ and noise $\xi' = \xi / (\omega_c \sqrt{T})$, the following stochastic equation emerges

$$\partial_\tau w(k, \tau) = -z(k) w(k, \tau) + \xi'(k, t),$$

with $z(k) = qk^2 - \gamma_0$, where the noise and other statistical properties are now time-independent. The shape of $z(k)$ shows an important phenomenon of the HCS, i.e. its instability for small k, as it changes sign when k is small enough. The instability signals the breakdown of the homogeneous field, and the emergence of macroscopic shear modes whose fate cannot be analysed in the linear framework. Simulations show the appearance of 'swarms' of granular particles, moving coherently in the same direction [106–108]. The steady-state autocorrelation function becomes:

$$C_\perp(k, \tau) = \frac{N}{2} \frac{qk^2}{z(k)} e^{-z(k)t}.$$

Driven granular systems introduce additional complexities. A study on inelastic grains driven by random velocity kicks derived a Langevin equation for the shear mode, incorporating both internal and external noise:

$$\partial_\tau U_\perp(k, t) = -\nu k^2 U_\perp(k, t) + \xi''(k, t),$$

with noise variance:

$$\langle \xi''(k, t)\xi''(k', t')\rangle = \delta_{k',-k}\delta(t - t')NT(2\nu k^2 + \zeta).$$

The corresponding autocorrelation is:

$$C_\perp(k, t) = \frac{N}{2}\frac{\zeta/2 + \nu k^2}{\nu k^2} e^{-\nu k^2 t},$$

which diverges for small k.

The situations changes if the external source of energy includes an external viscosity [75–110]. This model has been demonstrated to fairly describe the experimental behaviour of a quasi-2D system on a horizontal vibrating plate [111, 112], where roughness of the surface is modeled as a kick+dissipation mechanism which on average defines a 'bath temperature' T_b and an interaction time $1/\gamma_b$. Repeating the above arguments for this case, one gets

$$\partial_\tau U_\perp(k, t) = -(\nu k^2 + \gamma_b)U_\perp(k, t) + \xi'''(k, t), \tag{8.110}$$

with

$$\langle \xi'''(k, t)\xi'''(k', t')\rangle = \delta_{k',-k}\delta(t - t')2N(\nu k^2 T + \gamma_b T_b). \tag{8.111}$$

Equation (8.111) leads to the interesting result

$$C_\perp(k, t) = \frac{N}{2}\frac{\nu k^2 T + \gamma_b T_b}{\nu k^2 + \gamma_b} e^{-(\nu k^2 + \gamma_b)t}, \tag{8.112}$$

which turns back to (8.109) in the elastic case when the external bath is detached from the system ($\gamma_b = 0$). Equation (8.112) is interesting because the static structure factor $C_\perp(k, 0)$ defines a *finite* correlation length $\lambda_\perp = \sqrt{\nu/\gamma_b}$, a crucial difference with respect to the non-viscous bath, $\gamma_b = 0$. For length scales smaller than $\lambda_\perp$ the structure factor settles to the granular temperature T, while for larger length scales, it saturates to T_b. In a real experiment, the measure of $C_\perp(k, 0)$ is an effective way to access to T_b, an 'external temperature' not easy to be detected with other means.

In boundary-driven systems, such as those involving gravity, energy injection occurs at localized boundaries, and the system can be divided into layers. Each layer behaves like a homogeneous subsystem, with energy transferred from adjacent layers. Fluctuating hydrodynamics in such systems have been shown, in numerical simulations, to capture the dynamics effectively [113].

In conclusion, fluctuating hydrodynamics, while elegant and well-developed for equilibrium systems, presents significant challenges in nonequilibrium granular systems. Nevertheless, with appropriate assumptions and modifications, it remains a valuable tool for understanding fluctuation dynamics across a variety of complex systems.

8.5 Active matter

Parts of this section have been reprinted from [6] copyright (2015), with permission from Springer Nature.

Active fluids is a topic which, although newer than that of granular materials, reveals striking analogies in both the observed phenomena and the methods of investigation. Active fluids—or active matter—refer to a broad class of systems, often involving living organisms, where a large number of self-propelled constituents interact [114]. Examples include biofilaments and molecular motors *in vitro* or *in vivo* [115], collections of motile microorganisms such as algal blooms or biofilms, bird flocks [116], fish schools [117], and even chemical or mechanical imitations.

The main characteristic of active fluids is propulsion: the individual constituents or 'particles' move by walking, crawling, swimming, flying, and so on [118]. An active particle can transform energy from a reservoir into directed motion. Perhaps the most well-known example is that of molecular motors within cells, where propulsion is driven by chemical reactions involving the hydrolysis of ATP [119]. Across various scales, from bacteria to birds and fish, organisms exhibit a wide array of propulsion mechanisms, each with its own complexity and elegance. However, in the study of active fluids, the specific details of propulsion mechanisms are less important than understanding the collective behaviour of many particles under certain boundary conditions [120]. The presence or absence of a solvent fluid, for instance, determines the nature of interactions among particles, which can be either dissipative or conservative, and either contact-based or long-range. Self-propelled particles dispersed in a fluid, often possessing a well-defined polarity, are commonly referred to as 'swimmers'. These swimmers are typically classified into two categories: pushers and pullers, depending on whether the movement originates from the front or the rear, which in turn affects the flow generated in the surrounding fluid and the interaction with other particles.

A collection of self-propelled particles inherently exists out of equilibrium. Each particle injects kinetic energy into the system as it converts energy from an external reservoir. Viscous dissipation through the solvent, or, more rarely, non-conservative interactions with surrounding particles, balances this energy injection and may lead to a statistically steady state. This energy balance bears a resemblance to the energy cycle in shaken granular fluids, where grains acquire energy from collisions with the moving walls of a shaken container and dissipate it through further collisions. On closer examination, one might consider a granular fluid as an 'anti-active' fluid, where the smallest scale (the grain or particle) is responsible for dissipation, while in active fluids, it serves as the energy source. However, there are instances where the analogy between active and granular fluids is even more precise, such as in the thermostat model discussed in section 8.2.2.

One of the most common phenomena observed in active fluids is the occurrence of ordering transitions. These transitions, typically towards a polar (ferromagnetic) or nematic order, arise when parameters like propulsion velocity or particle density are varied. The elongated shape that allows for defining the 'direction' of a self-propelled particle is perhaps the primary distinction between active fluids and most granular systems. Nevertheless, the shear instability in cooling granular materials, where clusters of grains move like 'swarms', as discussed in section 8.2.1, is not dissimilar from collective behaviour observed in many models of active particles, such as bacterial suspensions and colonies [117].

Another landmark of active matter is the so-called motility-induced phase transition [121]. In this phenomenon, investigated in numerical and analytical models [122, 123], a solid–gas phase separation is observed which originates from the self-propulsion mechanism and not from attractive interactions. As a matter of fact, it can be observed even for non-attracting particles and it appears when the motility parameters (speed and persistence time) are increased.

The analogy between active and granular fluids has also been explored experimentally. In some studies, granular particles were shaped to break isotropy and placed on a vertically vibrated horizontal plate. The combination of anisotropy, subtle frictional mechanisms, and vertical vibration produced a self-propulsion effect. Several properties, including collective swarming-like behaviours, have been demonstrated [124].

Perhaps the most significant analogy with granular materials lies in the methods of investigation. Hydrodynamics is employed to describe many collective phenomena in both granular and active fluids [125]. The underlying principle is similar: a few slowly evolving observables are identified, and transport equations for these observables are constructed based on more or less rigorous kinetic theories. Symmetry arguments usually suffice to determine the basic structure of these equations, while more detailed (microscopic) calculations are required to assign values to the transport coefficients [126]. Even the qualitative form of hydrodynamic equations, along with approximate orders of magnitude for the transport coefficients, is often sufficient to capture the stable states and the transitions between them.

Active hydrodynamics, however, encounters challenges similar to those faced in granular hydrodynamics: the number of microscopic constituents is not vast, and the separation of scales between macroscopic and microscopic lengths or times is not always clear or guaranteed. In many cases, significant fluctuations are observed, necessitating a stochastic treatment of the macroscopic equations [127]. While fluctuating hydrodynamics for active fluids is still in its early stages [127], a few important strides have already been made in the context of granular systems.

Recent experimental, theoretical, and numerical investigations have revealed the presence of spontaneous equal-time velocity correlations in active matter systems, an intriguing emergent collective phenomenon. These velocity correlations represent a distinctive feature of active matter systems, and uncovering the underlying physical mechanisms behind them would be a significant breakthrough. In most cases, such correlations are attributed to velocity-aligning interactions, the fundamental ingredient behind the flocking transition seen in Vicsek-like models [120] and Toner–Tu hydrodynamics [117]. However, certain self-propelled particle systems display striking velocity and activity patterns even in the absence of these interactions [128]. Studies by Henkes *et al* [129] and Caprini *et al* [130] have demonstrated that spatial velocity correlations can emerge in systems of spherical particles without alignment forces. This phenomenon arises purely from the interplay between persistent active forces and steric repulsion. Caprini and collaborators examined two-dimensional active Brownian particles (ABPs) under high-packing conditions, developing a microscopic theory that predicts the exponential decay of spatial velocity correlations. The characteristic coherence length was determined based on model parameters, both with and without the inclusion of inertia. The basic principles of these theories are the same as discussed above for granular fluctuating hydrodynamics. These studies collectively suggest that the emergence of self-organized patterns in the velocity field is a fundamental property of active matter systems and may occur even in the absence of direct aligning interactions.

A fundamental property of self-propulsion is the existence of a finite persistence time τ_R: at small times, $t \ll \tau_R$, the particle goes roughly straight, while at time larger than this value, $t \gg \tau_R$, it tends to diffuse in random direction. This can be rationalized in terms of the mean-squared displacement which, for many simple active models (considering $d = 2$ for the sake of simplicity) reads

$$\langle |\mathbf{X}|^2(t)\rangle = [4D_T + v^2\tau_R]t + \frac{v^2}{2}\tau_R^2[e^{-2t/\tau_R} - 1],$$

where $D_T = k_B T/\gamma$ is the passive translational diffusion coefficient due to Brownian motion at the thermal temperature T, v is the typical velocity of self-propulsion, and τ_R is the mean re-orientation time, related to rotational diffusion (the rotational diffusion coefficient is τ_R^{-1}). The above expression for the mean-squared displacement is akin to that of a passive Brownian particle with inertia; however, the typical time separating ballistic to standard diffusion here is macroscopic (τ_R is in the range between micro-seconds and seconds) while for a passive Brownian particle is several orders of magnitude smaller, as it is related to the mean collision time with fluid molecules [131].

An interesting model for active particles—compatible with the above ballistic-diffusive behaviour—is the so-called active Ornstein–Uhlenbeck, where each active particle obeys the following equation of motion:

$$\begin{aligned} \dot{x}(t) &= v(t) \\ m\dot{v}(t) &= -\gamma v(t) + \sqrt{2\gamma T}\xi(t) - U'(x) + \gamma v_a(t) \\ \dot{v}_a(t) &= -\frac{v_a(t)}{\tau_R} + \sqrt{2\frac{D_{eff}}{\tau_R^2}}\xi_a(t), \end{aligned}$$

where $D_{\text{eff}} = v^2\tau_R$ is related to rotational diffusion, m is the particle's mass, $U(x)$ is an external potential and both $\xi(t)$ and $\xi_a(t)$ are Gaussian white noises with zero average and unitary variance.

In the literature, the model is usually presented in its overdamped limit [132], i.e. disregarding the timescales of order τ_m or smaller, that is equivalent to consider $\dot{x} \approx \frac{x(t+\Delta t) - x(t)}{\Delta t}$ with $\tau_R \gg \Delta t \gg \tau_m$. The overdamped equation reads:

$$\begin{aligned} \dot{x}(t) &= \sqrt{2\frac{T}{\gamma}}\xi(t) - \frac{U'(x)}{\gamma} + v_a(t) \\ \dot{v}_a(t) &= -\frac{v_a(t)}{\tau_R} + \sqrt{2\frac{D_{eff}}{\tau_R^2}}\xi_a(t), \end{aligned}$$

and it is usually studied in the athermal limit, i.e. by putting $T = 0$ because thermal noise is usually negligible for many purposes. This last assumption spoils the underlying time-reversal symmetry which is expected in the limit of vanishing self-propulsion ($D_{\mathrm{eff}} \to 0$) and has consequences for the definition of entropy production. In the case $T = 0$ it is possible to deduce a general approximation for the steady state distribution of this kind of particle in the presence of any potential, which goes under the name of unified coloured noise approximation and gives

$$P(\mathbf{x}) = \mathcal{Q}^{-1}\exp\left[-\frac{U(\mathbf{x})}{D_{\mathrm{eff}}} - \frac{\tau|\nabla U(\mathbf{x})|^2}{2D_{\mathrm{eff}}}\right]\left\|\mathbf{I} + \tau\nabla\nabla U(\mathbf{x})\right\|,$$

where $\mathcal{Q}$ is a normalization factor obtained by integrating over $\mathbf{x}$, $\mathbf{I}$ is the $dN \times dN$ identity matrix, $\nabla\nabla U$ is the Hessian of U and $\|\|$ stands for the absolute value of the determinant of a matrix [133, 134].

References

[1] de Gennes P G 1999 Granular matter: a tentative view *Rev. Mod. Phys.* **71** S374

[2] Andreotti B, Forterre Y and Pouliquen O 2013 *Granular Media* (Cambridge: Cambridge University Press)

[3] Savage S B 1984 The mechanics of rapid granular flows *Adv. Appl. Mech.* **24** 289

[4] Campbell C S 1990 Rapid granular flows *Annu. Rev. Fluid Mech.* **22** 57

[5] Ogawa S 1978 Multitemperature theory of granular materials *Proc. US–Japan Symp. on Continuum Mechanics and Statistical Approaches to the Mechanics of Granular Media* ed Media (Tokyo: Gakujutsu Bunken Fukyu-kai)

[6] Puglisi A 2014 *Transport and Fluctuations in Granular Fluids: From Boltzmann equation to Hydrodynamics, Diffusion and Motor Effects* (Berlin: Springer)

[7] Fan L T, Chen Y M and Lai F S 1990 Recent developments in solids mixing *Powder Technol.* **61** 255

[8] Faraday M 1831 On a peculiar class of acoustical figures; and on certain forms assumed by groups of particles upon vibrating elastic surfaces *Phil. Trans. R. Soc. London.* **52** 299

[9] Knight J B, Jaeger H M and Nagel S R 1993 Vibration-induced size separation in granular media: the convection connection *Phys. Rev. Lett.* **70** 3728

[10] Ehrichs E E, Jaeger H M, Karczmar G S, Knight J B, Kuperman V Y and Nagel S R 1995 Granular convection observed by magnetic resonance imaging *Science* **267** 1632

[11] Kuperman V Y, Ehrichs E E, Jaeger H M and Karczmar G S 1995 A new technique for differentiating between diffusion and flow in granular media using magnetic resonance imaging *Rev. Sci. Instrum.* **66** 4350

[12] Knight J B, Ehrichs E E, Kuperman V Y, Flint J K, Jaeger H M and Nagel S R 1996 Experimental study of granular convection *Phys. Rev.* E **54** 5726

[13] Knight J B 1997 External boundaries and internal shear bands in granular convection *Phys. Rev.* E **55** 6016

[14] Melo F, Umbanhowar P B and Swinney H L 1994 Transition to parametric wave patterns in a vertically oscillated granular layer *Phys. Rev. Lett.* **72** 172

[15] Melo F, Umbanhowar P B and Swinney H L 1995 Hexagons, kinks, and disorder in oscillated granular layers *Phys. Rev. Lett.* **75** 3838

[16] Umbanhowar P B, Melo F and Swinney H L 1996 Localized excitations in a vertically vibrated granular layer *Nature* **382** 793
[17] Metcalf T H, Knight J B and Jaeger H M 1997 Standing wave patterns in shallow beds of vibrated granular material *Physica* A **236** 202
[18] Kudrolli A and Gollub J P 1997 Studies of cluster formation due to collisions in granular material *Powders & Grains 97* (Rotterdam: Balkema)
[19] Kudrolli A, Wolpert M and Gollub J P 1997 Cluster formation due to collisions in granular material *Phys. Rev. Lett.* **78** 1383
[20] Falcon E, Fauve S and Laroche C 1999 Cluster formation, pressure and density measurements in a granular medium fluidized by vibrations *Eur. Phys. J.* B **9** 183
[21] Falcon E, Wunenburger R, Evesque P, Fauve S, Chabot C, Garrabos Y and Beysens D 1999 Cluster formation in a granular medium fluidized by vibrations in low gravity *Phys. Rev. Lett.* **83** 440
[22] Falcon E, Fauve S and Laroche C 2001 An experimental study of a granular gas fluidized by vibrations *An experimental study of a granular gas fluidized by vibrations Granular* **vol 564** ed T Pöschel and S Luding (Berlin: Springer)
[23] Olafsen J S and Urbach J S 1998 Clustering, order and collapse in a driven granular monolayer *Phys. Rev. Lett.* **81** 4369
[24] Eshuis P, van der Weele K, van der Meer D and Lohse D 2005 Granular Leidenfrost effect: experiment and theory of floating particle clusters *Phys. Rev. Lett.* **95** 258001
[25] Eshuis P, van der Weele K, van der Meer D, Bos R and Lohse D 2007 Phase diagram of vertically shaken granular matter *Phys. Fluids* **19** 123301
[26] Gnoli A, De Arcangelis L, Giacco F, Lippiello E, Pica Ciamarra M, Puglisi A and Sarracino A 2018 Controlled viscosity in dense granular materials *Phys. Rev. Lett.* **120** 138001
[27] Plati A, de Arcangelis L, Gnoli A, Lippiello E, Puglisi A and Sarracino A 2021 Getting hotter by heating less: how driven granular materials dissipate energy in excess *Phys. Rev. Res.* **3** 013011
[28] Folli V, Puglisi A, Leuzzi L and Conti C 2012 Shaken granular lasers *Phys. Rev. Lett.* **108** 248002
[29] Folli V, Ghofraniha N, Puglisi A, Leuzzi L and Conti C 2013 Time-resolved dynamics of granular matter by random laser emission *Sci. Rep.* **3** 2251
[30] Luding S, Clément E, Blumen A, Rajchenbach J and Duran J 1994 Studies of columns of beads under external vibrations *Phys. Rev.* E **49** 1634
[31] Warr S, Jacques G T H and Huntley J M 1994 Tracking the translational and rotational motion of granular particles: use of high-speed photography and image processing *Powder Technol.* **81** 41
[32] Warr S, Huntley J M and Jacques G T H 1995 Fluidization of a two-dimensional granular system: experimental study and scaling behavior *Phys. Rev.* E **52** 5583
[33] Warr S and Huntley J M 1995 Energy input and scaling laws for a single particle vibrating in one dimension *Phys. Rev.* E **52** 5596
[34] Pontuale G, Gnoli A, Reyes F V and Puglisi A 2016 Thermal convection in granular gases with dissipative lateral walls *Phys. Rev. Lett.* **117** 098006
[35] Kumaran V 1998 Temperature of a granular material 'fluidized' by external vibrations *Phys. Rev.* E **57** 5660

[36] Sunthar P and Kumaran V 1999 Temperature scaling in a dense vibrofluidized granular material *Phys. Rev.* E **60** 1951

[37] Wildman R D, Huntley J M and Parker D J 2001 Convection in highly fluidized three-dimensional granular beds *Phys. Rev. Lett.* **86** 3304

[38] Kudrolli A and Henry J 2000 Non-Gaussian velocity distributions in excited granular matter in the absence of clustering *Phys. Rev.* E **62** R1489

[39] Olafsen J S and Urbach J S 1999 Velocity distributions and density fluctuations in a 2D granular gas *Phys. Rev.* E **60** R2468

[40] Losert W, Cooper D G W, Delour J, Kudrolli A and Gollub J P 1999 Velocity statistics in vibrated granular media *Chaos* **9** 682

[41] van Noije T P C and Ernst M H 1998 Velocity distributions in homogeneous granular fluids: the free and the heated case *Granul. Matter* **1** 57

[42] Rouyer F and Menon N 2000 Velocity fluctuations in a homogeneous 2D granular gas in steady state *Phys. Rev. Lett.* **85** 3676

[43] Feitosa K and Menon N 2002 Breakdown of energy equipartition in a 2D binary vibrated granular gas *Phys. Rev. Lett.* **88** 198301

[44] Feitosa K and Menon N 2004 Fluidized granular medium as an instance of the fluctuation theorem *Phys. Rev. Lett.* **92** 164301

[45] Gallavotti G and Cohen E G D 1995 Dynamical ensembles in stationary states *J. Stat. Phys.* **80** 931–70

[46] Puglisi A, Visco P, Barrat A, Trizac E and van Wijland F 2005 Fluctuations of internal energy flow in a vibrated granular gas *Phys. Rev. Lett.* **95** 110202

[47] Naert A 2012 Experimental study of work exchange with a granular gas: the viewpoint of the fluctuation theorem *Europhys. Lett.* **97** 20010

[48] Mounier A and Naert A 2012 The Hatano-Sasa equality: transitions between steady states in a granular gas *Europhys. Lett.* **100** 30002

[49] Blair D L and Kudrolli A 2001 Velocity correlations in dense granular gases *Phys. Rev.* E **64** 050301

[50] Gradenigo G, Sarracino A, Villamaina D and Puglisi A 2011 Non-equilibrium length in granular fluids: from experiment to fluctuating hydrodynamics *Europhys. Lett.* **96** 14004

[51] Puglisi A, Gnoli A, Gradenigo G, Sarracino A and Villamaina D 2012 Structure factors in granular experiments with homogeneous fluidization *J. Chem. Phys.* **136** 014704

[52] D'Anna G, Mayor P, Barrat A, Loreto V and Nori F 2003 Observing Brownian motion in vibration-fluidized granular matter *Nature* **424** 909

[53] Gnoli A, Puglisi A, Sarracino A and Vulpiani A 2014 Nonequilibrium Brownian motion beyond the effective temperature *PLoS One* **9** e93720

[54] Plati A, Baldassarri A, Gnoli A, Gradenigo G and Puglisi A 2019 Dynamical collective memory in fluidized granular materials *Phys. Rev. Lett.* **123** 038002

[55] Plati A and Puglisi A 2022 Collective drifts in vibrated granular packings: the interplay of friction and structure *Phys. Rev. Lett.* **128** 208001

[56] Lucente D, Viale M, Gnoli A, Puglisi A and Vulpiani A 2023 Revealing the nonequilibrium nature of a granular intruder: the crucial role of non-Gaussian behavior *Phys. Rev. Lett.* **131** 078201

[57] Lohse D and Rauhè R 2004 Creating a dry variety of quicksand *Nature* **432** 689

[58] Costantini G, Puglisi A and Marconi U M B 2007 Brownian ratchet model *Phys. Rev.* E **75** 061124

[59] Eshuis P, van der Weele K, Lohse D and van der Meer D 2010 Experimental realization of a rotational ratchet in a granular gas *Phys. Rev. Lett.* **104** 248001
[60] Gnoli A, Petri A, Dalton F, Gradenigo G, Pontuale G, Sarracino A and Puglisi A 2013 Brownian ratchet in a thermal bath driven by Coulomb friction *Phys. Rev. Lett.* **110** 120601
[61] Brilliantov N V and Pöschel T 2004 *Kinetic Theory of Granular Gases* (Oxford: Oxford University Press)
[62] Javier Brey J, Moreno F and Dufty J W 1996 Model kinetic equation for low-density granular flow *Phys. Rev.* E **54** 445
[63] van Noije T P C, Ernst M H and Brito R 1998 Ring kinetic theory for an idealized granular gas *Physica* A **251** 266
[64] Goldshtein A and Shapiro M 1995 Mechanics of collisional motion of granular materials. Part 1. General hydrodynamic equations *J. Fluid Mech.* **282** 75
[65] van Noije T P C and Ernst M H 1998 Velocity distributions in homogeneous granular fluids: the free and the heated case *Granul. Matter* **1** 57
[66] Haff P K 1983 Grain flow as a fluid-mechanical phenomenon *J. Fluid Mech.* **134** 401
[67] Krook and Wu T T 1976 Formation of Maxwellian tails *Phys. Rev. Lett.* **36** 1107
[68] Blackwell D and Mauldin R D 1985 Ulam's redistribution of energy problem: collision transformations *Lett. Math. Phys.* **10** 149
[69] Ben-Naim E and Krapivsky P L 2002 Scaling, multiscaling, and nontrivial exponents in inelastic collision processes *Phys. Rev.* E **66** 011309
[70] Baldassarri A, Marconi U M B and Puglisi A 2002 Influence of correlations on the velocity statistics of scalar granular gases *Europhys. Lett.* **58** 14
[71] Ernst M H and Brito R 2002 Scaling solutions of inelastic Boltzmann equations with over-populated high energy tails *J. Stat. Phys.* **109** 407–32
[72] Bobylev A V, Gamba I M and Panferov V A 2004 Moment inequalities and high-energy tails for Boltzmann equations with inelastic interactions *J. Stat. Phys.* **116** 1651–82
[73] Villani C 2006 Mathematics of granular materials *J. Stat. Phys.* **124** 781–822
[74] Bobylev A V, Cercignani C and Gamba I M 2008 Generalized kinetic Maxwell type models of granular gases *Mathematical Models of Granular Matter* ed Lecture Notes in Mathematics (Berlin: Springer)
[75] Puglisi A, Loreto V, Marconi U M B and Vulpiani A 1998 Clustering and non-Gaussian behavior in granular matter *Phys. Rev. Lett.* **81** 3848
[76] Puglisi A, Loreto V, Marconi U M B and Vulpiani A 1999 Kinetic approach to granular gases *Phys. Rev.* E **59** 5582
[77] Kubo R, Toda M and Hashitsume N 2012 *Statistical Physics II: Nonequilibrium Statistical Mechanics* (Berlin: Springer Science)
[78] Bena I, Coppex F, Droz M, Visco P, Trizac E and van Wijland F 2006 Stationary state of a heated granular gas: fate of the usual H-functional *Physica* A **370** 179
[79] Bettolo Marconi U M, Puglisi A and Vulpiani A 2013 About an H-theorem for systems with non-conservative interactions *J. Stat. Mech. Theory Exp.* **2013** P08003
[80] De Soria M I G, Maynar P, Mischler S, Mouhot C, Rey T and Trizac E 2015 Towards an H-theorem for granular gases *J. Stat. Mech. Theory Exp.* **2015** P11009
[81] Plata C A and Prados A 2017 Global stability and h theorem in lattice models with nonconservative interactions *Phys. Rev.* E **95** 052121
[82] Sarracino A, Villamaina D and Costantini G 2010 Granular Brownian motion *J. Stat. Mech.J. Stat. Mech. Theory Exp.* **2010** P04013

[83] Puglisi A, Visco P, Trizac E and van Wijland F 2006 Dynamics of a tracer granular particle as a nonequilibrium Markov process *Phys. Rev.* E **73** 021301

[84] Risken H 1996 *Fokker-Planck equation* (Berlin: Springer)

[85] Sarracino A, Villamaina D, Gradenigo G and Puglisi A 2010 Irreversible dynamics of a massive intruder in dense granular fluids *Europhys. Lett.* **92** 34001

[86] Hansen J-P and McDonald I R 2013 *Theory of Simple Liquids: With Applications to Soft Matter* (New York: Academic)

[87] Sarracino A and Vulpiani A 2019 On the fluctuation–dissipation relation in non-equilibrium and non-Hamiltonian systems *Chaos* **29** 083132

[88] Smoluchowski M 1912 Experimentell nachweisbare, der üblichten Thermodynamik widersprechende Molekularphänomene *Phys. Zeitschr.* **13** 1069–89

[89] Feynman R P, Leighton R B and Sands M 1963 *The Feynman Lectures on Physics* (Reading, MA: Addison-Wesley)

[90] Costantini G, Bettolo Marconi U M and Puglisi A 2007 Granular Brownian ratchet model *Phys. Rev.* **75** 061124

[91] Cleuren B and Van den Broeck C 2007 Granular Brownian motor *Europhys. Lett.* **77** 50003

[92] Costantini G, Marconi U M B and Puglisi A 2008 Noise rectification and fluctuations of an asymmetric inelastic piston *Europhys. Lett.* **82** 50008

[93] Talbot J, Wildman R D and Viot P 2011 Kinetics of a frictional granular motor *Phys. Rev. Lett.* **107** 138001

[94] Cleuren B and Eichhorn R 2008 Dynamical properties of granular rotors *J. Stat. Mech. Theory Exp.* **2008** P10011

[95] Eshuis P, van der Weele K, Lohse D and van der Meer D 2010 Experimental realization of a rotational ratchet in a granular gas *Phys. Rev. Lett.* **104** 248001

[96] Gnoli A, Sarracino A, Puglisi A and Petri A 2013 Nonequilibrium fluctuations in a frictional granular motor: experiments and kinetic theory *Phys. Rev.* E **87** 052209

[97] Huang K 1988 *Statistical Mechanics* (New York: Wiley)

[98] Chapman S and Cowling T G 1970 *The Mathematical Theory of Nonuniform Gases* (Cambridge: Cambridge University Press)

[99] Ferziger J H and Kaper G H 1972 *Mathematical Theory of Transport Processes in Gases* (Amsterdam: North-Holland)

[100] Brey J J, Dufty J W, Kim C S and Santos A 1998 Hydrodynamics for granular flow at low density *Phys. Rev.* E **58** 4638

[101] Du Y, Li H and Kadanoff L P 1995 Breakdown of hydrodynamics in a one-dimensional system of inelastic particles *Phys. Rev. Lett.* **74** 1268

[102] Kadanoff L P 1999 Built upon sand: theoretical ideas inspired by granular flows *Rev. Mod. Phys.* **71** 435

[103] Goldhirsch I 1999 Scales and kinetics of granular flows *Chaos* **9** 659

[104] Landau L D and Lifchitz E M 1967 *Physique Statistique* (Éditions MIR)

[105] Mazenko G 2006 *Nonequilibrium Statistical Mechanics* (New York: Wiley)

[106] Orza J A G, Brito R, van Noije T P C and Ernst M H 1997 Patterns and long range correlations in idealized granular flows *Int. J. Mod. Phys.* C **8** 953

[107] van Noije T C P, Ernst M H, Brito R and Orza J A G 1997 Mesoscopic theory of granular fluids *Phys. Rev. Lett.* **79** 411

[108] van Noije T P C and Ernst M H 2000 Cahn-Hilliard theory for unstable granular flows *Phys. Rev.* E **61** 1765

[109] Gradenigo G, Sarracino A, Villamaina D and Puglisi A 2011 Fluctuating hydrodynamics and correlation lengths in a driven granular fluid *J. Stat. Mech. Theory Exp.* **2011** P08017

[110] Garzó V, Chamorro M G and Vega Reyes F 2013 Transport properties for driven granular fluids in situations close to homogeneous steady states *Phys. Rev.* E **87** 032201

[111] Gradenigo G, Sarracino A, Villamaina D and Puglisi A 2011 Non-equilibrium length in granular fluids: from experiment to fluctuating hydrodynamics *Europhys. Lett.* **96** 14004

[112] Puglisi A, Gnoli A, Gradenigo G, Sarracino A and Villamaina D 2012 Structure factors in granular experiments with homogeneous fluidization *J. Chem. Phys.* **014704** 136

[113] Costantini G and Puglisi A 2011 Fluctuating hydrodynamics in a vertically vibrated granular fluid with gravity *Phys. Rev.* E **84** 031307

[114] Ramaswamy S 2010 The mechanics and statistics of active matter *Annu. Rev. Condens. Matter Phys.* **1** 323

[115] Jülicher F, Kruse K, Prost J and Joanny J F 2007 Active behavior of the cytoskeleton *Phys. Rep.* **449** 3

[116] Cavagna A, Giardina I and Grigera T 2018 The physics of flocking: correlation as a compass from experiments to theory *Phys. Rep.* **728** 1–62

[117] Toner J, Tu Y and Ramaswamy S 2005 Hydrodynamics and phases of flocks *Ann. Phys.* **318** 170

[118] Bechinger C, Di Leonardo R, Löwen H, Reichhardt C, Volpe G and Volpe G 2016 Active particles in complex and crowded environments *Rev. Mod. Phys.* **88** 045006

[119] Jülicher F, Ajdari A and Prost J 1997 Modeling molecular motors *Rev. Mod. Phys.* **69** 1269

[120] Vicsek T, Czirók A, Ben-Jacob E, Cohen I and Shochet O 1995 Novel type of phase transition in a system of self-driven particles *Phys. Rev. Lett.* **75** 1226

[121] Cates M E and Tailleur J 2015 Motility-induced phase separation *Annu. Rev. Condens. Matter Phys.* **6** 219–44

[122] Fily Y and Marchetti M C 2012 Athermal phase separation of self-propelled particles with no alignment *Phys. Rev. Lett.* **108** 235702

[123] Speck T 2016 Collective behavior of active Brownian particles: from microscopic clustering to macroscopic phase separation *Eur. Phys. J. Spec. Top.* **225** 2287–99

[124] Deseigne J, Dauchot O and Chaté H 2010 Collective motion of vibrated polar disks *Phys. Rev. Lett.* **105** 098001

[125] Ramaswamy S, Liverpool T B, Prost J, Rao M, Marchetti M C, Joanny J F and Aditi Simha R 2013 Hydrodynamics of soft active matter *Rev. Mod. Phys.* **85** 1143

[126] Tiribocchi A, Wittkowski R, Marenduzzo D and Cates M E 2015 Active model h: scalar active matter in a momentum-conserving fluid *Phys. Rev. Lett.* **115** 188302

[127] Ramaswamy S, Simha R A and Toner J 2003 Active nematics on a substrate: giant number fluctuations and long-time tails *Europhys. Lett.* **62** 196

[128] Großmann R, Aranson I S and Peruani F 2020 A particle-field approach bridges phase separation and collective motion in active matter *Nat. Commun.* **11** 5365

[129] Henkes S, Kostanjevec K, Collinson J M, Sknepnek R and Bertin E 2020 Dense active matter model of motion patterns in confluent cell monolayers *Nat. Commun.* **11** 1405

[130] Caprini L, Marconi U M B, Maggi C, Paoluzzi M and Puglisi A 2020 Hidden velocity ordering in dense suspensions of self-propelled disks *Phys. Rev. Res.* **2** 023321

[131] Saragosti J, Silberzan P and Buguin A 2012 Modeling *E. coli* tumbles by rotational diffusion. Implications for chemotaxis *PLoS One* **7** e35412

[132] Marconi U M B, Gnan N, Paoluzzi M, Maggi C and Di Leonardo R 2016 Velocity distribution in active particles systems *Sci. Rep.* **6** 23297

[133] Jung P and Hänggi P 1987 Dynamical systems: a unified colored-noise approximation *Phys. Rev.* A **35** 4464

[134] Maggi C, Marconi U M B, Gnan N and Di Leonardo R 2015 Multidimensional stationary probability distribution for interacting active particles *Sci. Rep.* **5** 10742

Chapter 9

Stranger things

> Contradiction is not a sign of falsity, nor the lack of contradiction a sign of truth.
>
> Blaise Pascal

> ἔστι δὲ φῦλον ἐν ἀνθρώποισι ματαιότατον,
> ὅστις αἰσχύνων ἐπιχώρια παπταίνει τὰ πόρσω,
> μεταμώνια θηρεύων ἀκράντοις ἐλπίσιν[a].
>
> Pindar

Equilibrium systems are characterized by a time-symmetric dynamics, expressed by the detailed balance (DB) condition, and, as shown in the previous chapters, this implies some constraints on the allowed behaviours and on the observable phenomenology. For instance, equilibrium conditions exclude the possibility to extract work from a single thermal bath, or to induce a particle current in a system coupled to two reservoirs of particles at the same chemical potential. Also, in equilibrium, the response of an observable to an external perturbation is always proportional to the corresponding correlation function.

In this chapter we will present several counter-intuitive results that can be observed in nonequilibrium systems, when space and time symmetries are broken. In these more general cases, a more complex and surprising phenomenology can be observed, due to the central role played by dynamical and kinetic terms. We will focus on different examples where the presence of nonequilibrium sources are exploited to induce 'strange' effects, and will discuss the underlying basic physical mechanisms.

[a] 'For there is among mankind a very foolish kind of person, who scorns what is at hand and peers at things far away, chasing the impossible with hopes unfulfilled.' Pindar. Olympian Odes. Pythian Odes. Ed. and trans. William H Race. 2012. Cambridge (MA) and London: Harvard University Press.

doi:10.1088/978-0-7503-6229-0ch9

9.1 Motion from friction

Friction is usually considered as a hindrance to the motion of an object. The common experience tells us that the greater the friction the greater the force necessary to move an object. However, we could not walk on a perfectly smooth surface, and the precise role of frictional forces hides non-trivial effects.

The study of friction represents a huge field of research, from theoretical aspects related to its microscopic origin to practical applications in engineering [1]. Here we will focus on the simplest model of friction, namely the Coulomb or dry friction: a constant force which is always opposite to the motion (namely, to the velocity) of the object. Since we are interested in systems where fluctuations play a non-negligible role, we will consider a Langevin equation where such a friction force is also present. The problem of the study of Brownian motion in the presence of dry friction was addressed for the first time by de Gennes in 2005 [2], who showed how the diffusive properties of the particle are modified by the presence of this term.

9.1.1 Langevin equation in the presence of dry friction

We consider the generalized Klein–Kramers equation in one dimension, describing the motion of a Brownian particle of mass $m = 1$, with position x and velocity v, in a spatial potential $U(x)$, in the presence of a friction term [3]

$$\begin{aligned} \dot{x}(t) &= v(t) \\ \dot{v}(t) &= -\gamma v(t) - \Delta \mathrm{sign}[v(t)] - U'[x(t)] + \eta(t), \end{aligned} \tag{9.1}$$

where $\eta(t)$ is a white noise, with $\langle \eta(t) \rangle = 0$ and $\langle \eta(t)\eta(t') \rangle = 2\gamma T \delta(t - t')$, and Δ represents the amplitude of friction. Note that here we consider the more general case where a velocity-dependent force is present.

The corresponding Fokker–Planck equation for the evolution of the density probability function $P(x, v)$ of the process described by equation (9.1) is

$$\begin{aligned} \frac{\partial P(x, v)}{\partial t} &= -\frac{\partial}{\partial x}[vP(x, v)] + \frac{\partial}{\partial v}\{[F(v) + U'(x)]P(x, v)\} \\ &+ \gamma T \frac{\partial^2}{\partial v^2} P(x, v) = -\nabla \cdot \boldsymbol{J}^{rev} - \nabla \cdot \boldsymbol{J}^{irr}, \end{aligned} \tag{9.2}$$

where $F(v) = \gamma v + \Delta \mathrm{sign}(v)$, $\nabla = (\partial_x, \partial_v)$ and we have introduced the reversible and irreversible probability currents [4], respectively,

$$\boldsymbol{J}^{rev} = \begin{pmatrix} vP(x, v) \\ -U'P(x, v) \end{pmatrix},$$

and

$$\boldsymbol{J}^{irr} = \begin{pmatrix} 0 \\ -F(v)P(x, v) - \gamma T \partial_v P(x, v) \end{pmatrix}.$$

The stationary condition reads

$$\nabla \cdot (\boldsymbol{J}^{rev} + \boldsymbol{J}^{irr}) = 0; \tag{9.3}$$

the more stringent DB condition is $\boldsymbol{J}^{irr} = 0$. Such a condition can be satisfied only in the following cases: (i) if the external spatial potential is zero, namely $U = 0$, or (ii) if the frictional force is linear, namely $\Delta = 0$. This can be shown as follows. From $\boldsymbol{J}^{irr} = 0$ onc obtains the relation

$$\partial_v P(x, v) = -\frac{F(v)}{\gamma T} P(x, v); \tag{9.4}$$

substituting this relation in the stationary condition $\nabla \cdot \boldsymbol{J}^{rev} = 0$, one obtains the following form for the equilibrium distribution

$$P(x, v) = p(v) e^{-\frac{U(x)}{\gamma T}\frac{F(v)}{v}}, \tag{9.5}$$

where $p(v)$ is an unknown function which only depends on v. Now, taking the derivative with respect to v in equation (9.5) and imposing the condition (9.4) one has the following constraint for $p(v)$

$$\frac{p'(v)}{p(v)} = \frac{U(x)}{\gamma T}\frac{F'(v)v - F(v)}{v^2} - \frac{F(v)}{\gamma T}. \tag{9.6}$$

Since in the left-hand side does not appear any x-dependence, this constraint can be fulfilled only if $U(x) = 0$ or if $F'(v)v - F(v) = 0$, namely if $F(v) \propto v$. In reference [5] it is discussed how to introduce a multiplicative noise in order to recover DB with general nonlinear velocity-dependent forces. Here we note that for a linear friction $F[v(t)] = \gamma v(t)$, the system is characterized by the equilibrium stationary state $P(x, v) \propto \exp\{-[v^2/(2T) + U(x)/T]\}$ at the temperature T. In the general case with $\Delta \neq 0$, the stationary distribution is not known.

Now we show that the presence of nonequilibrium conditions, when coupled with a spatial asymmetry in the system, can lead to directed motion, namely a finite average velocity of the Brownian particle (see figure 9.1, left). Therefore, we will

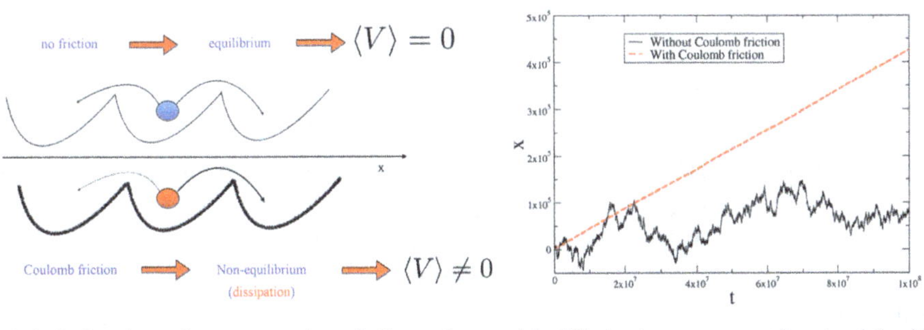

Figure 9.1. Left: schematic representation of a Brownian particle diffusing in an asymmetric potential, without (top) and with (bottom) Coulomb friction. Right: Time evolution of the position $x(t)$ for the model in equation (9.1) without Coulomb friction (black continuous lines), with parameters $\gamma = 0.05$, $\gamma T = 0.5$, and $\mu = 0.4$, and with Coulomb friction(red dashed line), with same parameters and $\Delta = 1$.

consider the presence of an asymmetric ratchet potential of the form usually studied in the literature of Brownian ratchets [6]

$$U(x) = \sin(x) + \mu \sin(2x), \tag{9.7}$$

where μ is an asymmetry parameter. Notice that, at variance with the other standard models of Brownian motors, such as flashing ratchets, rocking ratchets, and deterministically forced ratchets, in this system there are no external forces driving the motor, and the only source of dissipation is the presence of dry friction.

In figure 9.1, right panel, we show a trajectory of the Brownian particle described by equation (9.1), in the absence (continuous black line) and in the presence of dry friction (dashed red line). Only in the latter case can a net motor effect be clearly observed. This is due to the fact that, if only viscosity (i.e. a dissipative force linear in the velocity) is considered, the system is in an equilibrium state and the reversibility of the dynamics protects from any average directed motion.

9.1.2 The asymmetric Rayleigh piston in the presence of dry friction

The same mechanism can be exploited in a different setup to induce a directed motion of a probe of mass M and velocity V, in one dimension, interacting with particles of one or more gases with different parameters: mass m_i, temperature T_i, density ρ_i, coefficient of restitution r_i for the collisions between probe and gas molecules ($r_i \in [0, 1]$, with $r_i = 1$ for elastic collisions), where the index i represents different species. The motion of the tracer is also affected by dry friction, and we assume that the whole system presents an asymmetry in the dynamics as discussed below.

The dynamics of the probe is described by the following equation for the velocity probability $P(V, t)$

$$\begin{aligned}\frac{\partial P(V, t)}{\partial t} &= \int dV'[W(V|V')P(V', t) - W(V'|V)P(V, t)] \\ &+ \Delta \frac{\partial}{\partial V}\text{sign}(V)P(V, t),\end{aligned} \tag{9.8}$$

where Δ is the amplitude of the friction coefficient, and $W(V'|V)$ are the transition rates from V to V' due to collisions, and depend on M and the parameters (ρ_i, T_i, m_i and r_i), see chapter 7 for an explicit example. The asymmetry may be included in the model as follows

$$W(V'|V) = \begin{cases} W^+(V'|V) \text{ if } V' > V \\ W^-(V'|V) \text{ if } V' < V, \end{cases} \tag{9.9}$$

where $W^+(V'|V) \neq W^-(V'|V)$. This form of the transition rates introduces an asymmetry in the system but the condition (9.9) is not sufficient for a ratchet effect to be observed. Indeed, if $\Delta = 0$ and the transition rates satisfy DB with respect to the equilibrium distribution $P_0(V)$, then, denoting by $\langle ... \rangle_0$ the average over such distribution, one has $\langle V \rangle_0 = 0$ and $\langle V^2 \rangle_0 = T/M$, where T is the temperature of the thermal bath (in the case of several gases, $T_i = T$ for every i).

In this general model there are two different timescales: the mean collision times, that for large mass M, take the form $\tau_i \simeq \sqrt{m_i/T_i}/(\rho_i S_i)$, with S_i the scattering cross-section of the probe with the particles of gas i; and the stopping time $\tau_\Delta = V^*/\Delta$ due to friction, where V^* is the average post-collision velocity.

A formal general expression for the average velocity of the tracer can be obtained multiplying by V both members of equation (9.8) and integrating over V. In the stationary state, one has

$$0 = -\Delta\langle \text{sign}(V)\rangle + \langle \alpha(V)\rangle, \tag{9.10}$$

where $\alpha(V) = \int (V' - V)W(V'|V)dV'$ is the jump moment, depending on M through the rates W, and $\langle ...\rangle$ denotes a stationary average over $P(V)$.

Assuming the mass of the tracer large enough with respect to the largest mass among those of the gas particles m, we can perform an expansion around $V = 0$, as described in chapter 6. We get

$$\alpha + \alpha'\langle V\rangle + \frac{1}{2}\alpha''\langle V^2\rangle - \Delta\langle \text{sign}(V)\rangle \simeq 0, \tag{9.11}$$

where α' and α'' represent the first and second derivatives of α with respect to V, respectively, and all coefficients are computed for $V = 0$. The physical interpretation of these quantities is the following: $|\alpha'|^{-1}$ represents the thermalization time τ_{th} of the probe with the gas, in the absence of friction, and is related to the collision time, $\tau_{th} \sim \tau_i M/m_i$ [7]. The coefficients α, α' and α'' are functions of M, and can be expanded in powers of M^{-1}, taking into account that $\langle V^2\rangle \sim \mathcal{O}(M^{-1})$. Equation (9.11) then gives

$$\begin{aligned}\langle V\rangle &= -\frac{1}{\alpha'}\left[\alpha + \frac{1}{2}\alpha''\langle V^2\rangle\right] + \frac{\Delta}{\alpha'}\langle \text{sign}(V)\rangle \\ &= -\frac{A}{\alpha'}[T_k - T] + \frac{\Delta}{\alpha'}\langle \text{sign}(V)\rangle,\end{aligned} \tag{9.12}$$

where in the second line we have neglected thermal gradients (if present) so that one can define a reference temperature T, and we have defined the kinetic temperature $T_k \equiv M\langle V^2\rangle$ (assuming $\langle V\rangle^2 \ll \langle V^2\rangle$) and a general asymmetry A

$$\alpha \simeq -TA, \qquad \alpha''\langle V^2\rangle \simeq 2AT_k. \tag{9.13}$$

This general structure for the coefficients α and α'' can be found in many models [8–12], and follows from equation (9.9) (with the form of A depending on the specific system). See for instance the example discussed in chapter 7. Equation (9.12) makes clear that there are two mechanisms that can induce a ratchet effect: the first one is related to the quantity $D_{hf} \equiv -\frac{A}{\alpha'}[T_k - T]$, which is proportional to the temperature difference $T_k - T$, namely to the heat flux exchanged between the probe and the thermal bath, due to collisions; the second one is related to the friction and is proportional to the average of the frictional force: $D_\Delta \equiv \Delta\frac{M}{\alpha'}\langle \text{sign}(V)\rangle$. Notice that the first channel of dissipation can be at work due to the presence of *reservoirs* at different temperatures or dissipative collisions, but it is also modified by the presence of friction. Indeed, even if one considers elastic interactions and equilibrium baths, a net heat flow can still be induced by frictional forces.

For $\Delta = 0$, or when the thermalization time is much smaller than the stopping time, namely $\tau_{th} \sim 1/|\alpha'| \ll \tau_\Delta$, and therefore friction can be neglected, only the term D_{hf} plays a role. Then the ratchet effect occurs if and only if the transition rates are asymmetric (i.e. $A \neq 0$) and do not satisfy DB, so that $T_k \neq T$. In the opposite case, namely when friction dominates and cannot be neglected, both mechanisms are at work and the two contributions may produce nonmonotonic behaviours in the drift.

We now illustrate the above general structure in the specific case of the generalized Rayleigh piston [7], in the presence of Coulomb friction. We consider a piston of mass M, interacting with two different gases of elastic particles of mass m_r (at right) and m_l (at left), see figure 9.2. The two gases are at the same temperature T and have the same densities $\rho_i = \rho$, with $i = r, l$, so that the pressure on both sides of the piston is the same. The elastic collisions with the (right and left) gas particles are described by the rule $V' = V + \frac{2}{1+M/m_i}(v - V)$, where V and V' are the piston velocities before and after the collision, respectively. The particles velocities are drawn from the Maxwell–Boltzmann distribution $p_i(v) = \rho\sqrt{\frac{m_i}{2\pi T}}\exp(-\frac{m_i v^2}{2T})$, where Boltzmann's constant $k_B = 1$.

The (asymmetric) transition rates are give by [7]

$$\begin{aligned} W^+(V'|V) &= \left(\frac{M+m_l}{2m_l}\right)^2 (V' - V) p_l \left(\frac{M+m_l}{2m_l}V' - \frac{M-m_l}{2m_l}V\right), \\ W^-(V'|V) &= \left(\frac{M+m_r}{2m_r}\right)^2 (V - V') p_r \left(\frac{M+m_r}{2m_r}V' - \frac{M-m_r}{2m_r}V\right). \end{aligned} \tag{9.14}$$

These satisfy DB with respect to the Gaussian distribution

$$P_0(V) = (2\pi T/M)^{-1/2}\exp(-MV^2/2T). \tag{9.15}$$

The coefficients appearing in equation (9.12) are given by [13]:

$$\begin{aligned} \alpha &= \rho T[(M+m_l)^{-1} - (M+m_r)^{-1}] \\ &= -\rho T(m_l - m_r)/M^2 + \mathcal{O}(M^{-3}), \end{aligned} \tag{9.16}$$

$$\alpha' = -2\rho\sqrt{\frac{2T}{\pi}}\left[\frac{\sqrt{m_l}}{M+m_l} + \frac{\sqrt{m_r}}{M+m_r}\right], \tag{9.17}$$

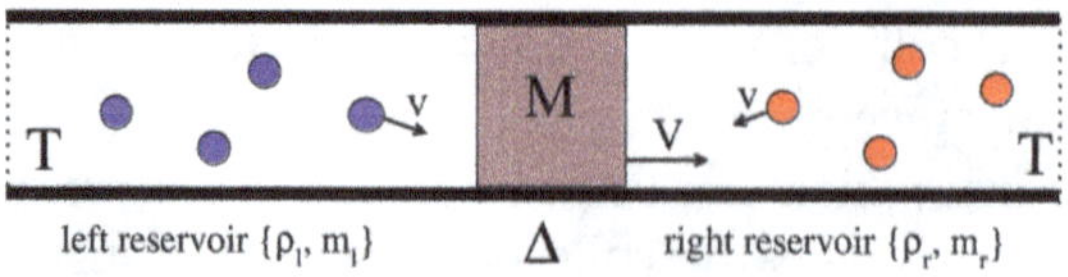

Figure 9.2. The asymmetric Rayleigh piston with Coulomb friction. Reprinted figure with permission from reference [13], Copyright (2013) by the American Physical Society.

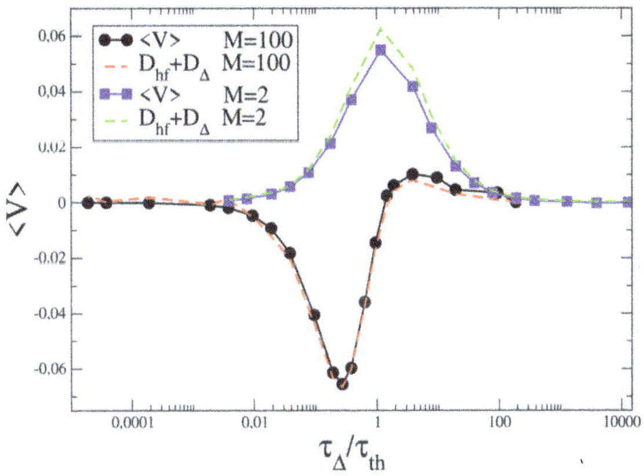

Figure 9.3. DSMC results for $\langle V \rangle$, with $T = 10$, $\rho = 0.5$, $m_r = 2$ and $m_l = 1$, as a function of τ_Δ/τ_{th}, for $M = 100$ (black dots) and $M = 2$ (blue squares). Equation (9.12) is represented by red (for $M = 100$) and green (for $M = 2$) dashed lines. Reprinted figure (adapted) with permission from reference [13], Copyright (2013) by the American Physical Society.

$$\begin{aligned} \alpha'' &= 2\rho[m_l/(1 + m_l/M) - m_r/(1 + m_r/M)]/M \\ &= 2\rho(m_l - m_r)/M + \mathcal{O}(M^{-2}). \end{aligned} \tag{9.18}$$

According to the relations (9.13), from equations (9.16) and (9.18) one has $A \simeq \rho(m_l - m_r)/M^2$. The timescales are $\tau_\Delta = V^*/\Delta = \sqrt{T/M}/\Delta$, due to the elastic collisions, and $\tau_{th} = 1/|\alpha'| \simeq \sqrt{\pi/(2T)}\,M/[2\rho(\sqrt{m_l} + \sqrt{m_r})]$.

The behaviour of the model can be studied with numerical simulations of the process in equation (9.8) with transition rates (9.14), using a direct simulation Monte Carlo (DSMC) algorithm [14]. The velocity v of a gas particle is drawn from $p_i(v)$, $i = r, l$ with probability 1/2, and then the collision with the piston with velocity V occurs with probability $\propto |v - V|$. In figure 9.3 we show $\langle V \rangle$ (black dots for $M = 100$ and blue squares for $M = 2$) as a function of the ratio τ_Δ/τ_{th}, varied by changing Δ. A net drift is observed in a wide range of parameters. Note that in this system the motor effect is entirely driven by dry friction, because the two gases are in equilibrium. The nonmonotonic behaviour of the drift can be described by the rhs of equation (9.12), represented in figure 9.3 by red (for $M = 100$) and green (for $M = 2$) dashed lines. The quantites $\langle V^2 \rangle$ and $\langle \mathrm{sign}(V) \rangle$ are computed in DSMC.

Similar models, also known as kinetic ratchets, have been studied in the literature, exploiting other sources of nonequilibrium conditions, such as granular interactions, and different geometries [8–12], see also the discussion in chapter 8.

9.2 Getting more from pushing less

Nonequilibrium conditions allow for other counter-intuitive phenomena, which can be investigated in terms of the response of the system to an external perturbation. Usually, when an equilibrium state is perturbed, one expects that if the external drive is increased the induced current will increase as well. This occurs for instance when the increase of the temperature leads to an increase of the energy, as represented by a positive specific heat. However, such a behaviour can be reversed in nonequilibrium conditions, where the presence of barriers or obstacles can trap the system in particular states, allowing for a negative specific heat (where an increase in the temperature of a thermal bath induces a decrease of the average energy of the system), a negative differential conductivity or a negative differential mobility, depending on the considered model: in general, a nonmonotonic behaviour of the current-forcing relation. This means that there exists an optimal value of the applied force at which the maximal current is obtained.

9.2.1 Simple models

First we discuss two simple models that can be exactly solved and that illustrate the two basic ingredients responsible for these nonmonotonic behaviours: nonequilibrium conditions and trapping effects.

Negative specific heat

We first consider the toy model described in reference [15]. It consists of a system with three possible states $\alpha = 0, 1, 2$ with energy levels, $E_0 = 0$, $E_1 = \varepsilon_1$, $E_2 = \varepsilon_2$, with $\varepsilon_2 > \varepsilon_1 > 0$. Instead of considering the system coupled to a single thermal bath at temperature T, one can introduce nonequilibrium conditions assuming for instance that the transitions $0 \leftrightarrow 2$ and $1 \leftrightarrow 2$ depend on the coupling with different thermal baths at temperatures T_x, and T_y, and forbidding the transition $0 \leftrightarrow 1$. Denoting by δ the energy level difference, $\delta = \varepsilon_2 - \varepsilon_1$, one has for the transition rates $W(0 \to 2) = \exp(-\varepsilon_2/k_B T_x)$ and $W(1 \to 2) = \exp(-\delta/k_B T_y)$. The stationary distribution is then given by

$$P_0^{st} = \frac{e^{-\delta/k_B T_y}}{Z}, \; P_1^{st} = \frac{e^{-\varepsilon_2/k_B T_x}}{Z}, \; P_2^{st} = \frac{e^{-\delta/k_B T_y} e^{-\varepsilon_2/k_B T_x}}{Z}, \tag{9.19}$$

where $Z = \exp(-\varepsilon_2/k_B T_x) + \exp(-\delta/k_B T_y) + \exp(-\delta/k_B T_y)\exp(-\varepsilon_2/k_B T_x)$. From these expressions one can obtain the average energy of the system

$$U = \sum_\alpha P_\alpha^{st} E_\alpha = \frac{e^{-\varepsilon_2/k_B T_x}\varepsilon_1 + e^{-\delta/k_B T_y} e^{-\varepsilon_2/k_B T_x}\varepsilon_2}{Z}, \tag{9.20}$$

and the specific heat

$$C_x = \frac{\partial U}{\partial T_x} = (\varepsilon_1 + e^{-\delta/k_B T_y}\varepsilon_2)\frac{e^{-\delta/k_B T_y} e^{-\varepsilon_2/k_B T_x}\varepsilon_2}{k_B T_x^2 Z^2} \tag{9.21}$$

$$C_y = \frac{\partial U}{\partial T_y} = [e^{-\varepsilon_2/k_B T_x}\varepsilon_2 - \varepsilon_1(1 + e^{-\varepsilon_2/k_B T_x})]\frac{e^{-\delta/k_B T_y} e^{-\varepsilon_2/k_B T_x}\delta}{k_B T_y^2 Z^2}. \tag{9.22}$$

While the first expression is always positive, the second one gives a negative specific heat if $T_x < (\varepsilon_2/k_B)/\log(\delta/\varepsilon_1)$. This means that an increase in the temperature T_y of one of the thermal baths will cause a decrease of the average energy of the system.

The mechanism responsible for this phenomenon can be illustrated as follows. The coupling with the baths allows the systems to populate the different energy levels. In the case of only two levels, there is only one pair of transition rates, ruled by a single bath with an effective temperature, and the system is in equilibrium. Then, if the temperature is increased, the relative occupancy of the higher level will be higher, leading to a positive specific heat. Considering a system with three energy levels, there are in general three pairs of transition rates, so that the system may be coupled to three different thermal baths. In the minimal model considered above, one can forbid the transition between the lower pair of levels, so that we only have two baths. In this case, if one increases the temperature of the bath coupled to the $1 \leftrightarrow 2$ transitions, the population P_1 will be depleted in favour of P_2. Since the $0 \leftrightarrow 2$ transition is coupled to a different bath, any increase in P_2 will produce an increase in P_0, and, if this bath is very cold, the increase in P_0 will be much larger than the increase in P_2. This mechanism produces a change in the populations from P_1 to P_0, even though the thermostat coupled to the $1 \leftrightarrow 2$ transition has been turned up. The fact that the $1 \leftrightarrow 0$ transition is forbidden in the model introduces a 'barrier' in the dynamics. As we will see in other examples, the presence of such 'obstructions', together with nonequilibrium conditions, represents the key ingredient to the display of negative response.

Negative differential conductivity

Consider a random walk in one dimension, under the action of an external force in the positive direction. Denote by p the transition rate to jump from position x to position $x + 1$ and by q the transition rate to jump from position x to position $x - 1$. As seen in chapter 5, according to the DB condition, we can set

$$\frac{p}{q} = e^{\beta W}, \tag{9.23}$$

where W is the work done by the force and β the inverse temperature. As we know, this condition does not univocally determine the transition rates p and q. We have also to set the escape rate $\gamma(W) = p + q$, which in general can depend on the work W. For instance, we can consider the random walk as a model to describe the motion of a Brownian particle in a channel, where some geometrical constraints are present, such as barriers or obstacles [16, 17]. When the particle is pushed against the obstacles by the external force, it can be trapped, and escaping from the trap can require a time which is increasing with the force. For instance, assuming the escape rate $\gamma(W) \sim e^{-\beta W}$, we have for the average current J

$$J = p - q = \gamma(W)\frac{1 - e^{-\beta W}}{1 + e^{-\beta W}}. \tag{9.24}$$

From the above expression, one can immediately observe that at large values of the force (large W), the current becomes decreasing in W, leading to a negative differential conductivity (see figure 9.4). This is a non-dissipative, kinetic effect, due to the waiting time distribution (which describes time-symmetric fluctuations). In the following, we will discuss more complex models where the same effect occurs.

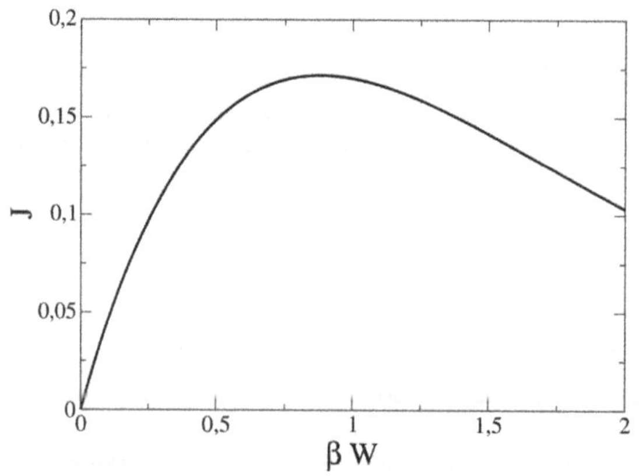

Figure 9.4. Current J as a function of the delivered work W in a model of a random walk with geometric constraints, described by equation (9.24).

9.2.2 Negative differential mobility

Here we consider the response of the velocity of a tracer particle to an external force F in a complex environment. We present results for two model fluids: a lattice gas described by a master equation, and an inertial tracer advected by a laminar flow described by a Langevin equation. In both cases we show that, quite surprisingly, in a certain range of parameters, when the external field is increased, the tracer stationary velocity V decreases. This means that if we denote by $\mu = V/F$ the mobility of the particle, we have $dV/dF < 0$, a phenomenon known as negative differential mobility (NDM). As we will discuss in detail, in analogy with the negative specific heat, this effect is due to the presence of barriers, or obstacles, that can trap the tracer: the interplay between the typical trapping time and other characteristic timescales in the model can lead to the nonmonotonic behaviour of the velocity–force relation.

Lattice gas model

Lattice gas models are discrete systems where particles can jump on the sites of a lattice, with excluded-volume interactions (two particles cannot occupy the same site at the same time). We consider a tracer particle (TP) subjected to an external force and 'bath particles' performing a random motion. The dynamics is the following: each bath particle waits for an exponentially distributed time with mean τ^* and then jumps in a random direction with probability $1/2d$, where d is the lattice dimension; if the target site is empty at this moment the move is accepted. Similarly, the TP waits an exponential time with mean τ and then chooses to jump in the direction ν ($\nu \in \{\pm 1, \ldots, \pm d\}$) with probability

$$p_\nu = \frac{e^{(\beta/2)\boldsymbol{F}\cdot\boldsymbol{e}_\nu}}{\sum_\mu e^{(\beta/2)\boldsymbol{F}\cdot\boldsymbol{e}_\mu}}, \tag{9.25}$$

where β is the inverse temperature, $\boldsymbol{e}_\mu$ are the corresponding $2d$ base vectors of the lattice, the lattice step has been taken equal to one and we denote $\boldsymbol{F} \equiv F\boldsymbol{e}_1$ the external force. The form of equation (9.25) satisfies the generalized DB condition [18, 19]. This condition is compatible with other choices that can lead to a different phenomenology [20–22].

Denoted by $P(\boldsymbol{R}_{TP}, \eta; t)$ the joint probability of finding at time t the TP at the site $\boldsymbol{R}_{TP}$ with the configuration of obstacles η (where $\eta(\boldsymbol{R}) = \{1, 0\}$ represents the instantaneous occupation of the site at position $\boldsymbol{R}$), the model is described by the following master equation

$$\partial_t P(\boldsymbol{R}_{TP}, \eta; t) = \frac{1}{2d\tau^*}\sum_{\mu=1}^{d}\sum_{r\neq \boldsymbol{R}_{TP}-\boldsymbol{e}_\mu, \boldsymbol{R}_{TP}} [P(\boldsymbol{R}_{TP}, \eta^{r,\mu}; t) - P(\boldsymbol{R}_{TP}, \eta; t)]$$
$$+\frac{1}{\tau}\sum_{\mu=1}^{d} p_\mu\{[1 - \eta(\boldsymbol{R}_{TP})]P(\boldsymbol{R}_{TP} - \boldsymbol{e}_\mu, \eta; t) - [1 - \eta(\boldsymbol{R}_{TP} + \boldsymbol{e}_\mu)]P(\boldsymbol{R}_{TP}, \eta; t)\}, \quad (9.26)$$

where $\eta^{r,\mu}$ is the configuration obtained from η by exchanging the occupation numbers of sites $\boldsymbol{r}$ and $\boldsymbol{r} + \boldsymbol{e}_\mu$. The stationary velocity $V(F)$ along the field direction is then given by

$$V(F) \equiv \frac{d\langle \boldsymbol{R}_{TP} \cdot \boldsymbol{e}_1\rangle}{dt} = \frac{1}{2d\tau^*}(A_1 - A_{-1}), \quad (9.27)$$

where the coefficients A_ν $(\nu = \pm 1, \dots, \pm d)$ are defined by the relation $A_\nu \equiv 1 + \frac{2d\tau^*}{\tau} p_\nu(1 - k(\boldsymbol{e}_\nu))$. Here, $k(\boldsymbol{e}_\nu) \equiv \sum_{\boldsymbol{R}_{TP},\eta}\eta(\boldsymbol{R}_{TP} + \boldsymbol{e}_\nu)P(\boldsymbol{R}_{TP}, \eta)$ represents the stationary density profile around the TP. The evolution equation for this quantity involves higher order correlations so that a closure scheme is necessary. This is obtained by introducing a decoupling approximation

$$\langle \eta(\boldsymbol{R}_{TP} + \boldsymbol{\lambda})\eta(\boldsymbol{R}_{TP} + \boldsymbol{e}_\nu)\rangle \approx \langle \eta(\boldsymbol{R}_{TP} + \boldsymbol{\lambda})\rangle\langle \eta(\boldsymbol{R}_{TP} + \boldsymbol{e}_\nu)\rangle, \quad (9.28)$$

which assumes that the occupation of the site just in front of the TP, and of a site some distance λ apart from it, becomes statistically independent. This represents a mean-field-like approximation and it is physically motivated in the observation that a fluctuation in the occupancy of the sites in the vicinity of the tracer does not affect the dynamics far from the tracer itself [23, 24]. The details of the computations can be found in [25, 26]. The accuracy of this analytical approach is demonstrated by the comparison with Monte Carlo simulations, as reported in figure 9.5, where a nonmonotonic behaviour of the force–velocity relation, corresponding to the effect of NDM, is clearly visible in a certain region of model parameters.

The physical mechanism underlying NDM can be illustrated in the low density limit, assuming a strong external force, so that $p_1 \simeq 1 - \varepsilon$, $p_{-1} = O(\varepsilon^2)$ with $\varepsilon = 2\exp(-\beta F/2)$. In this limit, the mean velocity in the absence of obstacles can be written as $(1 - \varepsilon)/\tau$. In the presence of obstacles, it is then given by the mean distance $1/\rho$ (being ρ the particle density) travelled by the TP between two obstacles divided by the mean duration of this excursion, which is the sum of the mean time of free motion $\tau/[\rho(1 - \varepsilon)]$, and of the mean trapping time τ_{trap}. Now, note that the

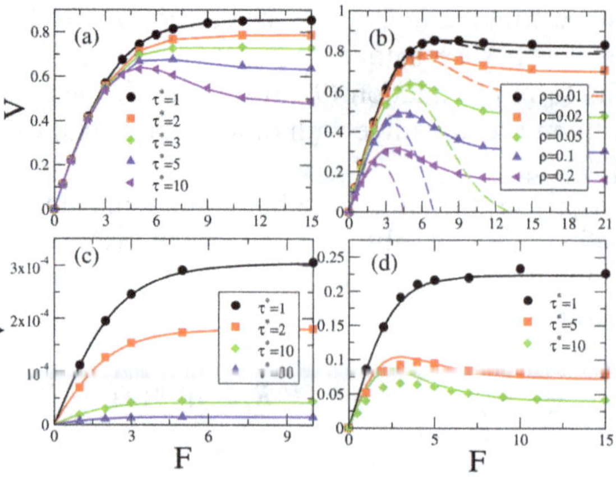

Figure 9.5. $V(F)$ for $d = 2$ and $\beta = 1$: (a) $\rho = 0.05$, $\tau = 1$ and different τ^*, analytic prediction (lines) and numerical simulations (symbols); (b) $\tau = 1$ and $\tau^* = 10$, analytic prediction (continuous lines), numerical simulations (symbols) and linearized solution (dashed lines); (c) high density limit, $\rho = 0.999$, with $\tau = 1$ and different τ^*, analytic prediction (lines) and numerical simulations (symbols); (d) $\rho = 0.5$, $\tau = 1$ and different τ^*, analytic prediction (lines) and numerical simulations (symbols). Reprinted figure with permission from reference [25], Copyright (2014) by the American Physical Society.

escape from a trap induced by surrounding obstacles results from two alternative independent events: the TP steps in the transverse direction with respect to the force (with rate ε/τ) or the obstacle steps away (with rate $3/(4\tau^*)$, for $d = 2$). This leads to $1/\tau_{\text{trap}} = 3/(4\tau^*) + \varepsilon/\tau$, and eventually

$$V(F) \simeq \frac{1 - \varepsilon}{\tau + 4\rho(1 - \varepsilon)\dfrac{\tau^*}{3 + 4\varepsilon\tau^*/\tau}}. \tag{9.29}$$

From this formula, it can be viewed that V is decreasing with F at large F (i.e. small ε), and therefore nonmonotonic with F, as soon as $\tau^* \gtrsim \tau/\sqrt{\rho}$, see figure 9.6. This unveils the physical origin of NDM, which is the result of two competing effects. Indeed a large force reduces the travel time between two consecutive encounters with obstacles, but it also increases the escape time from the traps. Eventually, for τ^* large enough, such traps are sufficiently long-lived to slow down the TP when F is increased.

Inertial tracer in a laminar flow

We now study the response of a tracer particle to an external field F in a different fluid model. We consider a stationary laminar flow in two dimensions and an inertial tracer, the dynamics of which is described by the following two-dimensional Langevin equation [27]

$$\dot{x} = v_x, \qquad \dot{y} = v_y \tag{9.30}$$

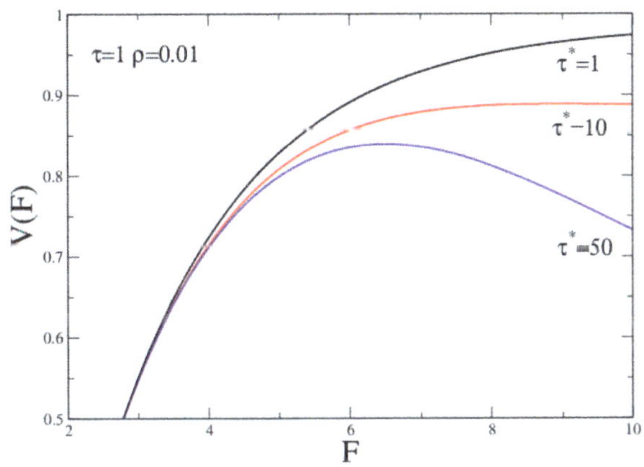

Figure 9.6. Equation (9.29) evaluated for different values of the typical time of the bath particles τ^*.

$$\dot{v}_x = -\frac{1}{\tau}(v_x - U_x) + F + \sqrt{2D_0}\xi_x \tag{9.31}$$

$$\dot{v}_y = -\frac{1}{\tau}(v_y - U_y) + \sqrt{2D_0}\xi_y \tag{9.32}$$

$$U_x = \frac{\partial\psi(x, y)}{\partial y}, \qquad U_y = -\frac{\partial\psi(x, y)}{\partial x}. \tag{9.33}$$

Here (x, y) and (v_x, v_y) denote the spatial coordinates and the velocity components, respectively, the force F is along the x direction, the divergenceless flow (U_x, U_y) corresponds to a two-dimensional convection and is derived from the stream function

$$\psi(x, y) = \frac{LU_0}{2\pi}\sin(2\pi x/L)\sin(2\pi y/L), \tag{9.34}$$

while ξ_x and ξ_y are uncorrelated white noises with zero mean and unitary variance.

It is important to observe that this system is out of equilibrium even for $F = 0$, because of the steady velocity field represented by the non-gradient forces of equation (9.33). Length and time are measured in units of L and L/U_0, respectively, where $U_0 = 1$ and $L = 1$, which defines a typical timescale of the flow $\tau^* = L/U_0 = 1$. We also consider the presence of microscopic noise with molecular diffusivity D_0, which guarantees ergodicity, and is related to the temperature T of the environment by $D_0 = T/\tau$.

In figure 9.7 we report the results of numerical simulations for $\langle v_x\rangle$ and the mobility $\mu = \langle v_x\rangle/F$ (inset) as a function of F, for $\tau = 10$ and different values of D_0. At small forces one has a linear regime, characterized by a constant mobility depending on D_0. The same happens for large enough forces: the effect of the velocity field $\boldsymbol{U}$ is negligible and, irrespective of D_0, the simple linear behaviour $\langle v_x\rangle(F) = \tau F$ is recovered. In constrast, at intermediate values of the force, a more

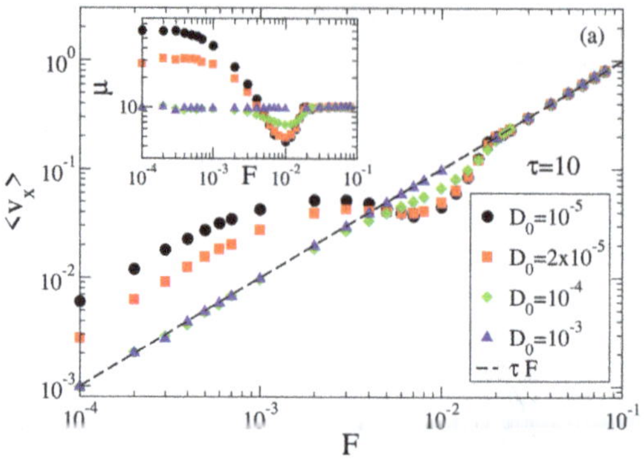

Figure 9.7. Force–velocity relation $\langle v_x \rangle(F)$ and mobility μ (inset) for different values of D_0, in the case $\tau = 10$. NDM is observable for $D_0 = 10^{-5}$ and $D_0 = 2 \cdot 10^{-5}$, around $F \sim 4 \cdot 10^{-3}$. Reprinted figure with permission from reference [27], Copyright (2016) by the American Physical Society.

complex nonlinear scenario emerges, characterized by a nonmonotonic behaviour, which corresponds to the phenomenon of NDM.

Even in this case, the physical origin of this kind of effect can be traced back in the formation of traps that slow down the motion of the tracer. Indeed, the interplay between the different timescales present in the systems (Stokes time τ, and characteristic time of the velocity field τ^*), coupled with the external force, leads the tracer to visit for long times regions where the underlying field is opposite to the driving force, resulting in a decrease of the stationary velocity. This argument will be developed in the following section, where an even more surprising phenomenon taking place in the same system is discussed.

9.3 Turn right to go left

Intuitively, the most efficient way to induce a finite velocity of a particle in a given direction, is to apply an external force in the same direction. However, quite surprisingly, there are particular cases where the stationary particle velocity turns out to be opposite to the external force, producing a negative response. In the context of fluids, this means for instance that the particle mobility $\mu < 0$, leading to the phenomenon known as absolute negative mobility (ANM). This kind of behaviour, defined by a negative induced current, has been experimentally observed in different systems, from colloidal particles in a microfluidic setup to strongly correlated electrons and Josephson junctions [28–30], and has been studied in several theoretical models, from single particles in a constrained geometry [31, 32] to many-particle interacting systems [33]. See also references in [33] for other examples[1].

Before presenting detailed results in the two models considered in the previous section, it is useful to discuss the simplest general mechanism that makes ANM possible when the considered system can show NDM. This also helps to clarify the

[1] There are also very particular cases where ANM can take place at equilibrium: in systems with negative temperatures [34], in a one-dimensional model with non-local interactions [35] or in Hamiltonian systems with coupled currents [36].

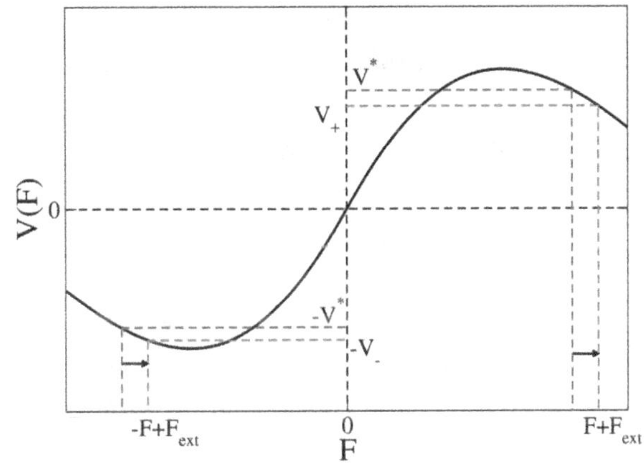

Figure 9.8. Schematic nonmonotonic force–velocity relation.

relation between these two intriguing behaviours. Let us consider a force–velocity relation with a nonmonotonic behaviour, and set the model parameters in the region where the differential mobility is negative, namely on the right of the maximum, see figure 9.8. Denote by V^* the stationary velocity corresponding to this value of the force F. Since the system is symmetric, to the opposite value of the force there corresponds the velocity $-V^*$ (see blue dashed lines in the figure). Therefore, if one considers an alternating force, taking positive and negative values, this will result in a zero average velocity. If now a further external force is applied on the system in the positive direction (red dashed lines), this will result in the change $V^* \to V_+$ and $-V^* \to -V_-$, with $|V_-| > V_+$, so that on average the stationary velocity will be negative, yielding an ANM.

The mechanisms to introduce an alternating force can be different: oscillating fields, presence of active forces that randomly change direction in time, underlying velocity fields, etc. In the following we will illustrate this phenomenon in two specific models where some of these conditions are realized.

9.3.1 Absolute negative mobility of an active tracer in a lattice gas model

We consider the same lattice gas model introduced above, where now the tracer particle is active, namely it is driven by an 'internal' force F_A that changes direction randomly with a characteristic time τ_α. We are interested in the force–velocity relation when a further constant external force F_E is applied. Denoting by $P_\chi(\boldsymbol{R}, \eta; t)$ the probability that the tracer is at position $\boldsymbol{R}$, the active force in direction $\boldsymbol{e}_\chi$ and the environment in configuration η at time t, the master equation reads

$$2d\tau^*\partial_t P_\chi(\boldsymbol{R}, \eta; t) = \mathscr{L}_\chi P_\chi - \alpha P_\chi + \frac{\alpha}{2d-1}\sum_{\chi' \neq \chi} P_{\chi'}, \tag{9.35}$$

where $\alpha = 2d\tau^*/\tau_\alpha$ is a dimensionless rate of reorientation of the active force, and $\mathscr{L}_\chi$ is the evolution operator when the active force is in direction $\boldsymbol{e}_\chi$:

$$\mathscr{L}_\chi P_\chi = \sum_{\nu=1}^{d} \sum_{r \neq R - e_\nu, R} [P_\chi(\boldsymbol{R}, \eta^{r,\nu}; t) - P_\chi(\boldsymbol{R}, \eta; t)] \\ + \frac{2d\tau^*}{\tau} \sum_\mu p_\mu^{(\chi)} [(1 - \eta_R) P_\chi(\boldsymbol{R} - \boldsymbol{e}_\mu, \eta; t) - (1 - \eta_{R+e_\mu}) P_\chi(\boldsymbol{R}, \eta; t)]. \tag{9.36}$$

As in the previous case, the first term in equation (9.36) describes the jumps of bath particles, while the second term accounts for the jumps performed by the tracer. The tracer jumps in direction $\boldsymbol{e}_\mu$ if the target site is empty, with rate $p_\mu^{(\chi)}/\tau$, where

$$p_\mu^{(\chi)} = \frac{\exp[(F_A \boldsymbol{e}_\chi + F_E \boldsymbol{e}_1) \cdot \boldsymbol{e}_\mu / 2]}{Z}, \tag{9.37}$$

and Z is a normalization factor.

This master equation can be treated analytically along the same lines described in the previous section, so that an expression for the active particle velocity can be obtained as a function of the model parameters (see reference [33] for details.) In figure 9.9 the behaviour of the velocity as a function of the external force F_E is reported: symbols represent numerical simulations, while curves are analytical predictions. One can observe that for large enough values of τ_α the phenomenon of ANM occurs.

We focus on the physical mechanism underlying this effect, that can be illustrated with a reasoning similar to that developed in the previous section. Assuming a low density ρ of bath particles, the obstacles are expected to diffuse independently, and for a given orientation of the active force χ, the effective jump rate for the tracer is $\tau + \rho\tau_p^{(\chi)}$, where $\tau_p^{(\chi)}$ is the mean time that the tracer spends with bath particles on

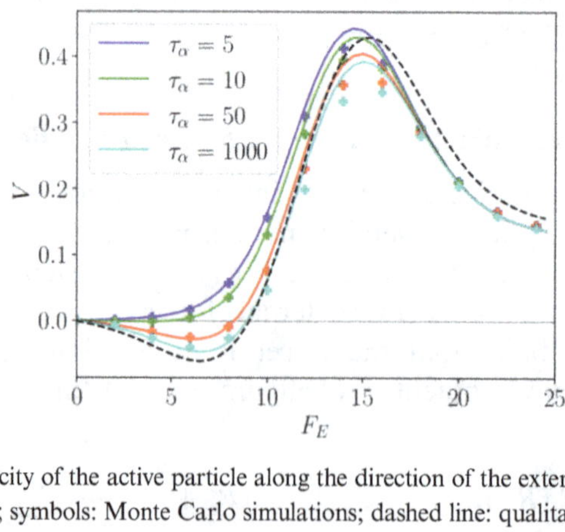

Figure 9.9. Stationary velocity of the active particle along the direction of the external force F_E on a 2D lattice. Lines: analytical prediction; symbols: Monte Carlo simulations; dashed line: qualitative argument in the limit of infinite persistence (equation (9.39) for $\tau_\alpha = \infty$). The other parameters are $\rho = 0.1$, $\tau^*=30$, $\tau = 1$, $F_A = 12$. Reprinted figure with permission from reference [33], Copyright (2023) by the American Physical Society.

one of its neighbouring sites. This timescale allows for a quantitative characterization of the 'trapping' effect due the bath particles and can be evaluated by considering that, when the tracer is at a given site $\boldsymbol{R}$ and is blocked by an obstacle at site $\boldsymbol{R} + \boldsymbol{e}_1$, the tracer can move forward if one of these three independent events occur: (i) the obstacle moves in a transverse direction with characteristic time $\frac{2d\tau^*}{(2d-2)}$; (ii) the active force changes direction with characteristic time τ_α; (iii) the tracer moves in a direction transverse to the direction of the obstacle with characteristic time $\tau/(1 - p_1^{(\chi)} - p_{-1}^{(\chi)})$. The mean trapping time therefore follows an exponential law of characteristic time $\tau_p^{(\chi)}$ given by

$$\frac{1}{\tau_p^{(\chi)}} = \frac{(2d-2)}{2d\tau^*} + \frac{1}{\tau_\alpha} + \frac{(1 - p_1^{(\chi)} - p_{-1}^{(\chi)})}{\tau}. \tag{9.38}$$

The tracer velocity is then estimated as an average over the directions of the active force χ:

$$V \simeq \frac{1}{2d}\sum_\chi \frac{p_1^{(\chi)} - p_{-1}^{(\chi)}}{\tau + \rho\tau_p^{(\chi)}}, \tag{9.39}$$

and the condition for the existence of absolute negative mobility is given by $\frac{\mathrm{d}V}{\mathrm{d}F_E}|_{F_E=0} < 0$. In figure 9.9 the dashed line represents the prediction of equation (9.39). This argument shows how the trapping effect of the tracer by slow obstacles can result in ANM when its activity (τ_α) is strong enough.

9.3.2 Absolute negative mobility of an inertial tracer in a laminar flow

The same phenomenon of a negative mobility can be observed in the model of an intertial tracer driven by an external force and advected by a steady laminar flow, introduced in the previous section. In this case, the complexity of the system prevents any analytical treatment. We report results of numerical simulations in figure 9.10, where there appears a small region of the parameter space where the effect of ANM occurs.

From a close inspection of typical trajectories of the tracer, one can observe that the motion is realized along preferential 'channels' characterized by $\langle v_x \rangle < 0$ or by $\langle v_x \rangle > 0$. Both inertia and noise induce random transitions between these channels and the force introduces a bias in such transitions, yielding an average $\langle v_x \rangle \neq 0$. In some cases an increase of the positive force may enhance the probability of transitions from channels where $\langle v_x \rangle > 0$ to channels where $\langle v_x \rangle < 0$, leading to NDM or even ANM. For particular values of the parameters, due to the coupling between the intertial dynamics, the underlying velocity field and the external force, the particle gets trapped in regions where the field is strong and opposite with respect to the force, resulting in the ANM. This phenomenon is quite robust, and has been observed in several similar systems [37, 38].

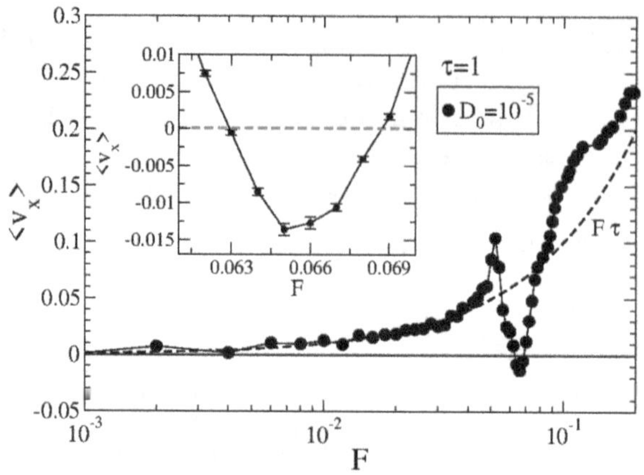

Figure 9.10. Force–velocity relation $\langle v_x \rangle (F)$ in the case $\tau = 1$ for $D_0 = 10^{-5}$. Notice the negative peak, corresponding to ANM, observed in a range of forces near $F \sim 6.5 \cdot 10^{-2}$, which is magnified in the inset. Reprinted figure with permission from reference [27], Copyright (2016) by the American Physical Society.

9.3.3 Generalized Einstein relation and the role of non-entropic forces

We conclude the analysis of this phenomenon with a discussion on the connection between negative response and fluctuation–dissipation theorem (FDT). Indeed, as illustrated in chapter 5, the response of a system is always related to the spontaneous fluctuations between some observables in the unperturbed dynamics. Therefore, it is natural to wonder how the occurrence of NDM and ANM can be explained in terms of correlation functions.

We start by considering a Langevin equation for a particle with mass m in two dimensions

$$\begin{aligned} \dot{\mathbf{x}}(t) &= \mathbf{v}(t) \\ m\dot{\mathbf{v}}(t) &= -\gamma m \mathbf{v} + \mathbf{F}(\mathbf{x}, \mathbf{v}) + \sqrt{2m\gamma T}\,\eta(t), \end{aligned} \tag{9.40}$$

where $\mathbf{F}$ is an external force, γ is the friction coefficient of the thermal bath, T its temperature and η an uncorrelated Gaussian noise. Applying the FDT derived in chapter 5, it can be shown that the following generalized Einstein relation (GER) can be derived [39]

$$\mu_{ij} = \frac{1}{T} D_{ij} - \lim_{t\to\infty} \frac{1}{2\gamma m T} \int_0^t ds \left\langle \frac{[\mathbf{x}(t) - \mathbf{x}(0)]_i}{t}; F_j[\mathbf{x}(s), \mathbf{v}(s)] \right\rangle, \tag{9.41}$$

where

$$\mu_{ij} = \lim_{t\to\infty} \frac{1}{t} \frac{\partial \langle [\mathbf{x}(t) - \mathbf{x}(0)]_i \rangle^f}{\partial f_j} \Bigg|_{\mathbf{f}=0} \tag{9.42}$$

is the mobility defined with respect to a further perturbation $\mathbf{F}(\mathbf{x}, \mathbf{v}) \to \mathbf{F}(\mathbf{x}, \mathbf{v}) + \mathbf{f}$,

$$D_{ij} = \lim_{t\to\infty} \frac{1}{2t} \langle [\mathbf{x}(t) - \mathbf{x}(0)]_i; [\mathbf{x}(t) - \mathbf{x}(0)]_j \rangle \tag{9.43}$$

is the diffusion coefficient, and $\langle A; B\rangle = \langle AB\rangle - \langle A\rangle\langle B\rangle$. With respect to the equilibrium expression, one observes an extra term which involves the correlation between the forcing and the displacement. This term comes from the time-symmetric part of the action [39].

We can apply the above formula to the system described in the previous section, where the velocity field $\boldsymbol{U}$ plays the role of the external force. Then, considering the x dircction, thc GER explicitly reads

$$\lim_{F\to 0} \mu(F) = \frac{1}{T}[D_{xx}(F = 0) - C_{x\Phi}(F = 0)] \tag{9.44}$$

$$C_{x\Phi}(F) = \lim_{t\to\infty} \frac{1}{2Tt}\langle [x(t) - x(0)]; \Phi(t)\rangle_F, \tag{9.45}$$

where

$$\Phi(t) = \int_0^t U_x[x(s), y(s)]ds, \tag{9.46}$$

and $\langle\cdots\rangle_F$ denotes an average measured at force F. Computing in numerical simulations the nonequilibrium contribution due to the coupling with the field Φ one can check the validity of the GER (9.45). Therefore, from equation (9.45) one obtains that ANM can be interpreted as the consequence of $C_{x\Phi}$ becoming larger than D_{xx}.

The same analysis can be carried out in the discrete model represented by the active particle in a lattice gas introduced above. Indeed, the expression of the mobility of the tracer in the limit of small external force takes the form

$$\begin{aligned} &\lim_{F_E\to 0} \frac{V}{F_E} = \\ &D_0 - \frac{1}{2d\tau}\left[(p_1 - p_{-1})v_1 + \frac{2(2d - 1)\tau^*}{\alpha\tau}v_1^2 + 2\sum_{\varepsilon\in\{-1,1,2\}} p_\varepsilon\left(\tilde{g}_\varepsilon - \frac{dk_\varepsilon}{dF_E}\right)(1 + (2d - 3)\delta_{\varepsilon,2})\right], \end{aligned} \tag{9.47}$$

where $D_0 \equiv \lim_{t\to\infty} \frac{1}{2}\frac{d\langle X_t^2\rangle}{dt}$ is the diffusion coefficient of the active tracer without external force [40], $p_\varepsilon = p_1^{(\varepsilon)}$, $k_\varepsilon = k_1^{(\varepsilon)}$, $v_1 = \sum_{\varepsilon=\pm 1}\varepsilon p_\varepsilon(1 - k_\varepsilon)$, and we introduced the cross-correlations between the position of the tracer and the occupation of the sites in its vicinity $\tilde{g}_\varepsilon = \langle (X_t - \langle X_t\rangle_\varepsilon)\eta_{X_t + e_1}\rangle_\varepsilon$. The second term in equation (9.47) represents again the correlations between the displacement of the tracer and the dynamics of its environment: its value rules the emergence of ANM and, if it exceeds D_0, then one has ANM.

These examples show the central role played by the correlations between degrees of freedom in the characterization of nonequilibrium behaviours. As discussed in chapter 5, the involved quantities are related to time-symmetric dynamical fluctuations and have a non-dissipative origin. They are indeed connected with the kinetic properties of the model, for instance through the dependence of the escape rate on the forcing (see references [17, 41] for an extended discussion).

References

[1] Vanossi A, Manini N, Urbakh M, Zapperi S and Tosatti E 2013 Colloquium: modeling friction: from nanoscale to mesoscale *Rev. Mod. Phys.* **85** 529

[2] de Gennes P G 2005 Brownian motion with dry friction *J. Stat. Phys.* **119** 953–62

[3] Sarracino A 2013 Time asymmetry of the Kramers equation with nonlinear friction: fluctuation-dissipation relation and Ratchet effect *Phys. Rev.* E **88** 052124

[4] Risken H 1996 *Fokker-Planck equation* (Berlin: Springer)

[5] Dubkov A A, Hänggi P and Goychuk I 2009 Non-linear Brownian motion: the problem of obtaining the thermal Langevin equation for a non-Gaussian bath *J. Stat. Mech. Theory Exp.* **2009** P01034

[6] Reimann P 2002 Brownian motors: noisy transport far from equilibrium *Phys. Rep.* **361** 57–265

[7] Alkemade C T J, van Kampen N G and Macdonald D K C 1963 Non-linear Brownian movement of a generalized Rayleigh model *Proc. R. Soc. Lond. Ser. A: Math. Phys. Sci.* **271** 449–71

[8] Meurs P, Van den Broeck C and Garcia A 2004 Rectification of thermal fluctuations in ideal gases *Phys. Rev.* E **70** 051109

[9] Cleuren B and Van den Broeck C 2007 Granular Brownian motor *Europhys. Lett.* **77** 50003

[10] Costantini G, Bettolo Marconi U M and Puglisi A 2007 Granular Brownian ratchet model *Phys. Rev.* E **75** 061124

[11] Gnoli A, Petri A, Dalton F, Pontuale G, Gradenigo G, Sarracino A and Puglisi A 2013 Brownian ratchet in a thermal bath driven by Coulomb friction *Phys. Rev. Lett.* **110** 120601

[12] Talbot J and Viot P 2012 Effect of dynamic and static friction on an asymmetric granular piston *Phys. Rev.* **85** 021310

[13] Sarracino A, Gnoli A and Puglisi A 2013 Ratchet effect driven by Coulomb friction: the asymmetric Rayleigh piston *Phys. Rev.* E **87** 040101

[14] Bird G A 1994 *Molecular Gas Dynamics and the Direct Simulation of Gas Flows* (Oxford: Oxford University Press)

[15] Zia R K P, Praestgaard E L and Mouritsen O G 2002 Getting more from pushing less: negative specific heat and conductivity in nonequilibrium steady states *Am. J. Phys.* **70** 384–92

[16] Baerts P, Basu U, Maes C and Safaverdi S 2013 Frenetic origin of negative differential response *Phys. Rev.* E **88** 052109

[17] Maes C 2017 *Non-Dissipative Effects in Nonequilibrium Systems* (Berlin: Springer)

[18] Lebowitz J L and Spohn H 1999 A Gallavotti-Cohen-type symmetry in the large deviation functional for stochastic dynamics *J. Stat. Phys.* **95** 333–65

[19] Maes C 2021 Local detailed balance *SciPost Phys. Lect. Notes* **32** 1–17

[20] Basu U and Maes C 2014 Mobility transition in a dynamic environment *J. Phys. A: Math. Theor.* **47** 255003

[21] Baiesi M, Stella A L and Vanderzande C 2015 Role of trapping and crowding as sources of negative differential mobility *Phys. Rev.* E **92** 042121

[22] Bénichou O, Illien P, Oshanin G, Sarracino A and Voituriez R 2016 Nonlinear response and emerging nonequilibrium microstructures for biased diffusion in confined crowded environments *Phys. Rev.* E **93** 032128

[23] Burlatsky S F, Oshanin G, Moreau M and Reinhardt W P 1996 Motion of a driven tracer particle in a one-dimensional symmetric lattice gas *Phys. Rev.* E **54** 3165

[24] Bénichou O, Cazabat A M, De Coninck J, Moreau M and Oshanin G 2000 Stokes formula and density perturbances for driven tracer diffusion in an adsorbed monolayer *Phys. Rev. Lett.* **84** 511

[25] Bénichou O, Illien P, Oshanin G, Sarracino A and Voituriez R 2014 Microscopic theory for negative differential mobility in crowded environments *Phys. Rev. Lett.* **113** 268002

[26] Bénichou O, Illien P, Oshanin G, Sarracino A and Voituriez R 2018 Tracer diffusion in crowded narrow channels *J. Phys. Condens. Matter* **30** 443001

[27] Sarracino A, Cecconi F, Puglisi A and Vulpiani A 2016 Nonlinear response of inertial tracers in steady laminar flows: differential and absolute negative mobility *Phys. Rev. Lett.* **117** 174501

[28] Ros A, Eichhorn R, Regtmeier J, Duong T T, Reimann P and Anselmetti D 2005 Absolute negative particle mobility *Nature* **436** 928

[29] Levitov L and Falkovich G 2016 Electron viscosity, current vortices and negative nonlocal resistance in graphene *Nat. Phys.* **12** 672–6

[30] Nagel J, Speer D, Gaber T, Sterck A, Eichhorn R, Reimann P, Ilin K, Siegel M, Koelle D and Kleiner R 2008 Observation of negative absolute resistance in a Josephson junction *Phys. Rev. Lett.* **100** 217001

[31] Eichhorn R, Reimann P and Hänggi P 2002 Brownian motion exhibiting absolute negative mobility *Phys. Rev. Lett.* **88** 190601

[32] Machura L, Kostur M, Talkner P, Łuczka J and Hänggi P 2007 Absolute negative mobility induced by thermal equilibrium fluctuations *Phys. Rev. Lett.* **98** 040601

[33] Rizkallah P, Sarracino A, Bénichou O and Illien P 2023 Absolute negative mobility of an active tracer in a crowded environment *Phys. Rev. Lett.* **130** 218201

[34] Baldovin M, Iubini S, Livi R and Vulpiani A 2021 Statistical mechanics of systems with negative temperature *Phys. Rep.* **923** 1–50

[35] Cividini J, Mukamel D and Posch H A 2018 Driven tracer with absolute negative mobility *J. Phys. A: Math. Theor.* **51** 085001

[36] Wang J, Casati G and Benenti G 2020 Inverse currents in Hamiltonian coupled transport *Phys. Rev. Lett.* **124** 110607

[37] Cecconi F, Puglisi A, Sarracino A and Vulpiani A 2017 Anomalous force-velocity relation of driven inertial tracers in steady laminar flows *Eur. Phys. J.* E **40** 1–5

[38] B-q Ai, W-j Zhu, Y-f He and W-r Zhong 2018 Giant negative mobility of inertial particles caused by the periodic potential in steady laminar flows *J. Chem. Phys.* **149** 164903

[39] Baiesi M, Maes C and Wynants B 2011 The modified Sutherland-Einstein relation for diffusive non-equilibria *Proc. R. Soc. A: Math. Phys. Eng. Sci.* **467** 2792–809

[40] Rizkallah P, Sarracino A, Bénichou O and Illien P 2022 Microscopic theory for the diffusion of an active particle in a crowded environment *Phys. Rev. Lett.* **128** 038001

[41] Maes C 2020 Frenesy: time-symmetric dynamical activity in nonequilibria *Phys. Rep.* **850** 1–33

Chapter 10

Pedagogical appendices

> Science is built up with facts, as a house is with stones. But a collection of facts is no more a science than a heap of stones is a house.
>
> Henri Poincaré

For the sake of completeness we include three short appendices which can be useful, in particular for students and young researchers.

Appendix I: Basic facts of the atomic world from a clever use of statistical mechanics and a few experiments

Today atoms are recognized as a fundamental scientific idea; on the other hand, until about one century ago, their physical existence was a rather controversial topic. For instance, even eminent scientists, e.g. E Mach, P Duhem and W Ostwald, did not believe that atoms were the basic constituents of the world, and, surprisingly, Planck, one of the fathers of modern physics, changed his mind about the physical existence of atoms only at the end of the 19th century.

Nowadays, thanks to several technological improvements in experimental physics, it is possible to 'see' the atoms, and determine their properties, for instance their mass, size and so on. It is remarkable that already at the beginning of the 20th century, the correct order of magnitude was at hand for the basic quantities which characterize the atomic world: a clever use of statistical mechanics and some experiments were sufficient for their estimation. A chief example is the work of Loschmidt who, in the 1860s, when atoms and molecules were still hypothetical, was able to find the first reasonable estimate of the molecular size by means of kinetic theory only [1].

This appendix is devoted to summarizing these ideas based upon a few basic concepts: the speed of sound, the viscosity of gases, the physics of Brownian motion, barotropic law for colloidal particles, the virial expansion and the van der Waals equation for a dilute gas.

doi:10.1088/978-0-7503-6229-0ch10

A fast recap of important results

Sound speed and thermal speed

Perhaps the oldest, and simplest, way to establish a link between macroscopic features and the microscopic world is using the sound speed c_S. Elementary thermodynamics gives

$$c_s = \sqrt{\frac{1}{m}\frac{\partial p}{\partial \rho}},$$

where p is the pressure, $\rho = N/V$ the number density, and m the mass of the molecule. For an adiabatic perfect gas one has $p\rho^{-5/3} = \text{constant}$, and using the equation of state for the perfect gas $p = \rho k_B T$ one obtains:

$$c_s = \sqrt{\frac{5k_B T}{3m}}.$$

Such a result, although very simple, is rather interesting: the sound speed gives an estimate of the thermal speed of the molecules $\sqrt{k_B T/m}$.

Viscosity

In a gas of hard spheres of mass m and diameter $2R$ at temperature T, the viscosity is

$$\eta = \frac{\sqrt{mk_B T}}{6\pi^{3/2}R^2}.$$

Such a remarkable result had been found by Maxwell, and shows a rather counter-intuitive fact: it does not depend on the density. The experiment performed by Maxwell himself (in his house, with the help of his wife) confirmed the theory and gave support in favour of the validity of kinetic theory.

Brownian motion

In chapter 1 we already discussed the relevance of the study of the Brownian motion for the determination of the Avogadro number N_A; let us now briefly recall the main results. For the diffusion coefficient D of a spherical Brownian particle of radius r in a liquid with viscosity η and temperature T, we have

$$D = \frac{k_B T}{6\pi\eta r}. \tag{10.1}$$

In addition, following the intuition of Einstein, i.e. assuming that the same statistical laws hold both for molecules and for colloidal particles, one has the validity of the barotropic formula for grains of size of the order of micron and mass M (order 10^{-12}–10^{-11} g) in water at temperature T, in a gravitational field; for the case of high dilution one has:

$$\rho(z) = \rho(0)\, e^{-\frac{z}{\zeta}}, \ \zeta = \frac{k_B T}{Mg}. \tag{10.2}$$

The value of ζ depends on the mass M, and for colloidal particles ζ is small enough so that the behaviour of $\rho(z)$ can be observed in the laboratory. The experimental work of J B Perrin on the diffusion of Brownian particles and the barotropic formula for colloids, allowed for the determination of k_B, or equivalentely N_A.

The problem of the interaction potential

A typical problem in a statistical mechanics course is the following: in a box of volume V there are N particles whose Hamiltonian is:

$$H = \sum_{i=1}^{N} \frac{\mathbf{p}_i^2}{2m} + \sum_{i,j} U(|\mathbf{q}_i - \mathbf{q}_j|);$$

assuming that the system can be studied with the classical statistical mechanics, and that it is in equilibrium at a temperature T, determine the equation of state, the free energy, and so on. If we know $U(r)$, in the case of dilute gases it is possible to find a fair approximation for the equation of state:

$$\frac{p}{kT} = \rho + b_2(T)\rho^2 + b_3(T)\rho^3 + \dots , \tag{10.3}$$

where $\rho = N/V$, and the virial coefficients $b_2(T)$, $b_3(T)$, ... can be explicetely computed from $U(r)$: in particular it is rather easy to find $b_2(T)$,

$$b_2(T) = 2\pi \int_0^{\infty} r^2[1 - e^{-\beta U(r)}]dr. \tag{10.4}$$

In real cases, even if the temperature is high enough and classical statistical mechanics holds, it is not easy to find the potential $U(r)$ for a given system. The reason is that the potential has always a quantum origin and, therefore, $U(r)$ must be computed from the Schrödinger equation. On the other hand, only in a few particular cases, e.g. simple liquids with spherical molecules, is it possible to find a good approximation of $U(r)$ from first principles. In general, it is necessary to use an empirical protocol: one assumes a shape for $U(r)$ suggested by phenomenological arguments, and containing some parameters which must be found *a posteriori*, e.g. for simple liquids often one uses the Lennard-Jones potential:

$$U(r) = 4\varepsilon\left[\left(\frac{\sigma}{r}\right)^{12} - \left(\frac{\sigma}{r}\right)^{6}\right], \tag{10.5}$$

where ε and σ must be fitted from experimental data. The origin of the tail r^{-6} is rather clear and represents the dipole–dipole interaction; in contrast, the repulsive part r^{-12} is introduced only with phenomenological arguments.

The strategy to find the potential is the following:

(a) one assumes a suitable shape for $U(r)$, e.g. Lennard-Jones like, containing some parameters;
(b) from the potential $U(r)$ one computes the coefficient $b_2(T)$;
(c) the parmeters, e.g. ε and σ for Lennard-Jones, are determined comparing the virial coefficients with the experimental data.

In order to discuss how from equation (10.3), and experimental data, it is possible to find the basic features of $U(r)$, we consider a very simple model for $U(r)$: $U(r) = \infty$ if $r < r_0$, $U(r) = -\varepsilon$ if $r_0 < r < r_1$, and $U(r) = 0$ for $r > r_1$. The computation of b_2 is trivial:

$$b_2(T) = 2\pi \int_0^\infty r^2[1 - e^{-\beta U(r)}]dr = \frac{2}{3}\pi r_1^3 - \frac{2}{3}\pi\left(r_1^3 - r_0^3\right)e^{-\beta\varepsilon};$$

at large temperature $T \gg \varepsilon/k_B$ we have

$$b_2(T) \simeq b - a\beta, \quad b = \frac{2}{3}\pi r_0^3, \quad a = \frac{2}{3}\pi\left(r_1^3 - r_0^3\right)\varepsilon, \tag{10.6}$$

and the state equation is

$$\frac{p}{kT} \simeq \rho + (b - a\beta)\rho^2. \tag{10.7}$$

Instructive is a comparison of the previous result with the van der Waals equation

$$\frac{p + A\rho^2}{kT}\left(\frac{1}{\rho} - B\right) = 1; \tag{10.8}$$

for small ρ we have

$$\frac{p}{k_B T} \simeq \rho + (B - A\beta)\rho^2. \tag{10.9}$$

We have $a = A$ and $b = B$, and equations (10.9) and (10.7) are in agreement up to order ρ^2. The van der Waals equation, originally obtained with phenomenological argument, can be obtained in the framework of the statistical mechanics and, overall, the coefficients A and B have an interpretation in terms of the parameters of the potential, i.e. r_0, r_1 and ε. It is interesting, and reassuring, that the previous results do not change too much if we use the Lennard-Jones potential. At large temperature, i.e. $k_B T \gg \varepsilon$, for $r < \sigma$ one has $1 - e^{-\beta U(r)} \simeq 1$, while for $r > \sigma$, $1 - e^{-\beta U(r)} \simeq \beta U(r)$: therefore $b_2(t) \simeq (b - \beta a)$ with

$$b = \frac{2}{3}\pi\sigma^3, \quad a = \frac{16}{9}\varepsilon\sigma^3.$$

The above results are qualitatively similar to those in equation (10.6); therefore, using the value of $b_2(T)$ obtained from the experimental data, we can determine the order of magnitude of the parameters appearing in the potential $U(r)$. Remarkably, the choice of the model is not particularly relevant.

A simple experiment for high school students and conclusive remarks

We conclude mentioning a rather simple experiment, which can be performed even in high school laboratories, and allows one to determine the size of a molecule [2]. It is enough to use oleic acid, $C_{18}H_{34}O_2$ (see figure 10.1), which is a rather common

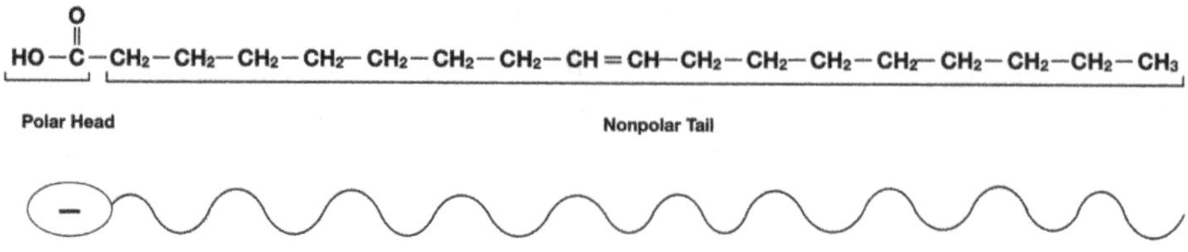

Figure 10.1. The molecule of the oleic acid

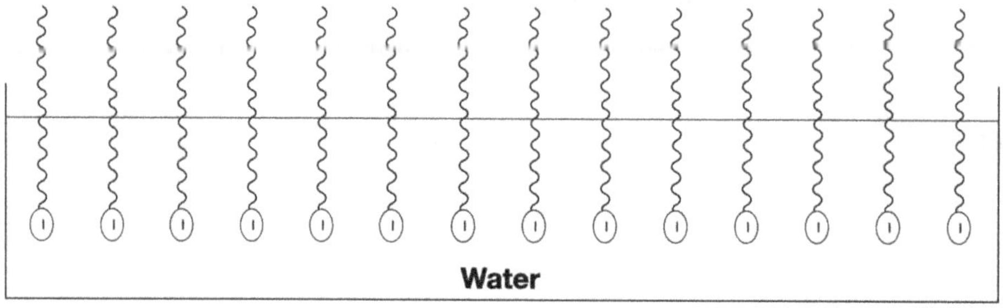

Figure 10.2. A monolayer of oleic acid in water.

material, e.g. it contributes to the 75% of olive oil. This fatty acid molecule has two ends. One is a nonpolar, hydrophobic carbon chain without oxygen atoms; let us call it the 'tail'. The other end is a polar, hydrophilic 'head', containing the two oxygen atoms. Oleic acid is not soluble in water, and therefore when placed in water, it tends to form a monolayer with the polar head sticking down towards the surface of the water and the nonpolar tail sticking straight up away from the surface of the water, see figure 10.2. The experiment is rather simple: one puts a small quantity, whose volume is known, of oleic acid on a water surface; then one measures the area of the surface film, so one has the thickness of the film. Of course the film must be at least one molecule thick; therefore, from the film thickness it is possible to estimate the length of an oleic acid molecule, which is $O(10^{-9})$ metres.

In the previous pages we showed how combining some theoretical results with experiments it is possible to obtain an estimate of microscopic quantities such as the Avogadro number N_A (or equivalently the Boltzmann constant $k_B = \mathscr{R}/N_A$), the mass m and the radius R of the molecules, the characteristic energy ε and distance σ of the interaction potential.

We can summarize as follows:

- from the diffusion coefficient of the Browinian particle, as well as the density of colloidal particles in a gravitational field, we can find N_A;
- from the viscosity in gases one has $\sqrt{m}/R^2$;
- comparing the virial coefficient of a suitable model for the potential with experimental data of the virial coefficent $b_2(T)$, at different temperatures we can determine σ^2 and ε;
- simple clever experiments allow us to estimate the size of the molecules.

Therefore, even without the modern technology, using in a smart way statistical mechanics and few experiments, one can have a correct scenario of the order of magnitude of the main quantities which characterize the microscopic world. Remarkably, in the second half of the 19th century, the above scenario was well known to Loschmidt, Boltzmann and the scientists who believed in the physical existence of the atoms (molecules).

Appendix II: Numerical methods for the Langevin equations

Let us now briefly discuss some numerical algorithms used for the numerical integration of stochastic differential equations. In order to introduce the basic ideas we start with the $1d$ Langevin equation, with an additive noise, of the form

$$dx = a(x)dt + \sqrt{2c}\, dW. \tag{10.10}$$

The simplest method for a numerical integration, called Euler method, is rather intuitive; we introduce a time discretization $t_n = n\Delta t$, with Δt small, and $x_n = x(t_n)$, so that

$$x_{n+1} = x_n + a(x_n)\Delta t + \sqrt{2c\Delta t}\, z_n, \tag{10.11}$$

where z_n are Gaussian independent processes with zero mean and unitary variance.

The above method is not very accurate, namely its precision is only $O(\Delta t^{1/2})$; following the general strategy used for the ordinary differential equations we can improve the precision using the so called Euler–Heun method. One introduces the first approximation given by the above result, i.e.

$$x'_{n+1} = x_n + a(x_n)\Delta t + \sqrt{2c\Delta t}\, z_n;$$

then we have a more accurate approximation:

$$x_{n+1} = x_n + \left[a(x_n) + a(x'_{n+1})\right]\frac{\Delta t}{2} + \sqrt{2c\Delta t}\, z_n. \tag{10.12}$$

The level of the accuracy of the previous methods can be understood studying the linear case

$$dx = -\frac{x}{\tau}dt + \sqrt{2c}\, dW. \tag{10.13}$$

With the Euler method one has

$$x_{n+1} = x_n - \frac{x_n}{\tau}\Delta t + \sqrt{2c\Delta t}\, z_n, \tag{10.14}$$

while the Euler–Heun method gives

$$x_{n+1} = x_n - \frac{x_n}{\tau}\Delta t + \sqrt{2c\Delta t}\left(1 - \frac{\Delta t}{2\tau}\right) z_n. \tag{10.15}$$

Using the results discussed in chapter 3 it easy to show that for the linear Langevin equation (10.12) one has the exact rule

$$x_{n+1} = e^{-\frac{\Delta t}{\tau}} x_n + \sqrt{2c\tau\left(1 - e^{-2\frac{\Delta t}{\tau}}\right)}\, z_n,$$

which holds for any Δt, not necessary small with respect to τ. An easy computation shows that for $\Delta t \ll \tau$, if we consider a precision $O(\Delta t^{1/2})$, one has the result in equation (10.14), while the Euler–Heun algorithm (10.15) is in agreement with the exact result with a precision $O(\Delta t^{3/2})$.

The treatment of the general case

$$dx_n = a_n(\mathbf{x})dt + \sum_j g_{nj}\, dW_j, \quad n = 1, 2, \ldots, N,$$

does not present particular additional difficulties. We can repeat the procedure in a straightforward way: of course now, at each step, one has to generated N independent Gaussian variables.

It is possible to generalize the above methods to the case of multiplicative noise, i.e. a Langevin equation of the form

$$dx_n = a_n(\mathbf{x})dt + \sum_j g_{nj}(\mathbf{x})\, dW_j.$$

In the following we'll adopt Ito's stochastic integration rule. Let us present the main result in the $1d$ case, only for simplicity of the notation:

$$dx = a(x)dt + g(x)\, dW.$$

At lower precision level one has the Euler algorithm:

$$x_{n+1} = x_n + a(x)\Delta t + g(x)\sqrt{\Delta t}\, z_n.$$

For a more precise approximation one can use the Milstein's algorithm, which can be derived from the the Ito formula

$$x_{n+1} = x_n + a(x)\Delta t + g(x)\sqrt{\Delta t}\, z_n + g(x)\frac{dg(x)}{dx}\Delta t(z_n^2 - 1),$$

for details see chapter 5 of reference [3]. In the Stratonovich integration scheme the Euler method gives the same result (therefore missing an important term of order Δt!), while for the Milstein's algorithm one has a slightly different result:

$$x_{n+1} = x_n + a(x)\Delta t + g(x)\sqrt{\Delta t}\, z_n + g(x)\frac{dg(x)}{dx}\Delta t z_n^2.$$

We conclude this short discussion on the numerical integration of stochastic differential equations mentioning that it is possible, in the spirit of Runge–Kutta method for ordinary differential equations, to obtain algorithms which improve the precision $O(\Delta t^{3/2})$, see e.g. reference [4].

An interlude on random numbers

We saw how in the numerical integration of the Langevin equation one needs to use a series of independent Gaussian variables; therefore we have to face the practical problem of the generation of random numbers. We can note that if we are able to obtain independent variables uniformly distributed in (0, 1), then it is easy to obtain independent variables distributed according to any probability distribution. For instance if x_1 and x_2 are uniformly distributed in (0, 1) then the variable

$$z = \sqrt{-2\ln x_1}\cos(2\pi x_2)$$

is a Gaussian variable with zero mean and unitary variance.

Strictly speaking, one can produce true random number sequences only using some non-deterministic physical phenomenon, for instance the decay of radioactive nuclei or the arrival on a detector of cosmic rays. A more practical way is to use an algorithm able to produce a 'random-looking' sequence of numbers, by means of a recursive rule. Let us call a pseudo-random number generator (PRNG) an algorithm, i.e. a deterministic system, designed to mimic a random sequence on a computer.

This issue is far from trivial; we cannot resist the temptation to recall a celebrated claim by von Neumann: *'Anyone who considers arithmetical methods of producing random digits is, of course, in a state of sin'* [5]. The reasons of this unavoidable sin are the following:

(a) since any algorithm must be deterministic, its Kolmogorov–Sinai (KS) entropy (h_{KS}) is finite. Therefore, the sequence $\{x(i)\}$ cannot be 'really random', i.e. with an infinite KS entropy. This limitation would be present also in a hypothetical computer able to work with real numbers.

(b) Since any deterministic system with a finite number of states must be periodic, any sequence produced by an algorithm working with discrete numbers (as happens with any computer) must also be periodic, possibly after a transient: therefore, not only $h_{KS} < \infty$, but also $h_{KS} = 0$. As consequence, we have that any computer-implemented system can be only 'pseudo-random'.

Let us comment on the problems one has to face in the use of a deterministic chaotic system such as a PRNG:

(1) The outputs $\{x(t)\}$ of a perfect PRNG, for small ε, have $h(\varepsilon) = \ln(1/\varepsilon)$. On the other hand, since in any deterministic d-dimensional system $h(\varepsilon) \simeq h_{KS}$ for small ε (i.e. $\varepsilon < \varepsilon_c$ with $\ln \varepsilon_c \sim -h_{KS}$), one should work with a very large h_{KS}. In this way the true (deterministic) nature of the PRNG becomes apparent only below the small scale ε_c.

(2) In order to actually observe the behaviour $h(\varepsilon) \sim -\ln \varepsilon$ (i.e. that of independent variables) for $\varepsilon \geqslant \varepsilon_c$ in a concrete deterministic algorithm, it is necessary that the time correlation is very weak.

A third point has to be added, dealing with the problem (**b**). Quantities like the h_{KS} and the ε-entropy have an asymptotic nature, i.e. they are related to large time behaviour. This allows the existence of situations where the system is, strictly speaking, non-chaotic ($h_{KS} = 0$) but its features appear irregular to a certain extent. Such property (denoted with the term pseudo-chaos [6]) is basically due to the presence of long transient effects. As already noted, the use of a computer discretizes the phase space of a dynamical system, cancelling (at least) its asymptotic chaotic properties. However, we may have confidence that, if the period of the realized sequence is long enough, the effects related to points (1) and (2) may survive as a chaotic transient.

According to this observation, a third request must be added:

(3) the period of the series generated by the computer (i.e. with a state-discretization of the deterministic system) must be very large.

Actually, in the daily activity, at practical level, there are no particular difficulties using PRNG, which are very efficient; see e.g. the well known book by Knuth [7]. On the other hand, even very popular generators are not perfect and can show some negative features; the interested reader can find in reference [8] a detailed discussion on the practical and conceptual problems one has to face in the use of deterministic systems such as PRNG.

Appendix III: Bibliographical suggestions

In this final appendix we offer a few bibliographical suggestions that are not usually mentioned in standard textbooks on statistical mechanics. Our aim is to draw the attention of the reader to some characters who played an important role in the development of mathematics, in particular stochastic processes applied to physics, in the second half of the last century, and who also devoted many efforts to the dissemination and to the conceptual analysis of the more general issues addressed in their studies. Indeed, we deem very important to explore, beyond the more technical works, the cultural melting pot shared during their lives and to get in touch with their achievements directly from their memories. The following is certainly a very partial and incomplete list of books (autobiographical or collecting reflections on sparse themes) of some authors close to our interests, and that in our opinion can be a source of inspiration for future generations.

The first books we mention are three autobiographies:

- Mark Kac, *Enigmas of Chance: An Autobiography*, 1985 [9];
- Stanislaw M Ulam, *Adventures of a mathematician*, 1983 [10];
- Freeman Dyson, *Disturbing the Universe*, 1979 [11].

Several links connect these three works. The most evident relies on the fact that in the introduction of his book, Kac mentions the other two books (together with *Heraclitean Fire* by Erwin Chargaff), providing his point of view on the personalities of their authors. In particular, Kac and Ulam, in their years of training, shared the cultural atmosphere of the 1930s in Lwów (the Polish name for Lviv, now in Ukraine; a city formerly in the Polish–Lithuanian Commonwealth). They both studied mathematics and were members of the Lwów School of Mathematics, founded by Hugo Steinhaus and Stefan Banach. The stimulating environment in which they developed their interests is widely attested in their autobiographies: meetings also took place outside the school, at the 'Scottish Café', where long hours were spent on mathematical problems[1].

The second link concerns the Second World War, that deeply influenced their lives. In particular, Kac and Ulam left their country for the United States (Kac in 1938 and Ulam in 1939, just a few days before the Germans invaded Poland; both their families, who remained in Poland, were later murdered by the Nazis). Ulam then joined the Manhattan Project in Los Alamos in 1943, giving important contributions to the development of the atomic bomb. Dyson also took a direct part in the Second World War as he was assigned to work in the Operational Research section of RAF Bomber Command. We mention that Kac was the co-chair of the Committee of Concerned Scientists and co-authored two letters, which publicized the case of persecuted scientists [12, 13].

Finally, as well as being a rich source of anecdotes regarding important figures in theoretical physics of the mid-twentieth century, these books bring the inspiring atmosphere of those years to life, revealing from within some of the mental and psychological processes underlying the results obtained by their authors.

- Gian-Carlo Rota, Mark Kac, Jacob T Schwartz, *Discrete thoughts*, Birkhäuser, 1992.

This is a collection of short essays on several topics, ranging from mathematics and statistics, to computer science and philosophy. Even if many papers were written in the second half of the 1970s of last century, they still address current issues in these fields, bringing to the fore central aspects, in particular in the relations among different disciplines, that represent heated topics nowadays. In particular, we would like to highlight the reflections on the role of mathematics in society and science (chapters 2, 3, 8 and 10), the discussions on the probabilistic approach in physics (Chapters 4 and 5), and on computer science and artificial intelligence (chapters 7, 9, 16 and 18), and the ever-present debate on the relationship between pure and applied research (chapter 13).

- Alfred Renyi, *Dialogues on Mathematics*, 1967 [14];
- Alfred Renyi, *A Diary on Information Theory*, 1984 [15].

Alfred Renyi represents another central figure in the development of probability theory with applications to information theory and statistical mechanics. In the first of the mentioned books, he addresses three fundamental issues concerning mathematics using the form of the dialogue. The first two dialogues are set in ancient Greece and the characters feature Socrates and Archimedes. They address the issue related to the *nature* of mathematics, namely about the objects of mathematics and

[1] These problems were collected in the 'Scottish Book'; in 1981, Ulam's friend R Daniel Mauldin published a version of this notebook.

their *existence*, and on its applications. In particular the reader can find a nice discussion on the role of models in science. The third dialogue is between Torricelli and Galileo and illustrates the modern scientific method and the indispensable role of mathematics in the description of the fundamental laws of Nature.

In the second work, he presents a very nice and readable introduction to information theory through the eyes of an imaginary (very diligent and acute) student. Using simple examples, the concepts of Shannon entropy, mutual information and information transmission through a noisy channel are discussed. In the second part of the book, beyond some applications of probability theory to game of chance and the theory of trees, one can find some interesting reflections on the general utility of teaching of probability theory in several disciplines.

- Shang-Keng Ma, *Statistical Mechanics*, 1985 [16].

Finally, we suggest a book on statistical mechanics which goes beyond the usual textbooks, and stands out due to the original discussions on several topics, such as the fundamental assumption of statistical mechanics and its logical foundation.

References

[1] Bader A and Parker L 2001 Joseph Loschmidt, physicist and chemist *Phys. Today* **54** 45–50
[2] King L C and Neilsen E K 1958 Estimation of Avogadro's number: an experiment for general chemistry laboratory *J. Chem. Educ.* **35** 198
[3] Pavliotis G A 2014 *Stochastic Processes and Applications: Diffusion Processes, the Fokker-Planck and Langevin equations* (Berlin: Springer)
[4] Tocino A and Ardanuy R 2002 Runge-Kutta methods for numerical solution of stochastic differential equations *J. Comput. Appl. Math.* **138** 219–41
[5] von Neumann J 1963 Various techniques used in connection with random digits *Collected Works* **vol V** ed A H Taub (London: Macmillan)
[6] Chirikov B V and Vivaldi F 1999 An algorithmic view of pseudochaos *Physica* D **129** 223–35
[7] Knuth D E 1981 *The Art of Computer Programming, Volume 2: Seminumerical Algorithms* (Reading, MA: Addison-Wesley)
[8] Falcioni M, Palatella L, Pigolotti S and Vulpiani A 2005 Properties making a chaotic system a good pseudo random number generator *Phys. Rev.* E **72** 016220
[9] Kac M 1987 *Enigmas of Chance: An Autobiography* (New York: Harper and Row)
[10] Ulam S 1983 *Adventures of a Mathematician* (New York: Charles's Sons)
[11] Dyson F 1979 *Disturbing the Universe* (New York: Harper and Row)
[12] Gottesman M, Kac M and Langer J 1980 A legacy and a hope *Phys. Today* **33** 102–3
[13] Kac M, Lebowitz J L and Plotz P H 1984 Yosif Begun *Science* **226** 114–6
[14] Renyi A 1967 *Dialogues on Mathematics* (San Francisco, CA: Holden-Day)
[15] Renyi A 1984 *A Diary on Information Theory* (Budapest: Akadémiai Kiadó)
[16] Ma S-K 1985 *Statistical Mechanics* (Singapore: World Scientific)

Index

doi:10.1088/978-0-7503-6229-0ch11

www.ingramcontent.com/pod-product-compliance
Lightning Source LLC
Chambersburg PA
CBHW080331240525
27014CB00005BA/22
* 9 7 8 0 7 5 0 3 6 2 2 7 6 *